RADIUM GIRLS

CLAUDIA
CLARK

RADIUM

Girls

WOMEN
AND
INDUSTRIAL
HEALTH
REFORM,
1910–1935

The
University
of North
Carolina
Press
Chapel Hill
and
London

Library of Congress

Cataloging-in-Publication Data

Clark, Claudia.

Radium girls: women and industrial

health reform, 1910–1935 / Claudia Clark.

p. cm.

Includes bibliographical references and index.

ISBN 0-8078-2331-7 (cloth: alk. paper). —

ISBN 0-8078-4640-6 (pbk.: alk. paper)

1. Watch dial painters—Diseases—

United States—History. 2. Radium paint—

Toxicology. 3. Consumers' leagues—United

States—History. 4. Industrial hygiene—United

States—History—20th century. I. Title.

HD6067.2.U6C55   1997   96-27358

363.11′9681114—dc20

CIP

05 04 03 02 01 6 5 4 3 2

THIS BOOK WAS DIGITALLY PRINTED.

CONTENTS

PREFACE

This is a story about illness and death and fairness. How is it that some deaths seem fair and some unfair? From one perspective, death is meted out one to each of us, inevitable and equitable, and thus indistinguishable and unremarkable. Yet some deaths seem "untimely," whereas others come "in the fullness of time." The issue here, however, is not just age of death. Although death by violence or accident is felt to be somehow more unfair the younger the victim is, still, violent deaths to elderly persons are thought to have unfairly deprived them of some of their allotted lifespans. Of American workers whose deaths in the early twentieth century were caused by radium, some died young, in their early twenties, whereas others succumbed to radium-induced diseases decades later, in their eighties. The early deaths seem perhaps the more deplorable, but we still sense something unfair about the later deaths, even if they occurred beyond the average lifespan.

Some deaths seem "senseless" and others "meaningful." A meaningful death might be one in the service of some ideal or some goal. Marie Curie discovered radium. She died from leukemia, likely caused by her long exposure to radium. She may be considered a martyr to science, and her death has meaning in the context of her remarkable scientific contributions. Sabin von Sochocky discovered a luminous paint formula that included radium and founded a company that sold luminous watch and clock dials. He died from aplastic anemia, also likely caused by radium exposure. He was perhaps a martyr to creative entrepreneurship as well as to science, and in both ways his death might seem meaningful. Some of the women who painted luminous numbers on watch dials died as a result of their work. What meaning do the deaths of the dialpainters hold for us? They seem senseless, unfair.

Marie Curie and Sabin von Sochocky were willing to risk the unknown to gain such things as knowledge, renown, financial profit. One way we might think about the fairness and meaning of death is by risk-benefit analysis. When people choose to face possible risks to gain possible benefits, often we admire their courage, and any harm to them seems meaningful in the context of their goals. It might be argued that the dialpainters stood to gain from their work, if in a small way, through their wages. In the nineteenth century, it was assumed that workers invested their good health in their jobs and stood to lose it in pursuit of an income. In the twentieth century, the "assumed risk" doctrine was overthrown in the courts in favor of employers' liability for the health of their employees. In part, this was because the size of the reward (wages) seemed incommensurate with the risks. From this perspective, poorly paid workers should not be expected to assume high risks, but highly paid workers might understandably risk their lives for great reward. The magnitude of the benefit affects our judgment about the appropriate level of risk.

The issue of choice is also relevant to the understanding of fairness. Twentieth-century courts have recognized that workers are not as free as scientists or capitalists to choose the risks they assume nor as powerful to mitigate those risks. Workers in modern society depend on their wages to survive and so are not completely free to leave dangerous jobs. Workers usually cannot control the conditions under which they labor. This powerlessness leaves them less able than others to act on rational calculations of risk versus benefit.

Most important, knowledge seems central to our understanding of fairness. Curie and von Sochocky may not have known of radium's long-term dangers, but as scientists, as pioneers, they understood the risks of the unknown. They also knew that radium emitted powerful radiations, that high energy was freed even by small quantities of the element, that radium's rays burned and destroyed living tissue. Their choice to work with such an element had qualities of informed consent. The dialpainters knew little of this. Without knowledge of the dangers of their work, dialpainters could not choose whether to face them or not, and their deaths seem less fair than others. In this book, the "sixty-four-dollar question," if you will, is whether businesses knew or had good reason to suspect the risk to their workers of close contact with radium. If they did, then they were not only negligent in providing an unsafe workplace. They were also guilty (in legalese) of the "fraudulent concealment" of danger, that is, they failed in their duty to inform workers of the possible risks they faced.

A last issue of fairness underlying this book concerns medical experts and public health officials. In recent memory, health workers have been assigned new ethical duties, such as to report possible child abuse or to educate patients enough to win informed consent for risky procedures. More traditionally, physicians have been charged with the duties inherent in their Hippocratic Oath—to do no harm and to care for the sick. Two vexing questions posed by the dialpainting experience are these: did researchers studying the dialpainters' illnesses have an ethical duty to publicize the dangers they discovered in workplaces? did they have an ethical duty to share their knowledge and skills with the dialpainters?

No definitive answer can be derived from this case study of radium poisoning. Rather, I hope to suggest two avenues of exploration in answering these historical/ethical questions. The first is to address the contemporary debate about these matters: at least some people recognized the ethical duties and ethical failings of health workers, whereas others defended health workers' inaction. Second, with knowledge of the health effects of radium and skills to measure those effects held by a very small number of scientists, we must consider whether those with knowledge and skills had a greater ethical duty to share them than if such knowledge and skills were widely available.

What meaning do the deaths of the dialpainters hold for us? As always, meaning is a matter of interpretation and mindset. As I began my work, the prevailing historical outlook was medical, and the discovery of radium poisoning was heralded as yet another victory of science over the unknown. I started from a different perspective, trying to imagine the experience of a dialpainter, sick, frightened—and angry, although given the restrictions of the female role in the early part of this century, often as scared of the anger as of the illness. I have traced in this book the steps the dialpainters took to find out what was killing them, and I hope I have successfully rendered what can only have been a furious frustration on their part that knowledge was withheld from them by their employers, by government officials, and by medical researchers. The dialpainters fought for recognition of their industrial disease, for compensation, for prevention; they sought allies to help them; they strategized various routes to confront all branches of government with their needs. The meaning of their story from this perspective is political, in that it reveals where power over occupational health lay and how it shifted as different interests collided over industrial radium poisoning.

ACKNOWLEDGMENTS

Because this is one of those books that started as a dissertation, my acknowledgments go back many years. I am grateful to the undergraduate professor who first suggested I might want to major in history, something that seemed a bizarre idea at the time. Perhaps it still is, but thank you, John Long, and also thank you for first recommending graduate school. Another undergraduate professor, Ava Baron, made the scholarly life seem important and plausible: thank you.

Many individuals helped me through my graduate school years at Rutgers University. I was most inspired by Suzanne Lebsock, most nurtured by Paul Clemens. My adviser, James Reed, stimulated me intellectually and supported me emotionally (and financially a few times), but I am most grateful for those times when his confidence in me exceeded my own. Gerald Grob taught me a lot about writing. Judy Walkowitz encouraged me to relish confusion as a source of good questions. Of course, my fellow students probably taught me the most. Thank you to Kathy Jones, Paula Baker, and Carol Green for holding my hand through my first year (and even now occasionally). John Rossi once told me he thought my work was important, and that has gotten me through some bad moments. David Schmitz drank beer and argued politics with me, even when I was a lowly first-year student. Most important, though, were the members of Rutgers's first Women's History Seminar, with whom I learned that not only the political but also the scholarly is personal: thank you to Polly Beals, Anna Clark, Carolyn Strange, and Laura Tabili. The most amazing thing is that we all have jobs!

The first person to suggest that the dialpainters might make a good dissertation was Dick Wolfe, who manages the archives at Harvard Medical School's Countway Library of Medicine. Mr. Wolfe has also helped track down many a document, and he introduced me to a network of scholars interested in industrial health. While at the Countway, I chanced to meet Dr. William Castle, a subject in this work, who kindly agreed to an interview, for which I am grateful. Countless other librarians and archivists have made this work possible. I particularly want to thank Aloha South, of the National Archives, who spent many hours finding some documents that proved to be very important to this book. Carl Lane at the New Jersey Historical Society was also very kind, as was Charles Cummings at the Newark Public Library and Barbara Irwin at the History of Medicine Collection, University of Medicine and Dentistry of New Jersey. The staff at Argonne National Laboratory were gracious, especially James Stebbings, then director of the Center for Human Radiobiology. Not only did Dr. Stebbings make it possible for me to undertake this research, but he also disagreed with my conclusions frequently enough to improve this work. This manuscript has also been much improved because of comments by David Rosner, Kathryn Kish Sklar, Barbara Sicherman, and the editors at the University of North Carolina Press.

This research has been supported by a number of grants and fellowships. I gratefully acknowledge the state of New Jersey for a Garden State Graduate Fellowship, the American Institute of the History of Pharmacy for a Grant-in-Aid, the College of Physicians in Philadelphia for a Francis C. Wood Fellowship, the New Jersey Committee for the Humanities for several grants, Argonne National Laboratory for a Thesis Parts appointment, Rutgers University for a University Dissertation Fellowship, and Emory University for a Mellon Post-doctoral Fellowship in the Humanities.

My family and friends helped see me through, too. My father Herbert Clark and my sister Nancy Clark may not have always understood what I was up to, but they were always polite about it and were there when I needed them. Betsy Watson and Ella Watson-Stryker gave me a life to return to when I wearied of scholarship. Thank you all.

RADIUM GIRLS

"POISONED! -- *as They Chatted Merrily at Their Work*
Painting the Luminous Numbers on Watches, the Radium Accumulated in Their Bodies, and
Without Warning Began to Bombard and Destroy Teeth, Jaws and Finger Bones, Marking
Fifty Young Factory Girls for Painful, Lingering,
But Inevitable Death"

*Through its newspaper empire, the Hearst Corporation distributed this drawing and
headline to notify Americans about the plight of New Jersey's "Radium Girls." (From*
American Weekly, *Sunday newspaper insert, February 28, 1926, p. 11. Drawing retouched
by Ronald Conner. Used by permission of the Hearst Corporation.)*

INTRODUCTION

When young American men went off to war in 1917, young American women went off to work. For working-class daughters, especially those of immigrant families, this meant, increasingly, employment in a factory. At this time a new factory job became available for a small number of working-class daughters in a few cities: dialpainting. World War I trench warfare had made the newly invented wristwatch crucial at the front and popular at home. Novel glow-in-the-dark paint soon illuminated the watches and instruments of both entrenched soldiers and trendy consumers. Painting luminous numbers on watch faces and other dials in "dialpainting studios" was sex-typed as women's work, and, like most white women who worked outside the home, the majority of dialpainters were young women, from their midteens to their early twenties, single or newly married and as yet childless. Dialpainting promised them relatively easy work and comparatively high wages.[1]

In the 1920s and 1930s some dialpainters and former dialpainters began to suffer from a variety of illnesses, often crippling and frequently fatal. The modern understanding of their illnesses is that radium, the ingredient in the paint that produced the energy required for luminosity, had been ingested from paintbrushes brought to a point between the lips and (perhaps) had been inhaled as dust from the air; radium further subjected the dialpainters to high levels of radiation in their workplaces. Most imbibed radium would have been excreted, but, because radium and calcium have similar chemical properties, a portion of each day's dose was plated into the dialpainters' bones. Radioactive emissions from radium fixed in bone could affect a dialpainters' health in a number of ways. Radiation could damage the constantly reproducing bone marrow, causing anemia. Radiation could weaken the bones, so that vertebrae and metatarsals (bones in the feet) might crush under normal pressure, or long bones in the legs or arms might snap spontaneously. Radiation could kill bone tissues so that they became infected easily, especially the jaw bones, which are open to infection through dental caries or gum diseases. Radiation constantly bombarded the bone and bone marrow, with bone cancer (osteosarcoma) or cancer of the marrow (multiple myelomas) possible outcomes. As radium decayed through its radioactive series, it produced radon, a radioactive gas, most of which would have been exhaled. If, however, some radon was retained within bone cavities in the skull—the nasal sinuses and the mastoids—its radiations and radioactive products might cause cancer (epithelial carcinoma) in the tissues lining these cavities.

The dialpainters were the first industrial victims of radium poisoning; indeed, they were among the first victims of any form of radioactivity. Their illnesses drove medical scientists to examine the effects of radioactivity on living tissues, yielding a new field of scientific study, human radiobiology, or health physics. Data from the dialpainters' cases shaped health standards for other industrial workers, for medical personnel and their patients, and for scientists, including those who worked on the Manhattan project during World War II. In part because of the significance of information about their illnesses, the "radium dialpainters" are well known as a case study among students of health physics, epidemiology, and occupational health. The episode is referred to as a "classic" case, one whose puzzling, multivariate symptoms yielded to the application of medical principles and logic.[2]

To historians the dialpainters' cases tell a different story. In rough outline, the account presented here is of women who knew as early as 1923 that they faced an industrial disease. For two years they sought medical and scientific corroboration, the last year with the collaboration of a middle-class women's reform group. By 1925 many scientists and most government officials conceded that the dialpainters' illnesses were caused by radium, but industry leaders still resisted this conclusion. Through remunerated scientific consultants, radium businesses were able to control knowledge about radium poisoning, concealing data that supported its existence and promoting opinions that put blame for the dialpainters' illnesses elsewhere. Radium businesses evaded liability for compensation and squashed efforts to regulate workshops. The dialpainters and their reformer allies continually fought the radium companies on all fronts, hoping to win recognition of their disease, compensation for its victims, and prevention of future illnesses. Although business managers changed work processes to make their job far safer, dialpainters still accumulated radium and contracted diseases, albeit diseases not officially recognized as radium-related.

The "discovery" of radium poisoning is considered here less as an event attributable to an individual than as a process through which society came to agree on an explanation for the variety of illnesses among radium-exposed workers; it is less a story of medical success than a story of political process. The "discovery" of radium poisoning as I mean it might be better called the "recognition" of radium poisoning, except that recognition was but one of three parts to the process of discovering this new industrial disease. As discussions about the dialpainters' illnesses proceeded among radium workers, business managers, government agency functionaries, medical specialists, and legal authorities, the recognition of radium poisoning was inextricably braided together with consideration of compensation for victims and prevention of further disease among new workers. Recognition, compensation, and prevention were part of one process because deciding how workers came to be poisoned on the job led inevitably to evaluation of production processes, which involved more than just identification of the industrial chemical responsible for the illnesses. Related issues included ascertainment of the techniques or working conditions through which workers were exposed, which implicated management and/or workers as determinants of those conditions, which, in turn, raised questions about negligence and liability. Other questions prompted by these considerations were what modifications to the production process would be necessary to make work safer? and what level of safety or, conversely, risk was acceptable — and acceptable to whom? Obviously, these are delicate questions, and radium poisoning must be understood as a social product born of political negotiation.[3]

From such negotiations may be abstracted a map of the shifting and relative political power and social authority of labor, management, government, and science. The discovery of radium poisoning, if compounded of interlocking conflicts about recognition, compensation, and prevention of industrial diseases, also partook of several larger political issues. The first involved disputes over workplace control, which had become by the 1920s an employer prerogative after decades of labor-management conflict. At issue in the 1920s was the role of government as arbiter of this conflict. Employees, having lost most union struggles to gain labor control of the shopfloor and most political battles to control government regulation of working conditions, acceded as a last resort to government's involvement as a neutral mediator of workplace disputes. Business leaders, having had positive experiences with government intercession in the workplace during World War I, were content to allow government to act as an allegedly neutral outside force. Thus by the twenties labor, management, and the state were primed for experiments in "corporatism."[4]

Corporatism is the second larger issue to which the discovery of radium poisoning relates. Corporatism was the belief that all parts of society were necessary to its harmonious functioning, and that therefore all parts should cooperate to see to the welfare of each part. The 1920s political philosophy that tried to build an "associative state" has also been called corporate liberalism. The state's role, according to this philosophy, was to remain a neutral partner and to advance the settlement of issues that maximally benefited all parties.[5]

The importing of scientists to debates over working conditions was an extension of the idea that neutral observers could contribute to peaceful solutions of complex political issues. In situations like industrial disease investigations, in which arose conflicting interpretations of events, scientists claimed the role of objective observer, capable of determining truth and worthy of legitimating truth. Corporatism in the 1920s included scientists as the fourth side of a labor/management/government table. Although often scientists worked for government or management, they still laid claim to independence and neutrality as features of their public service. In the discovery of radium poisoning, the U.S. Public Health Service (USPHS) ostensibly attempted to build consensus among business leaders and labor representatives about the cause and prevention of industrial radium poisoning, but decisions were ultimately ceded to scientists. Scientists did not always live up to their self-proclaimed neutrality, and they sometimes promulgated truth that leaned heavily on business interpretations.[6]

An interesting medical ethics quandary surfaces from consideration of the dialpainters' plight. The dialpainters needed influential advocates to help them win recognition of their industrial diseases, compensation for their injuries, and regulations to prevent future cases. Claims to impartiality prevented most physicians and government officials from aiding the dialpainters in their struggle, thus leaving largely unchallenged a status quo that benefited business. Generally it is difficult to prove that medical and governmental professionals choose their neutral observer role in deference to business interests, but in a few instances in the dialpainter history, we have evidence of professionals motivated to take allegedly objective stands because of consultantships directly with business or because of university research positions philanthropically supported by business. Some scientists refused to testify on behalf of the dialpainters; some scientists suppressed evidence of radium poisoning. Compounding the problem was the virtual monopoly that scientists and radium businesses held on the knowledge and equipment necessary to determine the radioactivity of dialpainters. The questions raised here are whether health researchers have an ethical duty to report knowledge about working conditions' adverse health effects and to bear witness to that knowledge in court.[7]

A third issue in which the discovery of radium became entangled was the long-standing debate about the social utility of commodity production and economic development. The radium poisonings and similar episodes raised explicit questions about the relative value of industrialization compared to its ill effects on the health of consumers, workers, or citizens in general. One well-explored aspect of this is workers' hesitation (perhaps at managers' insistence) to protest workplace toxicity and environmental pollution because of their dependence on manufacturing jobs. Conversely, movements for improved safety and for workers' compensation have also been widely studied.[8]

Industrial diseases, or industrial health, until recent years has been an almost neglected topic. Historians are now beginning to piece together a picture of shifting conditions in the workplace and varying responses to them by workers, managers, medical experts, government officials, and reformers. We are driven by recent events—the discovery of new industrial diseases, the creation of new government vehicles to attempt to control them, a convergence of labor and public concern over the effects of industrial chemicals—and by historiographical trends—attempts to wed social and political history, especially evident in efforts to bridge the gap between studies of workers in the workplace and studies of workers as social and cultural beings in the home and the community. Industrial health is largely a workplace phenomenon (as currently defined, although an archaic term, "workers' health," involved the study of all diseases of workers and considered the effects of poor housing, poor nutrition, crowded cities, in sum, poverty). Even in its current manifestation, however, industrial health can take the historian out of the workplace. The dialpainters drew on family and community connections to identify radium poisoning and to fight for its recognition, compensation, and prevention. Further, radium poisoning was not constrained to workers, but threatened scientists, medical patients, consumers of patent medicines, and, if it did not threaten, it at least frightened people living close to radium businesses. Industrial health bridges the history of labor and the social history of workers, the history of medicine and the social history of health and the environment, the traditional history of politics and the social history of politics. In the case of the dialpainters, we may also study the history of women. Industrial health is a good topic for those interested in breaking out of narrow historiographical traditions.[9]

What makes the dialpainters' cases an especially appealing vehicle for studying occupational health history is that individuals and institutions that played prominent roles at the national level of the industrial health debate appear in the dialpainters' chronicle. By 1910 a varied group of Progressive Era reformers, scientists, insurance experts, industry leaders, and government bureaucrats had created amid the many other reform efforts a movement for healthier working conditions. This industrial hygiene movement spanned the 1910s and the 1920s. This raises yet again the thorny question of what happened to Progressivism in the 1920s. These twenties and thirties episodes allow us to measure the successes and failures of a Progressive reform movement in industrial health. Efforts to prove the occupational origins of the dial-painters' illnesses, to win them compensation, and to prevent further radium exposure thus shed light on national occupational health developments.[10]

To understand how the dialpainters were treated within American institutions it is relevant to consider them not only as workers but also as women. Traditional labor history helps us little, as the dialpainters were not unionized and received barely token support from labor organizations. But because the workers were female, a middle-class women's reform organization, the Consumers' League, became involved, allowing us to study closely the actions of this group as it worked with and for the dialpainters and for industrial health generally.[11]

The Consumers' League was at the forefront of reform efforts in industrial health, for men as well as for women, and women reformers were effective in improving occupational health and safety in the United States. An evaluation of women reformers' goals and strategies exposes flaws in their analyses of the problems and in their proposed solutions, but it also reveals their creativity in rising to the challenge of the radium cases. Reformers devised means to influence decision makers in medicine and in legislative, administrative, and juridical law. Despite their mistakes and blind spots, the commitment and compassion of the Consumers' League reformers are unquestionable. Their efforts illuminate historians' understanding of how women forged (and sometimes forfeited) political and social power in the decade following their winning of the vote.[12] The industrial hygiene movement intensified in the 1920s, due mostly to the work of communities of women reformers, the Consumers' League the most prominent among them. The dialpainters' cases allow us to discover and acknowledge the work of women reformers throughout an era whose reform orientation is still debated.

The Consumers' League's interest in combating industrial diseases seems to have been sparked by Dr. Alice Hamilton, who is a central (if often a behind-the-scenes) figure in this study, in part because she operated within three groups central to occupational health. She was a reformer, she was a physician working in one of the earliest departments of industrial hygiene at Harvard University, and she was a government investigator of workplace hazards for both state and federal government agencies. In all three roles she exerted her influence to help the dialpainters.[13]

We gain in other fields as well. In terms of more traditional legal and political history, the discovery of radium poisoning necessitates consideration of the legal issues surrounding negligence, statutes of limitations, and equitable fraud during the 1920s. The history of industrial radium poisoning and, in comparison, the history of radium poisoning among medical patients also illuminate the stances taken by state and federal executive departments in struggles over workplace health.

Not least, the dialpainters' cases teach us about business attitudes toward worker health. The industry's intransigence in this instance endangered workers. The radium companies denied the dangers of imbibing radium well beyond the point at which consensus had been reached among most medical experts and government officials. Consultants were employed to bear scientific witness to radium's benign character. Medical and government experts were wielded in attempts to avoid or control the cost of compensation to injured employees. The industry united to withstand regulation aimed at preventing future cases of radium poisoning. An exploration of radium business leaders' perspectives on radium poisoning reveals that they had personal as well as financial motives for denying that radium was dangerous; nonetheless, their resistance to the scientific evidence resulted in more suffering.[14]

Information about the discovery of radiation poisoning is of contemporary interest. Data on the health and illnesses of the dialpainters were eventually used to devise recommendations for maximum radiation exposure; indeed, even today they underlie arguments over safe levels of radiation in soil, air, and water. People still suffer the effects of the early dialpainting industry: some workers survive, with variable morbid effects of radiation poisoning; industrial sites remain radioactive and demand expensive decontamination procedures; homes built on radioactive landfills (tailings from the radium extraction industry) now pose a risk from radon to their inhabitants.

The dialpainters' cases had significant ramifications within many circles, in medicine and science, law and government, business and insurance. Given their importance, it often escapes notice that the number of workers involved was small. Dialpainting was, numerically, an inconsequential occupation, and the percentage of workers definitely established to have suffered symptoms of radium poisoning was low. Dialpainting began in 1917. Over the next decade, perhaps 2,000 women were employed in the work, mostly in three locations: Orange, New Jersey; Waterbury, Connecticut; and Ottawa, Illinois. The names of 1,600 of these workers are known to researchers. The market for luminous dials decreased in the late 1920s, and only 441 dialpainters are known to have been employed between 1927 and 1940; perhaps 20 percent more actually worked. But World War II boosted the demand for dials that could be read in the dark, and during the 1940s another 1,350 women are known to have painted them; again, this is a minimum figure. Perhaps, then, as many as 4,000 women were dialpainters, with half of them working before 1927.[15]

Ascertaining the extent of mortality and morbidity among the dialpainters is difficult. Medical studies of these women have focused on the occurrence of cancers. Cancers definitely attributable to radium (the bone cancers and the "head cancers") thus far have been found only in dialpainters who worked before 1927; after that, safety precautions seem to have been effective enough to prevent those kinds of cancers. Of the 1,600 identified workers in the pre–1927 cohort, 63 contracted bone cancer and 23 more developed cancer specifically of the nasal sinuses or mastoid cavities—86 cancers in all, or an incidence rate of 5.4 percent. Data on other cancer deaths possibly caused by radium exposure are still controversial, but dialpainters may also have died from breast cancers, lung cancers, and bone marrow tumors (multiple myeloma).[16]

Early dialpainters died from illnesses other than cancer. The first deaths attended infections of dead bone and other tissues in the jaw ("necrosis" of the jaw) and anemia. Death from such causes seems to have been confined to the earliest dialpainters and has not been studied since the 1920s and 1930s. We have only limited evidence of how many dialpainters may have died from anemia and jaw necrosis—and no figures as to how many suffered from such maladies but survived. In 1929 the U.S. Department of Labor identified twenty-three fatalities attributed to radium poisoning, and cancer was implicated in none of these: cancer was first reported in two deaths among New Jersey dialpainters in 1929. By 1931 eighteen women in New Jersey were known to have died from radium poisoning, thirteen from radium diseases not involving cancer and five from cancers. By 1940 fifteen women in Connecticut were known to have died, six from illnesses other than cancer and nine from cancer.[17]

Noncancer deaths seem to have had a negligible impact on the overall mortality data. Of 1,235 workers employed before 1930, expected deaths by 1976 were 461 and actual deaths were 529 (before age eighty-five), or 68 excess deaths, or 5.5 percent. With the incidence of bone, sinus, and mastoid cancers among the 1,600 known dialpainters who worked before 1927 calculated at 5.4 percent, it is likely that deaths from noncancerous industrial diseases only minimally affected the population from an aggregate perspective. For the first six years following the discovery of radium poisoning, however, such effects as anemia and jaw infections caused most of the deaths and sparked most of the fears among dialpainters.[18]

Many women survived these afflictions without developing cancer but were disabled by bone lesions; without morbidity data we do not know how many became ill or disabled as a result of their work. Besides the bone afflictions, another possible effect of radium poisoning was infertility: there are data suggesting that fertility among dialpainters was lower than normal, but more research is needed before this can be substantiated. Although cancer caused most of the deaths among the dialpainters, these other effects should not be ignored and merit further medical study.[19]

We do know that many dialpainters accumulated dangerous levels of radium. Measurement and estimation of deposited radioactivity is a controversial science, as is the determination of dangerous levels of deposited radioactive substances, so the studies quoted here should be considered suggestive, not determinative. And although radioactivity is measured in a multitude of units, readers should keep in mind that cancer deaths attributable to radium dwindle to nil in populations with body-burdens below 10 microcuries of deposited radium, which roughly corresponds to 10 micrograms of radium (damage to the bones and illnesses still occur at body-burdens below this level, however). One study indicates that 18 percent of all dialpainters who worked before 1950 accumulated a significant dose (greater than 10 microcuries) of radium, enough to put them in a high-risk category. Most of the high-risk workers were in the World War I cohort. In a second study, of 693 women who worked before 1930 and who were measured for accumulated radium, 42 percent had an estimated initial body-burden of 5 microcuries or more. Sixteen percent had an initial body-burden of 50 microcuries or more. These women were at high risk for cancer deaths, jaw necrosis, anemia, and the undetermined levels of illnesses and handicaps caused by radium.[20]

The dialpainters have been exhaustively studied by medical scientists, but except as bodies seldom appear in historical accounts about the discovery of radium poisoning. I have tried to reconstruct dialpainter attitudes and efforts from newspaper interviews, court records, and one brief autobiography, with some success, and have quoted dialpainters directly as often as possible, to remind us that these were real people taking actions to help themselves and not just victims or patients. I had hoped to learn more about the dialpainters themselves through interviews of survivors. Unfortunately, I was unable to contact any survivors. The epidemiologists who controlled the data at the Human Radiobiology Center at Argonne National Laboratory were concerned that interviews might interfere with the dialpainters' cooperation with long-term studies of radiation's effects and so refused me access to the addresses of known survivors. My own efforts to find survivors failed, due partly to a lack of funds, partly to my promise not to use the Argonne data to contact dialpainters, and partly because I hesitated to repeat a daunting task already accomplished by investigators.

What follows is a (perhaps unwieldy) attempt to weld together a narrative history with an analytical one. I strive to tell not only the moving story of the discovery of radium poisoning and its effects on individuals and society but also to analyze the changing knowledge, attitudes, and institutions that affected the dialpainters' struggle for recognition, compensation, and prevention of radium poisoning. My strategy has been to move the narrative throughout while employing each chapter to focus on a particular sector of the intricate web of individuals and institutions implicated in industrial health history.

Chapter 1 discusses the New Jersey dialpainters' early years, their first suspicious illnesses, and their first year's work to determine the cause of their illnesses. I demonstrate that the dialpainters actively pursued recognition of the industrial origin of their diseases.

Chapter 2 considers the history of the use of radium as a medicine and early studies about the effects of small quantities of internally deposited radium. The literature on radium, I establish, originated largely from those with a financial stake in radium's beneficence.

In Chapter 3 the dialpainters' alliance with the Consumers' League is examined, and another year of effort to win recognition of the dialpainters' industrial disease is followed. I review the history of the industrial hygiene movement and the league's role in it.

Chapter 4 explores the scientific work that finalized for most the "discovery" of radium poisoning, although radium businesses' continued resistance meant that dialpainters still had to fight for recognition of their disease. I demonstrate that much of this effort relied on Consumers' League data and networks, and that therefore the league was crucial to the discovery of radium poisoning.

Chapter 5 follows dialpainters in Connecticut and New Jersey in their demands for compensation. The two states are compared, and the Consumers' League's intervention in New Jersey is shown to be decisive in explaining differences between the two states.

The federal government's responses to industrial radium poisoning are considered in Chapter 6 and are compared to their responses to iatrogenic radium poisoning in Chapter 7. As these accounts show, the federal government acted with more alacrity when consumers of radium medicines were threatened than when workers were endangered.

Chapter 8 follows the Illinois dialpainters through the late discovery of radium poisoning among them, through fights for compensation, and through struggles over prevention. Although after 1927 radium accumulation was low enough to prevent the necroses, anemia, and bone and head cancers of the early dialpainters, workers still accumulated significant levels of radium that some blamed for new disease outbreaks among them.

In conclusion, I suggest what might be learned from the dialpainters' experiences to help us improve workers' health in our own time. We need to know why so little progress has been made in ensuring workplace health after over a century of research and reform involving industrial diseases. The best estimate is that occupational diseases currently kill about 100,000 Americans each year and cause 400,000 new illnesses. Criticism abounds concerning these figures, which emanate from the National Institute for Occupational Safety and Health.[21] What is most significant about this criticism is that we do not know what risks are posed by industrial poisons and unhealthy working conditions. Beginning in the 1970s, a few states passed "Right to Know" laws, but earlier no workers had the right to know about the dangers inherent in the materials they used in their jobs. I worked six years in the chemical industry—the last three using toxic pesticides. Poisoning incidents involving myself and my coworkers left me curious about why workplace health was so poorly regulated, why the possible toxic effects of common chemicals were so poorly understood, why what was known was so narrowly disseminated: in sum, why so many workers' lives were and are at risk. These are the questions that motivate this study.

WATCH ALICE GLOW

The New Jersey Radium Dialpainters

The first illnesses linked to industrial radium poisoning came to light in Orange, New Jersey, beginning in 1922, among women who had been employed to apply luminous numbers on dials. To these women, the "dial-painters," must be attributed the "discovery" of radium poisoning, although others would claim that honor. Experts' diagnosis of radium poisoning followed a long campaign by the dialpainters for medical and legal acknowledgment of the existence and cause of their illnesses.

Among the first to make the connection between dialpainting and the dialpainters' illnesses was Katherine Schaub, a dialpainter herself. We know more of Schaub than of most dialpainters. She participated in a lengthy and notorious damages suit against her employer, which was covered, often melodramatically, in newspapers and magazines. Her lawyer's records, detailing her testimony, survive. An extensive medical record yields further details about this dialpainter and her life, and one physician, Harrison Martland, saved letters she wrote to him expressing her fears about her illnesses and treatments. Finally, Schaub wrote about her trials and triumphs, and a brief autobiographical fragment, published in a social reformers' magazine, is extant. As an individual, Schaub deserves a place among those influential in discovering and responding to radium poisoning. As one of a group about which, as a whole, we know relatively little, Schaub, in this account, also serves as the quintessential dialpainter, whose reactions and actions we must take as tokens of her coworkers' responses.[1]

Katherine Schaub was born in Newark, New Jersey, on March 10, 1902, the second child of William and Mary Rudolph Schaub. William Schaub's occupation is uncertain: on Katherine's birth certificate it appears in a handwritten scrawl that might read "baker"; a newspaper of 1928 lists him as a "Mason," but this might be a social affiliation, not a trade. Mary Schaub, according to her daughter's birth certificate, was an "H.K.," which must mean housekeeper. These occupations would fit the profile arrived at by an impressionistic review of various dialpainters' families. The dialpainters seem to have come from moderately well-off working-class families. Fathers or husbands worked at semiskilled or skilled trades, as plumbers, hatters, carpenters, and machinists. At least a few families owned their own homes. Like Katherine Schaub, most of the Orange dialpainters were born in the United States of American-born parents, many with German surnames, but about a third of the women were first-generation Americans whose families had immigrated most often from Ireland or Italy.[2]

Although there is no record of impoverishment among such families, there was enough financial need to send daughters out to work. Most of the dialpainters were young women from their midteens to early twenties. Generally, they were single and lived with their parents. Dialpainters quit work either on marrying or with their first pregnancy.[3]

Death and ill health were no strangers to workers or their families in the early twentieth century and often led to the formation of extended households. As a child Katherine Schaub lived with her parents, her two sisters and two brothers, a grandfather, and a cousin. The cousin was variously known as Regina, Virginia, or Irene Rudolph, but most often the latter. She was Katherine's age, born, like Katherine, in Newark in 1902. Rudolph had lost both parents to "consumption" (probably tuberculosis) and a sister to meningitis. In 1919 Mary Schaub would die of mastoiditis, an infection in a sinus behind the ear, which spread into her brain. Katherine's oldest sister, Josephine, then took over management of the household.[4]

Two years earlier, in 1917, Katherine Schaub and Irene Rudolph had started jobs in the newly opened "dialpainting studio" of the Radium Luminous Materials Corporation in nearby Orange, New Jersey. They were both fifteen years old. Schaub had become a lively young woman, known for her good looks—blue eyes, attractively bobbed fair hair, a "pretty little blond"—and her friendly, social disposition—she "like[d] the nightlife." No description or photographs of Rudolph survive. The cousins painted luminous numbers on watch and clock dials.[5]

The Radium Luminous Materials Corporation was begun by the Austrian immigrant Sabin von Sochocky, a medical doctor and physicist, with a partner who was also a physician; originally, both practitioners were interested in radium medicines. In 1915 von Sochocky developed a formula for a self-luminous paint. It contained but an infinitesimal quantity of radium, which, by releasing alpha particles, stimulated zinc sulfide in the paint to luminesce. This paint could be produced cheaply.[6]

The widest use for radium paint would be found in the luminous dial industry. World War I promoted both a new kind of timepiece and a need for luminous instrument faces. Before the war pocket watches had been the common personal timepiece for men, but soldiers found that these interfered with the creeping and slithering of trench warfare. They coped by transferring their time pieces to their wrists, thus the wristwatch. Synchronizing movements in the dark was also difficult, creating the need for a luminous wristwatch. Soldiers found the self-luminous wristwatch a most practical timepiece; so did Americans at home. The luminous wristwatch became a fad in America, first for men and then for women when ladies' wristwatches were introduced around 1920. The military also used radium paint on instrument dials in tanks, ships, planes, and other war machines, while those at home found domestic uses for luminous paint, such as glow-in-the-dark numbers for houses or theater seats and luminescent lamp-pulls. But the watch or clock face remained the most commonly luminized surface.[7]

Around 1917 the Radium Luminous Materials Corporation undertook a major expansion. The firm acquired radium mines in Colorado's Paradox Valley and bought an extraction plant in Orange, a city just west of Newark. A new dialpainting studio was built in Orange to New Jersey factory law specifications—offering "ideal working conditions," a company vice president bragged.[8]

The new plant was not without its problems, however, and the company's response to neighborhood grievances hint at how it would handle the dialpainters' later charges. In 1919 residents in the vicinity of the radium plant complained of harsh fumes. Laundry hung out to dry turned yellow, and it was so difficult to breathe that one morning gas masks were improvised. The local health department investigated, and legal proceedings were started to close the plant. When the company added equipment to its stacks to scrub out the acids causing the problem, the city dropped its suit. A company official later recounted that, although he was sure it was not the plant's fault, he had given a local woman five dollars in compensation for ruined laundry. But then, everyone else wanted money, too. People in this "poor, residential neighborhood" were "anxious to take advantage of the company." More compensation was not forthcoming. The new equipment did help decrease the acidity of the fumes, but the remaining fumes continued to raise protests in the neighborhood.[9]

The dialpainting studio was considered a good place for a woman to be employed. Applying the paint to watch dials and watch hands was delicate, exacting work that required a light touch and several weeks of training. Women, according to the prevailing stereotypes, were seen as particularly fitted to this work because it required little strength but much manual dexterity. As semiskilled laborers, dialpainters received relatively good pay, for women. An average dialpainter made about the same as the average woman worker elsewhere in New Jersey, but a dexterous dialpainter could earn much more. The median income for women around this time was about $15 dollars a week, with the better paid obtaining about $23. The average full-time employee at the Orange studio might draw about $20 a week painting about 250 watches a day, at one and a half cents per piece over a five-and-a-half-day week. A quicker worker could earn $24 or more during the same time. Of course, like most piecework jobs, dialpainting was not always regular, and dialpainters faced periods of no work or slack work; thus their average earnings were closer to the state average of $15 dollars a week. But during the war work was plentiful. In 1913 several small companies produced about 8,500 luminous watches; by 1919 a few larger, more centralized radium companies made 2.2 million luminous watches. In New Jersey the number of dialpainters rose from the 100 employed in 1916 in Newark to an average of 300 in Orange in 1917 and 1918. In 1919 about 200 women painted dials for the Radium Luminous Materials Corporation.[10]

The dialpainting studio, built on the second floor of one of the new buildings in Orange, was a pleasant place to work. Light streamed in through large windows. In front of the windows were long tables at which the dialpainters sat. The room was quiet and the young women gathered there could socialize. Most of the dialpainters were neighbors. They came from Orange and walked or took the trolley to work, or, like Schaub and Rudolph, they commuted from Newark by train. One former employee remembers them as "pleasant girls and young women, many of them high school girls wanting a good summer job." Another recalled: "We were a fine upstanding bunch of girls, even though not many of us had been to high school. We were always neatly dressed—in our shirtwaists and skirts—and we worked hard in the studio." Such enjoyable surroundings prompted the women to encourage friends and relatives to become dialpainters, and often families sent two or more daughters to the Orange studio. Schaub found the work "interesting and of far higher type than the usual factory job," and she had enticed her cousin Irene Rudolph to join her there.[11]

Besides the good pay and pleasant working conditions, the cousins had other reasons for liking their jobs. Many of the watches produced were sent overseas to soldiers, and Schaub noted that "I was pleased with the idea of a job which would engage me in war work." Some of the young women would scratch their names and addresses into these watches, and sometimes a lonely soldier would respond with a letter.[12]

Part of what made dialpainting an attractive job must have been the work with such a sensational product. The young women applied radium to their buttons, their fingernails, their eyelids; at least one, described by a friend as "a lively Italian girl," coated her teeth with it before a date, for a smile that glowed in the dark. After working in the studio, the women were covered with luminous powder. At night, at home, they would notice that their clothes luminesced, their fingers glowed, their hair shone in the dark. One young dialpainter, employed during the summer by a dialpainting company in Brooklyn, would return from work "with her hair and dress speckled with yellow. Sometimes we'd get the neighborhood kids to watch Alice glow in a dark closet. They'd get a good laugh out of that."[13]

The women worked at their own pace, rather than on an assembly line. They mixed the dry luminous paint powder with paste and thinner, carefully pointed a small brush with their lips before dipping it in the paint, and then meticulously filled in the numbers and other marks on the watchfaces arrayed on trays before them. This method was followed for water-based paint used on paper watchfaces; metal watchfaces required a paint thinned with turpentine, and the awful taste precluded the lippointing of brushes. But at Orange, most of the work must have been paper dials, because former employees always recall being taught to lippoint. Schaub herself became an "instructress" at the plant. "The method of pointing the brush with the lips was taught us, to give the brush an exceedingly fine point," she remembered. "I instructed them to have a very good point on the brush. . . . I instructed them to put the brush in their mouth to get the best point on it."[14]

The technique of lippointing was little remarked. The earliest dialpainters had been drawn from the china painting industry, where lippointing was employed, and they passed on the technique. Moreover, because radium was being sold by the Radium Luminous Materials Corporation for medicinal use, dialpainters' exposure to the element was considered good for their health. As an example of the corporate attitude, in 1920 the New Jersey company was selling the residue from radium ore at the end of the extraction process as sand for children's playboxes. Some parents expressed concern about the radioactivity remaining in the sand, as the extraction process was at best only 85 percent efficient. Dr. von Sochocky assured them that the sand was "most hygienic" for their children to play in; in fact, it was "more beneficial than the mud of world renowned curative baths."[15]

Only once did someone at Orange suggest a problem. According to one dialpainter, von Sochocky warned her that putting her paintbrush in her mouth might make her sick. She then "spoke to different people about it and talked to the woman in charge and she told me it wasn't so."[16]

Meanwhile, precautions with radium were taken in parts of the plant. Florence Wall was hired during the war as an assistant to von Sochocky. While working in his physics lab, she was required to wear protective clothing and to work behind screens. In 1920 the Electroscopic Laboratory, where radioactivity measurements were made, was moved a mile away from the extraction plant and dialpainting studio because the "open radium" at the plant contaminated the readings.[17]

At some point during her employment, Schaub—perhaps just a typical teenager—broke out in pimples. She consulted a doctor, who asked her if she worked with phosphorous. Phosphorous poisoning was well known among New Jersey's working class, and because the element was phosphorescent, it was logical to suspect its presence in the luminous paint. Schaub became alarmed at the idea that her work might be dangerous. She shared her worries with coworkers, who became concerned as well. As a result, both von Sochocky and the medical colleague with whom he had started his business lectured the dialpainters, convincing them that their work was not hazardous. For a while, small bowls of water were provided so the women could rinse their brushes before pointing, but these were soon discontinued because too much paint was being wasted in the rinsing.[18]

Katherine Schaub was not unfamiliar with the concept of industrial disease, since she suspected (with some prodding by her physician) that phosphorous poisoning had some relation to her outbreak of pimples. Chronic phosphorous poisoning became a notable disease with the invention of the "lucifer match," a strike-anywhere match that became popular in the 1830s. White (sometimes called yellow) phosphorous contributed to the matches' flammability. Often, women and children were employed in factories to coat matches with phosphorous, exposing them and adult male coworkers to phosphorous fumes. Chronic exposure to the fumes led to phosphorous poisoning, with three major symptoms. One symptom was anemia. A second was that the bones became so brittle that they were prone to easy fracturing. The final and most notable symptom was phossy jaw, whereby phosphorous fumes attacked the gums in mouths with carious teeth; symptoms included tooth loss, gum swelling, necrosis (decay) of the jaw bone, facial disfiguration, and terrible pain. Infection of the decaying tissues could kill. Because these symptoms were similar to those that radium would eventually produce in the dialpainters, phosphorous poisoning would be suspected—not illogically—by workers and their families and physicians as a cause of the dialpainters' illnesses.[19]

English physicians recognized phossy jaw and its cause by the 1860s, and by the 1870s European nations began to outlaw the use of white phosphorous in match formulas. This process accelerated after the discovery in 1897 of an alternative formula for matches. In 1906 the International Association for Labor Legislation called a conference at which eight European countries agreed to prohibit the production or importation of white phosphorous matches. In 1909 the American Association for Labor Legislation (AALL) undertook an investigation of the American match industry. Because the federal Bureau of Labor had been studying women and children employed in match factories, the two investigations were dovetailed to avoid duplication of labor. Despite manufacturers' assertion that phosphorous poisoning was not present in American factories, over one hundred cases were quickly uncovered. The AALL published its findings in the *Bulletin of the Bureau of Labor*, and ultimately, in 1912, the U.S. Congress passed a prohibitive tax on white phosphorous matches, thus effectively bringing an end to their manufacture.[20]

Physicians and workers in the industrial area of northern New Jersey were surely aware of phosphorous poisoning, for several reasons. For one, there was the publicity attending the American investigation and legislation. For another, of the sixteen U.S. match factories, two were in New Jersey, close to Orange, one in Passaic and one in Garfield. Further, the 1912 law did not end the threat of white phosphorous to workers, for it was still used in the manufacture of fireworks through the mid-1920s. Several small fireworks firms lodged briefly in Newark and the surrounding area before World War I, and afterward one of the three biggest fireworks manufacturers utilizing white phosphorous located in nearby Berkeley Heights. Cases of phossy jaw in fireworks producers were uncovered in the early twenties, including cases in New Jersey.[21]

Katherine Schaub, her family, and her friends may have known about another industrial hazard in this period, mercury poisoning. The "great mercury-using industry" of the early twentieth century, according to a 1911 study by the Women's Welfare Department of the National Civic Federation, was felt hat manufacture, and "the heart of the industry" was in the Newark-Orange district. To make felt, furs were first treated with a mercury solution in a process called "caroting," and in that and all subsequent operations workers were subject to mercury poisoning from inhalation of dusts and fumes or from absorption through the skin.[22]

Early symptoms of chronic mercury exposure are vague: malaise, depression, headaches, nervousness, insomnia, loss of appetite, loose teeth and ulcerated gums, sometimes a slight anemia. With continued exposure, the skin becomes dry and wrinkled. The mouth further ulcerates, teeth fall out, and even the bone can become carious. Copious quantities of saliva may be produced, and this is one sure way to diagnose the disease. Another telltale sign is a blue line of mercury visible across the gums. Anemia becomes more certain with higher or longer exposure, and the digestive tract is affected, alternating between constipation and diarrhea. The most well-known symptom of mercury poisoning, the "hatter's shakes," constitutes a second stage of chronic mercury poisoning and indicates an even greater exposure. The shaking starts in the hands and spreads so that the body and then the speech become uncontrollable. Melancholia is common, as are other psychological symptoms such as loss of memory or hallucination. If exposure continues, death may result from hemorrhage through the mouth, nose, and intestines. In women, mercury poisoning may be accompanied by amenorrhea; if pregnant, women may miscarry, and children born live may be mentally and physically impaired.[23]

Mercury poisoning among hatters had been known in Orange since the nineteenth century. In 1860 an Orange physician published "Mercurial Disease among Hatters" in the *Transactions* of the Medical Society of New Jersey. Another Orange physician was the author of "Hatting as Affecting the Health of Operatives," which appeared in 1878 in the *Annual Report* of the New Jersey Board of Health. "Hatter's Consumption," published in the *New York Medical Journal* in 1886 by a third Orange medical man, highlighted the high death rate among hatters from that scourge of the working class, which in this case may have been tuberculosis but could also have been a lung ailment associated with breathing in fur.[24]

Mercury poisoning was still widespread among hatters in the early twentieth century. Between consumption and mercurialism, the health of hatters was so precarious that some categories of hatters were barred from purchasing life insurance. So insidious was mercury poisoning that some workers did not recognize their illness until a ten-month "Great Strike" in 1909 demonstrated how much their health improved with a vacation from mercury.[25]

Numerous such incidents of occupational poisoning among New Jersey's workers could be recounted, all to the same point: knowledge of industrial disease was widespread among these workers and among physicians in their communities. Workers have always faced occupational poisons and have long recognized diseases that could be simply correlated with causes, like lead poisoning and silicosis. Although incidence rates have not been well documented, industrial diseases were so common that nineteenth-century workers referred to them colloquially: in addition to "hatter's shakes" from mercury, there were, for instance, "painter's colic" from lead in paint, "brass founder's ague" from metal oxide fumes, and "miner's asthma" from rock dust. Students of the field deplore the impossibility of estimating the number of industrial diseases contracted by workers, now and in the past, but one estimate betrays the severity of the problem in the late nineteenth century. A study of western hard rock miners estimated that 20 percent of them suffered from silicosis and that deaths from silicosis, at one thousand a year, exceeded deaths from accidents in the mines by 50 percent. Adding in deaths from other industrial diseases common among miners, such as lead poisoning and intoxication from toxic gases, led to the conclusion that far more workers suffered and died from health problems than from safety hazards in the mines.[26]

As statistics improved near the turn of the century, workers became aware of their typically poorer health compared to the general population. In the early twentieth century "industrial health" was conceived as a field encompassing all causes of poor health among workers, including not just specific industrial diseases but workplaces that were too hot, too cold, too dusty, too stuffy—and work at too high an intensity producing general fatigue. A major trend in industrial health has been its narrowing from the study of all diseases of workers to just those caused by industrial poisons. Tuberculosis was considered a workers' disease and so an industrial health problem. Although often not attributable to a specific cause like silicosis or byssinosis ("brown lung" from textile dust), tuberculosis had high rates among workers and contributed to significantly higher death rates among workers when compared to other classes. In 1906 the American Federation of Labor (AFL) recognized the high death rates from tuberculosis in fifty-three occupations. The 1908 census mortality reports showed that, of deaths among workers between the ages of twenty-five and thirty-four, tuberculosis was responsible for 49 percent of printers and compositors, 40–45 percent of glassworkers, 57 percent of hatters, 54 percent of leather workers, 41 percent of marble and stone cutters, and 40 percent of textile operators. In a 1934 study disparate death rates were documented by class: when the death rate among professional males was set at 100, the comparative rate among skilled workers was 116; among semi-skilled workers, 141; and among unskilled workers, 187.[27]

Other available facts and figures can only suggest the severity of the problem. During hearings in 1914 U.S. senators were told that, of the 34 million workers in America, 3 million a year became sick from an industrial illness. By 1918 life insurance companies would no longer sell policies to asbestos workers because of their high rate of death, presumably from asbestosis and tuberculosis.[28]

The problem of industrial diseases likely worsened over time as new methods and new materials were introduced into factories. Silicosis, for instance, increased as American foundries adopted sandblasting and power tools to replace the manual methods of the nineteenth century. During and after World War I, American industry, jarred by shortages of chemicals formerly purchased from the advanced German chemical industry, undertook new manufacturing processes, and chemical hazards of which workers and industry had been long aware, like lead poisoning and mercury poisoning, were joined by new problems. For example, to create the explosives demanded by wars, Americans built plants to capture and distill solvents formerly purchased from Germany, solvents created as by-products in the manufacture of coke from coal. These "coal tar distillates," such as benzene and toluene, were the basis of explosives like trinitrotoluene (TNT). As executives in the American chemical industry planned for conversion to peacetime production and sales, they considered new uses for the products of their chemical plants. The solvents had a number of applications: in paints, shellacs, and lacquers; in the production of rubber; in rubber cement and other glues; and in cleaning, particularly in the removal of fats and greases. But many of these solvents are toxic, in two ways: acutely, in large doses with an immediate effect, and chronically, in small doses with an accumulative affect. To illustrate, benzene—used in the early-twentieth-century production of footwear, sealed cans, straw hats, artificial leather, gasoline, varnishes and lacquers, and in dry cleaning—in acute poisoning cases represses respiration and in chronic poisoning cases affects the bone marrow, leading to anemia and hemorrhages. As the "chemicalization of industry" advanced, that is, as uses were increasingly found for the products of America's expanding chemical industry, more and more workers were exposed to dangerous new compounds.[29]

New Jersey workers were, unfortunately, in the forefront of these developments. In 1933 a physician in the state suggested that "the area within a 25 mile radius of Newark is the most important area in the world from the standpoint of occupational and health hazards." His observation was based on the coincidence of three industrial poisoning episodes in New Jersey all occurring during the 1920s, all accompanied by widespread publicity and public alarm. The three he cited were poisoning from benzene, tetraethyl lead (a gasoline additive), and radium.[30]

Benzene poisoning was identified only a few years after benzene's introduction into industry in the late nineteenth century, although its chronic toxicity took longer to be noticed than its acute toxicity. Chronic toxicity was studied most vigorously during the war years by Alice Hamilton, a woman who will figure largely in the radium account. Dr. Hamilton was an early contributor to industrial toxicology in the United States, and she often succeeded in mitigating, if not eliminating, chemical threats to workers in various industries. In part, her power came from her ability to balance three roles, as a respected university professor and researcher, as a government-appointed investigator, and as a crusading reformer. Her benzene work is the least studied of her wide-ranging career to date. As a federal inspector charged with protecting workers' health in the war industries during World War I, Hamilton uncovered an epidemic of chronic benzene poisoning. Her essay, "The Growing Menace of Benzene (Benzol) Poisoning in American Industry," appeared in the *Journal of the American Medical Association* in 1922. A National Safety Council study followed in 1924, and Hamilton was mollified when "many manufactures voluntarily gave it [benzene] up," prompting her to publish "The Lessening Menace of Benzol Poisoning in American Industry" in 1928. Still, she worried that other manufacturers persisted in its use and so "it continued to be a serious danger." In New Jersey the state Labor Department undertook an investigation of benzene use in the mid-1920s. After consultations with manufacturers, some self-regulation resulted in a decrease in the number of reported poisoning cases. New York State conducted its own investigation in 1926.[31]

Tetraethyl lead was a gas additive developed by General Motors in 1922. General Motors shared directors with the DuPont Chemical Company, and both of those firms combined with Standard Oil of New Jersey to produce tetraethyl lead, with leaded gasoline sales beginning in 1923. A few scientists warned that the leaded gas was a threat not only to workers' health but also to the public health. Workers' health became a subject of controversy in 1924, when illness and death struck refinery workers at Standard Oil's Elizabeth facility in northern New Jersey. Neurological symptoms included tremors, palsies, and hallucinations, leading to death in five instances, with thirty-five other workers affected. Leaded gas became known as "looney gas" among New Jersey workers. This case unfolded alongside the radium one, with many of the same people and institutions involved in both. The public health threat of leaded gas returned to controversy in the 1970s.[32]

Many of these cases concerned women. Although risks increased for all workers in chemicalized twentieth-century industries, women may have suffered a disproportionate amount of chronic industrial disease. As jobs opened in new industries, women were considered especially suitable for the "lighter work" in shoe factories, rubber factories, and tin can factories. Their jobs often involved glue or rubber dissolved in benzene, and thus many of the earliest victims of chronic benzene poisoning were women. In a 1929 study of fifty-five factories employing women in Massachusetts, approximately 6,000 of the 10,000 female workers were exposed to chemicals such as benzene, methanol, and carbon tetrachloride in the manufacture of such products as shoes, rubber, and candy.[33]

Early twentieth-century experts on industrial poisoning concluded that women were more susceptible than men to toxic chemicals. This has never been statistically established, but many factors other than sex might lead women to suffer more from industrial poisoning than men. Alice Hamilton listed three reasons for higher rates of lead poisoning among women. First, few women were organized into unions and so were denied the opportunity to redress poor working conditions through collective pressure. Second, women generally were poorly paid, resulting in inferiorly housed and inadequately nourished workers who were more susceptible to any disease, including industrial diseases. (Modern scholars have also considered women's family obligations on top of their employment duties—women's "double day"—as productive of greater fatigue among women than men, increasing women's susceptibility to disease.) And third, women often had to support families on their low wages, and the worry and strain of this might have left them more susceptible to disease. Elsewhere, Hamilton identified another crucial factor explaining higher rates of industrial diseases among women, the younger age on average of women workers. Men of all ages worked in the early twentieth century, but women tended to leave paid employment outside the home upon marriage; growing teenagers may be more injured by toxic chemicals than adults. Two additional factors may be considered. Because women are generally smaller than men, smaller doses may be sufficient to poison them. Moreover, women's physical role in reproduction—pregnancy and lactation—increases their chances of injury from chemicals because of the stress on their bodies and because of risks like miscarriage.[34]

Finally, women may have been employed more than men in the new industries using chemical substances in ways that might lead to chronic poisoning. Returning to the example of benzene during the World War I era: acute poisoning, caused by one massive exposure, occurred among men working in heavy industry as distillers or tank cleaners or pipe fitters; chronic poisoning, caused by long-term exposure to small quantities, tended to occur among women using rubber cement. Other light jobs involving the use of poisonous chemicals that came to be identified as women's work, included dyeing, dry cleaning, match manufacturing, felt hat making, pottery manufacturing, and developing tobacco products.[35]

Few statistics on comparative rates of industrial poisoning between men and women are available, but the Women's Bureau of the U.S. Department of Labor, considering the matter in 1924, cited mortality data from the International Labor Office showing that the rate of death among working women in Germany and Austria was higher than that among working men in the age groups in which women had the greatest employment. Morbidity was higher for women, too. And studies in the United States of industrial policyholders—workers and their families—revealed that although men in most age groups had higher mortality rates than women, the opposite was true in the 15-to-24 age group, when women died in greater numbers, coinciding with the period in women's lives when they were most likely to be employed outside the home. Whether greater exposure or greater susceptibility was at fault, the Women's Bureau held industry responsible.[36]

How did workers respond to these threats to their health? Two contradictory perspectives shape historical consideration of working-class attitudes and actions in this regard. First, some historians expected to find more working-class anger (or other reaction), given the toll of industrial diseases on workers' health. Their disappointment arises from union inaction. William Graebner, in a study of coal mining unions, asserted that health and safety were a "decidedly peripheral concern" of the unions. Three historians who studied byssinosis concluded that before World War II health and safety were not major organizing issues for textile workers' unions. Angela Nugent noted that "organized labor rarely sustained a coordinated campaign . . . to address the hazards of their trades." Archival evidence supports this view. For instance, apathy is suggested in this citation from a Western Federation of Miners organizer, referring to silicosis: "This deadly plague is so prevalent among the miners of Butte that it is looked upon with stoical indifference, the majority taking it for granted that it is the inevitable doom of all who remain long enough and work in the mines." In another example, the president of the Pennsylvania Federation of Labor in 1927 concluded that "for the most part" unions had ignored industrial health.[37]

Several explanations have been offered for this alleged dearth of working-class activism. Lowered expectations about life and health may have undercut workers' anger; in other words, many workers may have fatalistically accepted poor health and short lives as their lot. The struggle just to house, clothe, and feed themselves and their families may have dulled their demands for more than a living wage; short-term economic issues may have superseded long-term health issues in working-class consciousness. If family welfare were more important to workers than individual welfare, then setting aside personal health considerations might have been necessary in order to assure income adequate to care for the entire family. Definitions of illness also may have been a factor. European mercury workers, for instance, seem to have defined illness as the inability to work. They sought to counteract the pernicious effect of mercury with home remedies and ignored professional medical advice to leave their jobs; they avoided physicians altogether until too sick to work or unless new, potentially successful treatments became available. Finally, male workers' identification with the "cult of manly bearing" may have led them to discount symptoms of industrial disease in order to avoid appearing weak, the explanation offered for New Jersey hatters who in the 1870s refused to admit that they suffered from mercury poisoning. As Craig Donegan wrote, "it was a sign of manly bearing and solidarity . . . for laborers to ignore regulations even if they were meant to make work safer." [38]

From a second perspective, safety and health issues were central to nineteenth-century working-class dissatisfaction and organization. As David Montgomery has argued, workers' control issues lay at the heart of labor strife up to the 1920s as craftworkers resisted the deskilling and division of labor of the factory. Nineteenth-century workers' fears about their health were linked to their anger about increasing loss of control over their work. Working too many hours, too fast, in cramped, stagnant positions, in fetid air were aspects of industrialization that were condemned on grounds of health as well as indignity. As labor activists and social reformers of the nineteenth century pondered the transformations of industrial capitalism, health was redefined as a political and economic issue. Craig Donegan has observed that as early as the 1830s workers and labor leaders identified insufficient income, long hours, and poor working conditions as sources of working-class ill health. Workers' health was identified as an inherent right of labor. "Health is the Capital of the Laboring Man," workers claimed, and, as property rights were deemed worthy of protection by the individual and the state, so laborers claimed the right to fight for their health. David Rosner and Gerald Markowitz have argued in a review of Montgomery's work that health and safety issues demand center stage in synthesizing a general history of workers in America. [39]

According to this viewpoint, much of the history of laboring people's politics can be reinterpreted as part of the history of occupational health and safety. An early stage of worker organization, the formation of mutual assistance and benevolent societies, can be seen as a response to the insecurities inherent in high death rates among workers, caused in part by their working conditions. A second labor response was to limit the hours workers had to spend in dangerous conditions and to argue for the ten-hour and eight-hour day for reasons of health. Third, restrictions on the work of women and children were often promoted as a means to protect the health of those believed most susceptible to fatigue and illness; protective legislation had a significant health and safety component. Fourth, at least some unions negotiated for health and safety protection in contracts (and I suspect that as historians explore these concerns further, we will find that many workers organized to win healthier workplace conditions). Miners and railroad workers, in particular, have long histories of massive and radical strikes over health and safety issues, accounting, perhaps, for their early success in winning safety and health legislation. A fifth strategy was for workers to rely on their employers' paternalism to keep the workplace safe, and corporate welfare programs of the 1910s and 1920s were successful only so long as workers felt such plans were in their best interest.[40]

Using government was a final strategy. Workers urged the government to investigate threats to their health, sought regulations to control those threats, and supported legislation for "employers' liability" and then "workmen's compensation" (called in this study "workers' compensation") to make employers liable for their ill health. But workers were not united around this last strategy. Many, including AFL leader Samuel Gompers, considered the government an arm of capitalist power and thought workers would do better to rely on their own power developed through unified action in the workplace. Still, Gompers was persuaded to support some labor legislation, cooperating, for instance, with the National Civic Federation to win over labor to the idea of workers' compensation.[41]

New Jersey is a good case study of labor's strategy to seek government investigation and regulation of industry, because of its early industrialization and because of its strong progressive reform movements. In New Jersey, by the 1920s, two state organizations ostensibly protected the health of workers, the Department of Labor and the Board of Health with its affiliated local Boards of Health. When the dialpainters' story is resumed, it will demonstrate how poorly these institutions served workers' health interests. Here, we will see how they took shape before the 1920s.

Stuart Galishoff has provided the best examination of New Jersey health reform during the Progressive Era, a case study of public health in one city, Newark. In the nineteenth century Boards of Health were the first government vehicles to which workers turned. The New Jersey Board of Health undertook a few industrial health evaluations in the late nineteenth century, most notably a study of hatting in 1878. Then, in the 1890s, with the successes of bacteriology and urban housekeeping, emphasis switched to communicable diseases and research on industrial diseases waned. The state board's neglect of working conditions suited those serving on Newark's Board of Health, business and professional people with no desire to confront business interests. But after 1915 the board was composed of civil service appointees and, according to Galishoff, it became "an organ of progressive public health work." In 1920 it established a Division of Industrial Hygiene. The dialpainters, however, would receive little aid from state or local Board of Health offices.[42]

New Jersey's Department of Labor, which will figure more prominently in the dialpainters' story, offered little more help and certainly more hindrance than did the health boards. The department's history begins in the 1870s, when worker demand led to the establishment of the first Bureaus of Labor Statistics, in New Jersey as elsewhere. The first such bureau was formed in Massachusetts in 1869; by 1892 twenty-nine states and the federal government had labor statistics bureaus, which were often directed by union leaders or supporters. New Jersey's Bureau of Statistics for Labor and Industry was created in 1878. This bureau was markedly responsive to labor concerns, including health and safety. Beginning in 1883 it published annual reports that, in addition to details of wages, hours, and conditions of labor, included accounts by workers of the effects of employment on their health and a series of reports on the relation of occupation to longevity.[43]

So strident were early labor statistics bureaus, often publishing inflammatory accounts of working-class health, that employers campaigned to free them from working-class control. By the twentieth century the bureaus were headed by professionals like Carroll Wright, of Massachusetts, who sought to "disconnect [the Massachusetts bureau] from politics" (meaning the politics of class strife) by focusing on objective fact-finding, especially through dry statistical analyses. Labor often supported the new commissioners, however, because they frequently continued to document working conditions and endorsed legislation to improve workers' health.[44]

In New Jersey, conservative reform aiming to replace worker-sympathetic labor commissioners with ostensibly objective professionals at first outflanked rather than directly attacked the Bureau of Statistics for Labor and Industry. In 1885 the state had established a Bureau of Factory and Workshop Inspection after the New Jersey Federation of Labor successfully lobbied for regulation of women's and children's labor and factory safety and hygiene, both to be enforced through factory inspection. The number of inspectors steadily increased from one to twelve between 1885 and 1902, but factory inspection, Galishoff declared, "failed abysmally." Labor and progressive politicos blamed the chief inspector, known for his hostility to labor organization. After 1892 inspections virtually came to a halt. In 1902 the Federation of Labor complained to the governor that child labor was still common in the state and blamed the chief inspector's inadequate enforcement of child labor laws. The governor welcomed the charges and financed an investigation by a labor representative, which produced photographs of children working in South Jersey's glass industry, Trenton's potteries, and Passaic's and Paterson's textile mills. The chief inspector was suspended for ignoring such abuses. Labor leaders next brought complaints about high tuberculosis rates in the dusty trades and about deaths from factory fires due to the lack of fire escapes. In 1904 the governor, with support from some organized labor groups and a women's progressive reform group, the Consumers' League, secured a General Factory Law, which regulated workers' ages, hours, safety and health conditions, and which fixed administration of the law within a new Department of Labor. Responsibility for factory inspection was moved into the department, and specialized inspectors worked out of a number of bureaus, including a Hygiene and Sanitation Bureau.[45]

On the surface this all sounds progressive, but in part the moves were orchestrated to take power away from the labor-supported and Democratic-Party-identified Bureau of Statistics for Labor and Industry. Annual labor reports were removed from the jurisdiction of the still extant Bureau of Statistics, their publication now supervised by the new Department of Labor. The reports became much shorter and, in at least one respect, less responsive to labor concerns: data on job conditions and workers' life expectancy were no longer published. The Department of Labor completed absorption of the Bureau of Statistics in 1915, when antireform Republicans controlled the legislature. A reorganization destroyed the Bureau of Statistics of Labor and Industry and stripped its Democratic chief of his authority, transferring the bureau's responsibilities to the Department of Labor, headed by a Republican commissioner. The measure was approved over the Democratic governor's veto.[46]

The reorganization met the requirements of New Jersey's Manufacturers' Association, which had lobbied for it. Much of the association's support derived from its allegiance to the man appointed as the first labor commissioner of the new Department of Labor in 1904, Lewis Bryant. Bryant was a lawyer, a civil engineer, and a close friend of the governor who first appointed him. How this fitted him to be labor commissioner is unclear, but Bryant would serve from 1904 until his death in 1923, through many governors' terms, always with the support of the Manufacturers' Association. In Galishoff's study of public health in Newark, he concluded that Lewis Bryant "embodied much of what was best in Progressivism": moral indignation, a pragmatic and scientific approach, and faith in men and in democracy. With Bryant "at the helm," Galishoff continued, New Jersey became "a leader in industrial hygiene." My reading is different. Over the years of Bryant's service, he symbolized assurance that the interests of the New Jersey Manufacturers' Association would receive great consideration in the state Labor Department. To make this case, I must review the history of the workers' compensation movement in New Jersey.[47]

New Jersey was a leader in the workers' compensation movement, in part because of manufacturers' support for it, in part because workers and reformers also favored it. The first step down the road to workers' compensation had been the passage of an employers' liability law in 1909 with wide labor support. Throughout most of the nineteenth century employers nationwide had been protected from injured employees' suits by three defenses: "contributory negligence," which implied that the employee had caused his or her own injury; "the fellow-servant doctrine," which suggested that a fellow employee was responsible; and "assumed risk," which meant that the employee had accepted dangerous conditions with his or her job. In all three cases the employer was not held liable for damages. Around 1894 the AFL and other unions began to support employer liability laws to overturn one or more of these common-law defenses. By 1907 labor had won employer liability laws in twenty-six states.[48]

If New Jersey workers lagged behind the nation with their 1909 employers' liability law, they were quick to move to the next stage, workers' compensation. Once businesses became liable for employees' injuries and illnesses, they sought means to contain those costs; thus they supported adoption of various insurance schemes to administer and limit the awards to workers. Employees also supported workers' compensation because it promised them awards more surely and quickly than might be possible in court, even though the awards would be far less than court-ordered restitution. In 1911 workers' compensation was passed in New Jersey—the first state to adopt it. In 1916 new legislation sponsored by the American Association for Labor Legislation began the transfer of contested compensation decisions from the courts to an administrative board within the Department of Labor; this transfer was complete by 1918. The move confirmed the AALL's preference for administrative law over legislative and juridical law; its social experts thought that decisions about workplace accidents and diseases were best left to other social experts on compensation administration boards. The Manufacturers' Association concurred. The transition to administrative law was problematic for workers, however. Workers' compensation was a business strategy to limit payments to injured workers: when workers started winning liability suits, their employers sought compensation programs, and when workers won compensation decisions from juries, employers pressed for a "neutral" arena less sympathetic to destitute workers. Physicians and other professionals seeking increased social authority on the basis of their objective professionalism stood ready to assume decision-making power. These administrators were less likely than juries to support workers' rights to compensation.[49]

Industrial diseases were not included in the early compensation laws.

Galishoff wrote that in 1912 the New Jersey labor commissioner began an "exhaustive inquiry" into worker health and safety, examining the effects of fifty poisons as well as dusts and fumes. Perhaps this was instigated by Alice Hamilton's studies that year of lead poisoning among potters in Trenton and Perth Amboy, in which she discovered that one of every seventeen or eighteen potters had lead poisoning. At any rate, 1912 found New Jersey passing an occupational disease–reporting law, requesting physicians to inform the Bureau of Hygiene and Sanitation of any poisonings from lead, phosphorous, arsenic, or mercury, and illnesses from anthrax (a common industrial disease in leather workers) or from compressed air ("the bends" that tunnel and bridge workers suffered from high-pressure ventilation). The law, Galishoff admitted, was "poorly enforced."[50]

In 1911 the labor commissioner had been authorized to compel ventilation installation to dispel dusts and fumes; in 1914 he gained the authority to administratively promulgate industrial health and safety codes. An example of how the longtime Republican labor commissioner, Lewis Bryant, established health and safety codes attests that the growing power of the Department of Labor did not necessarily undermine employers' control of working conditions. New Jersey, together with its neighbors Pennsylvania and New York, was home to "dry color manufacturing," which involved particularly dangerous lead compounds. In creating codes to monitor these substances, the labor commissioner and his staff consulted closely with the manufacturers themselves, both privately, soliciting individual critiques of proposed regulations, and publicly, sponsoring conferences to discuss revamping the codes. Businesses in New York and Pennsylvania were consulted as well, and similar measures were proposed to those state governments in order to avoid unfair competitive advantages from differing health regulations. No workers were included in these deliberations, because, the New Jersey Department of Labor claimed, the workers were not unionized and so no representatives were readily available, and because the workers were illiterate immigrants and could not represent themselves. The Labor Department shouldered responsibility for seeing to the welfare of the dry color workers. Thus codes were shaped largely according to the manufacturers' needs. In addition, the department considered the ability of each manufacturer to afford the stipulated health and safety codes. Again, reform was dictated by the industry.[51]

In 1923 Lewis Bryant died, giving Governor George Silzer the chance to appoint a new labor commissioner and thus influence future management of the Labor Department. Silzer was a Democrat, an old Woodrow Wilson supporter, who sought to restore Progressive values to the governor's office. Hoping, he claimed, "to put government back into the hands of the people," he presented himself as pro-labor and pro-reform. He backed protective legislation, including a Consumers' League bill barring night work for women. He strove to improve workers' compensation. He revived efforts to regulate industrial health. When state Progressives had initiated an Industrial Diseases bill in 1915, it had been buried because of manufacturers' resistance. Now, under Silzer's leadership, a commission to study industrial diseases was created. Consisting of representatives of labor, manufacturers, and state government, as well as a physician and a chemist, the commission proposed a law to make industrial diseases compensable. The measure passed in 1923 and became effective on January 1, 1924.[52]

The form in which industrial diseases were added to the compensation bill, however, was highly conservative. The only diseases to be covered were the nine specifically listed in an attached "schedule." A five-month statute of limitations in the original compensation act was left in effect for the diseases listed. Such a short period in which to file for compensation might have been reasonable for most accidents, where injury was usually immediately apparent, but it would bar compensation for the many industrial diseases that developed over time.[53]

One might have expected Silzer to name a labor commissioner who was strongly supportive of labor. Instead, he nominated Andrew McBride, who was confirmed by the Republican-controlled legislature. McBride had a background in public health: he had been a county physician for eleven years and had served as president of the Paterson Board of Health for five years. Both these appointments ended with his election as mayor of Paterson in 1908. McBride held office during the 1913 Paterson silk strike, an episode known, among other infamies, for the city's use of violence against the strikers. As mayor, McBride controlled the police who had attacked these workers. He had also attempted to interfere with the strikers' tactic—to gather sympathy for their position—of seeking temporary homes for their children elsewhere. McBride insisted that the city could take care of its own children.[54]

McBride's actions in Paterson made him an odd choice for a supposedly labor-sympathetic governor. Perhaps the governor was paying a political debt—McBride was a longtime member of the New Jersey Democratic Committee. Or perhaps Silzer wished to propose someone with a good chance of being confirmed by the Republican-controlled legislature. Whatever the reason for his nomination, McBride fit in with the evolving anti-labor, pro-business stance of the New Jersey Department of Labor. It was Andrew

McBride who, as commissioner of the Labor Department, would be responsible for the state's response to discoveries of illness among dialpainters.[55]

The Radium Luminous Materials Corporation was the largest dialpainting firm during the war but not the only one. Many small companies also existed. By the early 1920s dialpainting was concentrated in three areas of the country. The first was in Orange, New Jersey, where in 1921 the Radium Luminous Materials Corporation became the U.S. Radium Corporation, the victim of what we today would call a corporate takeover. Dr. von Sochocky and his partner were ousted from the company after the partner sold a large share of his stock to the corporate treasurer, Arthur Roeder, who became the new president. A second center developed in Waterbury, Connecticut. Around 1919 the Radium Luminous Materials Corporation began to curtail its dialpainting, eventually maintaining only a few dialpainters in Orange. The reasons for this are unknown to us. Public demand for luminous watches was high, so there was no postwar slump in business. Perhaps the radium company felt that it would generate sufficient income by selling its luminous paint powder to others. The Orange company helped its largest customers, including Waterbury's watch and clock manufacturers, set up their own dialpainting facilities. New Jersey dialpainters traveled to Waterbury and other locations to demonstrate the proper mixing of paint and correct application methods, including lippointing. A third dialpainting center developed in Illinois, home to U.S. Radium's biggest competitor, the Radium Dial Company, a branch of the Standard Chemical Company.[56]

Dialpainting waned in Orange as it waxed in Waterbury. In 1920 only a hundred women were employed in Orange and in 1921, only twenty-five. The dialpainter Katherine Schaub left U.S. Radium in 1920 for a clerical job at a roller bearing manufactory. She enjoyed this job as she had her U.S. Radium work: the other women were sociable, and after work she would play the piano while they sang and worked on articles for their hope chests. Schaub's cousin Irene Rudolph also left for a job as a clerk. In late 1920 Schaub returned to U.S. Radium at the company's request, working for about six months. In late 1921 she found another dialpainting position at the Luminite Company in Newark. This job lasted no longer than had the return to U.S. Radium, and eventually Schaub took a position as a parcel wrapper at Bamburger's Department Store, most likely in Newark.[57]

The two cousins, then, had worked as dialpainters for about three years each, leaving for other work in 1920 and 1921. In 1922 they were both twenty years old. That spring Irene Rudolph began to have trouble with her mouth.

A cheek swelled, and she had a tooth extracted. The socket never healed properly and became increasingly inflamed and sore. By July the swelling extended along her jaw, and Rudolph began a series of treatments: more teeth were removed, and incisions opened her gum to permit the drainage of pus and removal of what could now be seen as rotting bone. Her oral surgeon, Dr. Walter Barry, recalled the outcome of her first operation:

> I made an incision along the lower jaw and cleaned out the pus and infection with a curette, removed all the decayed and diseased bone until I got down to the good healthy area; then I packed the wound and sutured, partially closed, leaving it open sufficiently for drainage. . . . For the first few visits, up to within a week or ten days, the wound seemed to be getting along very nicely and seemed to be doing as well as it could possibly be expected to do and then after a period there was a recurrence of the same condition. . . . Then I opened the wound again and again cleaned it out thoroughly.

Her dentist, Dr. James Davidson, who was Barry's partner, had recently examined some women poisoned from working with phosphorous in a fireworks plant in nearby Berkeley Heights. Their symptoms were similar to Rudolph's: severe infections of the jaw bone accompanied by anemia. Davidson, Katherine Schaub, and her parents became convinced that Rudolph suffered from phossy jaw. By December Rudolph had been admitted to a local hospital for more treatment of the necrosis (dead tissue) and accompanying infection spreading through her mouth, and for the anemia that was beginning to debilitate her. She reported at the hospital that she knew of another dialpainter experiencing similar symptoms.[58]

This was probably Hazel Vincent, a dialpainter who had worked with Rudolph and Schaub from 1917 until 1920. She left her job in 1920 because of her dental troubles. In 1920 and 1921 Vincent was seen by a doctor, who suspected phossy jaw from the black discharge with its "garlic odor" exuding from her nose and mouth and because concurrently she was anemic. Vincent also consulted Dr. Davidson, Rudolph's dentist.[59]

Or perhaps the case Rudolph referred to was that of Amelia Maggia. Maggia was twenty-one when she started dialpainting, one of at least five of the seven Maggia sisters who worked at the Orange studio. Amelia was known as an unusually productive worker. In 1921 she began to have tooth trouble: a toothache led to an extraction that never healed. She quit her job when her illness worsened in early 1922; within several months her lower jawbone and the surrounding tissue had so deteriorated that her dentist lifted her entire mandible out of her mouth. She died in September 1922. Her death certificate listed the cause of death as "ulcerative stomatitis" — inflammation and tissue destruction of the mouth — with syphilis given as a contributory cause. Maggia's immigrant background may have inspired the assumption of venereal disease; however, later autopsies on her exhumed remains found evidence of radium deposition in her bones and no evidence of syphilis. Hers is the earliest known suspicious death of a dialpainter; as a quick worker she may have been more exposed to the luminous paint than other, slower colleagues.[60]

Rudolph's doctor reported Irene's case to Newark health officials in December 1922, the month she was first hospitalized. The local experts turned to the state Department of Labor, which organized an inspection of the factory and an analysis of the paint. Lillian Erskine, chief of the Bureau of Industrial Statistics in the Labor Department, whose office was in the Industrial Safety Museum in Jersey City, visited the nearby Orange plant twice but could find nothing that conflicted with state factory laws. She recommended further study.[61]

As part of this first investigation by New Jersey's Health and Labor Departments, a consulting chemist, Dr. M. Szamatolski, was asked to analyze the paint used by the dialpainters. He found no phosphorous in the paint: the luminous paint glowed because of the action of radioactive particles emanating from radium or a radium isotope. Szamatolski was the first person to suggest that radium might be behind the illnesses of the dialpainters. He stated: "As you know, radium has a very violent action on the skin and it is my belief that the serious condition of the jaw has been caused by the influence of radium. I would suggest that every operator be warned . . . of the dangers of getting this material on the skin or into the system, especially into the mouth." Nevertheless, after reviewing his report and that of Erskine and a Newark Department of Public Health physician discussing Rudolph's health and Maggia's death, the state Labor Department took no action.[62]

Irene Rudolph died in July 1923 following what Katherine Schaub called "a most terrible and mysterious illness" of a year and a half. That week Schaub personally visited Department of Health and Department of Labor offices in Newark to file complaints about the cause of her cousin's death, which was listed as phosphorous poisoning on her death certificate. Schaub informed officers in Newark of two similar cases among former employees: one had already died and the other was failing. These must have been Amelia Maggia and Hazel Vincent (now Hazel Vincent Kuser since her marriage). The state departments asked Leonore Young, a health officer in Orange to investigate.

She looked up the death records and referred the matter back to the Labor Department. Again, no action was taken.[63]

In January 1924 the Department of Labor contacted Leonore Young about a *third* complaint about former employees of the dialpainting studio, this time concerning Hazel Vincent Kuser. By now, Kuser's condition had deteriorated to the point where parts of her upper jaw had to be removed and her anemia required blood transfusions. Her father had spent all of his retirement savings on her treatments, and her husband had mortgaged their home.[64]

Health officer Young investigated and reported in February on five suspicious cases: three women already deceased and two ill women. In addition to the deaths of Maggia and Rudolph, Young uncovered a third death. Helen Quinlan had started working for the radium company in Orange in 1918, when she was eighteen, and left several years later. In the fall of 1921 she began to suffer from a sore throat and a swollen face. Soon her teeth and gums also bothered her. She was treated for her aching teeth and ulcerated gums but continued to fail until her death on June 3, 1923. At the time, the cause of death was determined to be "Anemia, with contributory: Vincent's angina ["trench mouth"] and secondary: Pneumonia."[65]

The ill women in Young's report were Kuser and a new case, Marguerite Carlough. Carlough had begun treatment in January with Rudolph's oral surgeon, Dr. Walter Barry. Carlough had started dialpainting at the age of eighteen and had worked for over four years, until her illness forced her to quit during 1923. Fatigue and weakness had begun in 1921; in October 1923 her face swelled, and in December she had two molars extracted. A piece of decayed jawbone came out with the extractions, suggesting necrosis of the jaw.[66]

Barry, who was treating other dialpainters as well, was becoming convinced that an industrial disease was at the root of their problems. One patient, not discussed by health officer Young, was Irene Rudolph's cousin, Katherine Schaub. Schaub later wrote:

> It was in November 1923 I began to have trouble with my teeth. My dentist [probably Davidson] advised that I have two teeth extracted by the same dentist [oral surgeon Barry] who had treated Irene. After the teeth were taken out, I continued to go to the dentist for treatment, but failed to get any relief from pain. I kept thinking about Irene, and about the trouble she had with her jaw. I felt that there was some relationship between Irene's case and mine; I cannot explain just why, but I did.

At the time that he removed Schaub's teeth, Barry noted: "Bone structure does not look to be of good consistency. Curetted out thoroughly. Patient has been employed in Radium works in Orange[,] same place as Miss Rudolph . . . who died of necrosis of jaw and septicemia [the spread of a localized infection into the blood stream; blood poisoning]." Barry was also treating Genevieve Smith, Marguerite Carlough's best friend, who had been employed as a dialpainter since 1917. Smith had two sisters working at the Orange studio, Josephine and Anna; Genevieve and Josephine consulted with Barry, who told them to quit their jobs or he would no longer continue their dental treatments. Genevieve did leave.[67]

These informal networks among friends, families, and physicians coalesced into at least one and maybe two formal meetings, at the offices of Drs. Barry and Davidson, in which the dialpainters agreed that an industrial disease stalked them. Altogether, Barry and Davidson had treated at least six dialpainters: Barry attended Irene Rudolph, Marguerite Carlough, Katherine Schaub, and Genevieve and Josephine Smith; in addition, Davidson treated Hazel Vincent Kuser. The doctors' and dialpainters' growing fears led to a January 19, 1924, meeting of Barry with some of the affected dialpainters; a second meeting may have been held as well. Although no one was sure of the exact cause of their illnesses — the idea of phosphorous poisoning still clung despite the absence of phosphorous in the luminous paint's formula — this group talked of dialpainting as the source of the problems.[68]

Meanwhile, Leonore Young discovered that, in December 1923, the Office of Industrial Hygiene and Sanitation in the U.S. Public Health Service had published a report, "Preliminary Note on Observations Made on Physical Condition of Persons Engaged in Measuring Radium Preparations," based on an eighteen-month study of persons employed at the U.S. Bureau of Standards. The bureau measured and certified practically all the radium sold in the United States. The study was prompted by new medical evidence that radioactivity affected the blood. Briefly, the report reviewed medical findings of radium's harmful effects, including tissue damage and anemia. It then stated that "no serious defects" had been found among Bureau of Standards personnel even though two cases of skin erosion and one case of anemia were noted among the nine technicians studied. Finally, the report suggested safety precautions to be undertaken by those handling radium. Young suggested that the New Jersey Department of Labor contact the U.S. Public Health Service for more information.[69]

After notifying the New Jersey Labor Department of her findings, Young bowed out of any further role in the dialpainting investigations. Later she offered several explanations for this. First, she claimed, "I did not have the means and time to investigate." Second, "the field was not one of general public health but one involving workers at a particular plant." Furthermore, she said, because she had to investigate complaints about fumes emitted by the Orange plant and the city of Orange was involved in another suit against the corporation, "there would be no cooperation should she turn to the company" for information.[70]

The dialpainters' employers were aware of the suspicions of local dentists as well as their own employees' fears. On January 10, 1924, a vice president of the U.S. Radium Corporation wrote to its insurance company: "There recently have been rumors and comments made by individuals, particularly dentists, in which they claim work in our application department is hazardous and has caused injury and poor health to a former operator of ours and they are advising that other of our operators should discontinue being in our employ. We do not recognize that there is any such hazard in the occupation." Although the company did not believe the accusations of occupational poisoning, it did begin investigating them. U.S. Radium asked its insurance company to study the matter, but there is no evidence that the insurer ever complied with the request. A medical firm specializing in service to industry, the Life Extension Institute, was hired to examine a half dozen current employees; in March it reported no evidence of occupational poisoning, although one of the women examined would die a year later from radium poisoning. Shortly afterward corporation officials interviewed Drs. Barry and Davidson. Both dentists voiced their suspicions of phosphorous poisoning and recommended that the dialpainting procedures be discontinued.[71]

Their advice was not followed: production of luminous dials continued in Orange. The Labor Department took no steps, either. In replying to Leonore Young's report, the department stated that it did not have enough evidence to contact the Public Health Service. Nor did it pursue further study.[72]

By early 1924 the dialpainters were stymied in efforts to prove the occupational origins of their illnesses to those with power over their working conditions and medical compensation. Although a Labor Department chemist had made a connection between the dialpainters' symptoms and their exposure to radium, no one had as yet communicated the idea of radium as culprit to the women involved. U.S. Radium, for reasons discussed in the next chapter, rejected the idea that dialpainting was inimical to health. The state Departments of Health and Labor, although staffed with sympathetic investigators, were by political design subservient to the interests of business in New Jersey; they failed in three attempts to substantiate the dialpainters' conclusions that the deaths and illnesses among them were job related.

THE UNKNOWN GOD

Radium, Research, and Businesses

Post–"bomb" generations need no convincing that radioactivity is dangerous, but a different set of assumptions prevailed at the beginning of the century, when radioactive substances were hailed as therapeutic agents. Although radioactive emissions are still used to destroy unwanted tissues, a far different application was popular in the 1910s and 1920s: radium was employed as an internal medicine. It was promoted as an up-to-date panacea by enthusiastic radium entrepreneurs, including many still active in radium corporations when luminous paint became suspected of doing harm.

Scientific progress is not sufficient to explain the eventual change in the status accorded internally deposited radium. The qualities of radium cited as harmful by the mid–1920s were known by 1917, when the Orange dial-painters began their work, but qualities that scientists deemed dangerous in 1925, radium promoters had considered therapeutic a decade earlier. The resistance on the part of the U.S. Radium Corporation and other businesses to suggestions that radium was causing the dialpainters' illnesses was linked to this promotional literature. Except for the Waterbury clock and watch companies, dialpainting was undertaken by firms also producing radium for medicinal uses. These firms and others in the radium medicine business were the originators of most of the research data indicating that radium was not just safe but beneficial to imbibe.

Americans concerned about their health in the early twentieth century faced a confusing medical world. Life expectancy had increased dramatically in the late nineteenth century, a trend that would continue over the first half of the twentieth. Improved public health rather than new treatments for disease was primarily responsible. Although the advances of scientific medicine had improved diagnosis of diseases, few new ways of curing them had yet been discovered. The twentieth-century citizen could now be optimistic about a longer life but must rationally adopt a medically sound pessimism about curing life-threatening diseases. In such contradictory circumstances, great hope was attached to the discovery of any new medical treatment, and new treatments were sought with great fervor. One outcome of this climate was an astounding increase in the sale of patent medicines. Thus, rising expectations and concomitant frustrations with modern medicine affected interpretation of radium's properties.[1]

The years around the turn of the century also brought fundamental changes in the physical and chemical understanding of the world. The first step was the discovery of the X ray in 1895. X rays can penetrate some opaque matter like light penetrates glass or water, and they were immediately put to work to explore the interior of the human body. Their destructive effects were also quickly noted. Patients, their physicians, and other X-ray experimenters suffered skin destruction varing from inflammation to ulceration to complete destruction of the skin resembling serious burns. Hair fell out; the eyes burned and blurred. The first death from X-ray–induced cancer was in 1904. Such devastation, as frightening as it was, suggested a use for the technology beyond diagnostic applications, and X rays were employed as a medical treatment to destroy unwanted tissues from birthmarks to cancers. Lead shielding and standardization of doses made these devices safer over time, although absolute safety has never been achieved.[2]

All of the properties of X rays are explicable through an understanding of their physical nature. X rays belong at the high-energy end of the radiation spectrum, which begins at the low-energy end with radio waves and proceeds upward through microwaves, infra-red waves, light waves, and ultra-violet waves to X rays. X rays with other radiation perch on the boundary between energy and matter, with properties similar to waves and properties inherent in particulate matter. Radiation may be thought of as packets of energy, called photons, and X rays can be considered particularly high-energy photons, or photons in fast motion. X rays' destructive power is relative to their energy or speed. When an X ray collides with an atom, it breaks the atom into pieces of uneven charge. X rays are therefore a form of "ionizing radiation" in that they create charged particles, called ions. Charged particles are unstable and reactive; in living tissues, they can disrupt the chemical and physical bonds necessary to the healthy functioning of cells. An X ray may pass through matter without crashing into an atom, as atoms leave much empty space between them (X rays might pass through atoms, which also contain much empty space); in this case, the X ray's journey will cause no damage. Denser materials will, of course, stop more X rays, casting shadows that allow X-ray photographs of bones and such to be made.[3]

X rays do exist in nature, but in the early part of the century they were characterized as artificially created radiation, caused by shooting fast-moving electrons at a metal target. Radioactivity, on the other hand, was discovered as ionizing radiation emitted naturally by radioactive elements. Uranium, a known element, was found to be radioactive by Henri Becquerel in 1896. Two years later Marie Curie detected radioactivity in a uranium sample in excess of what could be explained by the presence of the uranium. She suspected the presence of a small amount of an unknown, strongly radioactive element, leading her and her husband, Pierre Curie, to identify two new radioactive elements, polonium and radium. The Curies first isolated a radium sample in 1902.[4]

Isolated radium had astounding properties. It was spontaneously luminous—it glowed in the dark! It continuously radiated heat. Both properties seemed to violate the principle of the conservation of energy, considered at the time a fundamental law of nature. Radium was also quickly found to spontaneously emit helium, a gas. This was the first evidence of the transmutation of elements, or one element turning into another, a goal of scientists since medieval alchemists yearned to turn lead into gold.

These properties are a function of the nature of radioactivity. Radioactivity is found in many of the largest atoms. Their size makes them unstable, rather as a house of cards becomes less stable the bigger it gets. Radioactive elements fall apart in a series of steps until a stable atom of lead is formed. Each step is initiated by the spitting-out of a subatomic particle accompanied by a release of energy as heat or light or other radiation. At each step the atom rests as a new element, smaller than the first. The atom may rest at each step for less than a second or for more than centuries, depending on the imbalances within each particular element.

Atoms are themselves bundles of smaller particles, some with positive charge, some with negative charge, and some with no charge at all; an atom is usually balanced with equal numbers of particles of opposite charges and thus overall is neutrally charged. The nucleus of an atom is made up of positively charged protons and neutral neutrons and is surrounded in most atoms by a cloud of much smaller negatively charged electrons. In an uncharged or neutral atom the number of protons and electrons are the same, thus canceling out each others' charges.

Two different particles may be emitted in the radioactive transformation of one element into another. Alpha particles are positively charged particles, with the structure of the nucleus of the helium atom. Alpha particles are comprised of protons and neutrons, two of each, with a positive charge of two. The alpha particle's large size (relative to other subatomic particles) means that a short time after penetrating into matter it is likely to bump into another particle. Also, its charge means that the alpha particle can disrupt matter by proximity alone, by stealing negatively charged electrons. Although it does not travel far, the alpha particle's size and charge make it very efficient at disrupting other atoms. Adding two electrons to an alpha particle yields ordinary helium.

Beta particles are nothing more than electrons, although extremely high-speed electrons, and their source is not the electron cloud orbiting most atoms but the nucleus: in an unstable atom a neutron may be transformed into a proton, with the emission of a high-energy newly created electron as well. Beta particles, too, have mass and tend to bump into other atoms rather quickly, although not as quickly as the larger alpha particles, and beta particles, with a negative charge, can also disrupt atoms from proximity alone.

Gamma radiation is indistinguishable from X rays, albeit very high-energy X rays, except for its source in radioactive transformations. Gamma radiation can penetrate far into various materials, including human tissue. But alpha and beta radiation, if their source is close at hand, can do much more damage because of their size and their charge.[5]

Soon after its discovery, radium was found to be dangerous in ways similar to X rays. The ability of radium rays to burn skin was first observed in

1901, when Henri Becquerel, who identified uranium's radioactivity, carried a radium sample in his vest pocket for several hours and later suffered a severe burn on his abdomen. Pierre Curie, together with Marie Curie the discoverer of radium, then intentionally produced a burn on his arm. As X rays were being used to burn off skin diseases and cancers, the Curies loaned some radium to a Parisian hospital for experiments. In his 1903 Nobel address Henri Becquerel mentioned that radium was being tried to treat skin lesions caused by lupus and by cancer.[6]

The amazing properties of X rays and radium soon led to speculation about their medicinal qualities, and by 1903 radiation was being tested in the treatment of just about any disease. First X rays and then radium became medical fads, celebrated as modern panaceas. An editorial in the *Boston Medical and Surgical Journal* in 1903 illustrates the high expectations for radium: "In view of the extraordinary impetus which has been given to the therapeutic action of various little understood rays and emanations by the introduction of the x-ray, we have no right to assume any other attitude [toward radium] than one of expectant anticipation."[7]

Two kinds of medical uses developed for radium. Like X-ray radiation, radium's radiation could destroy tissue. Physicians employed radium and X rays to attack serious medical complaints like cancer, more benign medical problems like warts, and cosmetically unattractive features like birthmarks, moles, and even unwanted body hair. Radium's second medical use was as an internal medicine, as the rest of this chapter outlines.

Early experimenters believed that radium did not just burn away tissue; they suggested that it selectively destroyed morbid cells and even more amazingly promoted the growth of healthy cells. This misinformation has been credited to Marie Curie herself. Another early claim—that radium was a bactericide—led the prominent physicist Frederick Soddy in 1903 to suggest the inhalation of radioactive gas as a treatment for tuberculosis. His was perhaps the first suggestion that radioactivity might have application in internal medicine.[8]

Medical researchers quickly began to use radium in a host of experiments, but a confusion early arose over how to manufacture "radium water"—a water solution made radioactive by exposure to radium—for medicinal purposes. Radium water might be manufactured by dissolving radium salts in water, but radium was too expensive and rare at first to use up in experiments. Radon was employed most often instead. Radon is a "daughter" of radium, a product of radium's series of transmutations into other elements; radium constantly emits radon in steady quantities. In early years called "radium emanation," radon is a gas, itself radioactive. It can be dissolved in water, thus rendering a solution radioactive. An insoluble radium compound can be brought into contact with water, and a radioactive solution results through the dissolution of radon, while the radium is seemingly unchanged. Or, radon can be drawn off from a radium source and bubbled through water to produce the radio-active solution without the radium ever contacting the water. Early in this century some experimenters mistakenly thought that water molecules could become radioactive through exposure to radium, and at least a few of them produced allegedly radioactive "radium water" by suspending a sealed tube of radium in water; this, of course, could not work because the radiation re-leased by radium cannot make water molecules radioactive and because the radon that might produce a radioactive solution was sealed intact with the radium. Much of the earliest literature on the effects of radioactivity, then, is suspect because just what form of radium water was employed is unclear.[9]

Increasing the difficulties in reconstructing these earliest experiments, re-searchers also used different forms of "radium" in their medicines, which varied in their effects. An element is determined by the number of protons in its nucleus; each element has a specific number of protons. But the number of neutrons can vary, with the various forms of an element called "isotopes." Radium, for instance, with 88 protons, can have 135, 136, 138, or 140 neu-trons (these are the common forms of radium; altogether, twenty-five isotopes of radium have been recognized). Isotopes are identified by the combined number of protons and neutrons; the common forms of radium are radium-223, radium-224, radium-226, and radium-228. Isotopes vary in their stability. Earlier I compared a radioactive element to a house of cards, less stable the bigger they get. The analogy was less than precise, for although of all the elements the bigger atoms tend to be the radioactive ones, within radioac-tive elements stability is not dependent on size. For instance, radium-226 is the longest lived isotope of radium; radium-223, the shortest one. "Stability" among radioactive elements is measured in "half-lives," that is, the length of time it takes for one-half of any given amount of a radioactive substance to transform itself one step or more down the radioactive series. (Note the pecu-liarities of half-lives: a substance loses one-half of its atoms during the first period of the half-life, one-half of what remains during the second period, another one-half of that remainder during a third period, and so on, so that the substance remaining at the end of each half-life is $\frac{1}{2}$, $\frac{1}{4}$, $\frac{1}{8}$, $\frac{1}{16}$, and so on. Theoretically, the original substance never completely disappears. It makes no difference how much material one starts with, a ton or a milligram, one-half of each will disappear over one half-life of the substance, one-half ton in the first case and one-half milligram in the second.) Radium-226 has a half-life of 1,620 years, a very long time. Radium-228 has a half-life of 5.7 years; radium-223, 11.7 days; and radium-224, 3.6 days. Usually, when "radium" is written, it refers to radium-226, the longest-lived isotope. But early researchers were not always precise, and the composition of their medicines varied.

Radium water, of whatever manufacture, was quickly reported to be germicidal and "antifermentive," and its use was proposed for many diseases including diphtheria, typhoid, and malaria; liver diseases, diabetes, and Bright's disease (kidney malfunction). In its role as a promoter of healthy cells, radium water was shown to stimulate the blossoming of lilac buds and accelerate the growth of tadpoles. It was even suggested as a means to promote the function of the reproductive organs.[10]

The discovery in 1903 that many health spas' naturally occurring mineral waters were radioactive may have first suggested, or at least promoted, the health benefits of radioactivity. Health spas were popular in the early twentieth century. Germany, the leading nation—with its neighbor Austria—in the transformation of medicine by scientific method, was the home of the most popular spas and of much of the initial research on the internal use of radioactivity. The springs at Gastein were the first to be studied. Eventually, the radioactivity of spa waters was attributed to radon, released into mineral water from the rocks through which water percolated on its way to the surface, although in a few cases radium itself is present in radioactive mineral waters. In 1904 a Gastein researcher placed uranium pitchblend residues in a bathtub for fourteen hours (some of the radon given off by radium in the pitchblend would have dissolved in the water), thus preparing "artificial Gastein waters," perhaps the first artificially prepared radium water. He reported favorable results following its use to treat rheumatism, arthritis, and neuralgia, some of the mainstays of the spa trade. Further studies were conducted at the hydrotherapeutic clinic of the University of Berlin, in a Berlin medical clinic, and in a publicly supported Viennese clinic. Waters were either drunk or bathed in, but new research suggested other applications. In 1906 scientists announced that workers in the uranium mine near the spa at St. Joachimsthal suffered amazingly low rates of arthritis and rheumatism because of their inhalation of radon gas. Folk medicine in the area included the use of radioactive pads or compresses. Physicians began to experiment with radon "emanatoria"—apparatus or rooms that allowed the inhalation of radon fumes—and with the use of radioactive pads, creams, and implements for pain relief. Again, German medicine led the way. Thus was born the "Berlin school" of radium therapy.[11]

In the early years there was also a "French school" of radium medicine.

The Laboratoire Biologique du Radium was established in 1906 with support from an industrialist who owned a radium factory. He also founded the first journal to disseminate information on the medical use of radium: *Le Radium* was published from 1904 to 1907. Like researchers in other radium institutes organized in the decade following radium's discovery — in London and Manchester, Vienna and Trieste, Stockholm, and New York and Baltimore — French scientists focused on the use of radium's rays in treating skin lesions and cancers, but they dabbled in the study of the internal use of radium solutions. Most important, they introduced a new method of radium therapy: the injection of solutions containing radium salts or radon.[12]

By the 1910s radium had evoked some interest among American medical experts, both in its ability to destroy unwanted tissue and in its potential as an internal medicine. As for the latter, American physicians often made pilgrimages to Germany during the early twentieth century to imbibe the spirit of scientific medicine there. Others kept in touch with German research through abstracts of studies published in U.S. medical journals. As early as 1905 at least one American physician was convinced of the value of radioactivity, writing, "Leading physiochemists are now quite ready to admit that the hitherto unaccountable properties of certain of the more famous spring waters of the world . . . depend . . . on their infinitesimal radium content or influence." An English translation of the French school's treatise, Louis F. Wickham and Paul Degrais's *Radium Therapy*, became available in 1910. An added English introduction admitted that "the notion of radioactivity has inflamed some minds to a degree far beyond anything for which the facts so far ascertained afford justification" but praised "the judicial temper" of Dr. Wickham, promoter of the injection of radium solutions for infections and physiological malfunctions.[13]

American practitioners showed little interest in radium as an internal medicine until after 1910, but by 1913 and 1914 radium medicines had been legitimated through research and publication. Research and publication most commonly emanated from three centers: the radium industry, the spa industry, and physicians in private practice who utilized radium. None of these may be considered completely objective sources, as all three hoped to use their findings to promote the sale of radium products.

In 1911 Joseph Flannery, president of the Standard Chemical Company of Pittsburgh, became interested in radium for two reasons: first, his sister was suffering from cancer, and there was not enough radium in America to attempt the new radium treatment for cancer; and second, the ore utilized in his vanadium refining business also contained radium that might be profitably extracted. But when the company began research to develop a process to extract radium from the low concentration ores available in the United States, it also faced the necessity of developing a domestic market for radium. Thus far, the country's best grade ores had been shipped to Europe for extraction, and domestic demand was so low that less than half a gram returned to the United States between 1911 and 1913. Part of the reason may have been the expense, the very thing that made radium an attractive industrial product. A price per gram as high as $180,000 was reported during the years 1912 through 1914. When Standard Chemical started producing radium in 1913, it tried to establish $120,000 a gram as the normal price. Between 1914 and 1921 actual prices varied from $89,000 to $125,000 a gram.[14]

To stimulate the U.S. market, Flannery borrowed a few tricks from the French radium producers. First, he opened a medical clinic and laboratory in 1913, touted as the "first free radium clinic in America." William Cameron, a physician, was the director of the medical clinic. He was assisted by the physicist Charles Viol, who had been with Standard Chemical since 1912 and became its director of research in 1913. Patients at the clinic were promised no less than thirty hours in the emanatorium and thirty liters of radioactive water to drink.[15]

The Standard Chemical Company clinic pioneered in the use of radium in internal medicine in the United States. A few isolated American physicians had experimented previously with small amounts of radium, but Cameron, Viol, and their staff used relatively large amounts systematically on a large number of patients and began to publicize their results immediately. In 1913 Cameron published three articles detailing the use of radon and radium in the treatment of arthritis, gout, neuritis, diabetes, neurasthenia, and pyorrhea. He also reported that an intravenous injection of radium chloride—"which procedure, I believe, originated at the clinic at Pittsburgh"—led to a decrease in blood pressure. Another prodigious researcher and writer was Frederick Proescher, who worked in the medical clinic's Pathological Research Laboratory. In 1913 and 1914 he published four articles on the intravenous injection of soluble radium salts, finding this treatment especially efficacious for high blood pressure, pernicious anemia, and leukemia. One can trace in Proescher's research the use over time of larger and larger amounts of radium. In his first article, he gave "a brief resume of my experience with intravenous injections of larger doses of radium bromide for therapeutic purposes." These had not been tried before, he claimed, but were probably safe: subcutaneous injections of 10 to 20 micrograms (a microgram is $\frac{1}{1000}$th of a milligram, which is $\frac{1}{1000}$th of a gram) of soluble radium salt had shown no harmful effects. Various amounts of radium bromide were tried intravenously—from less than 100 micrograms to 500 micrograms (half a milligram) of radium element—in cases of pernicious anemia, uterine cancer, arthritis, and neuritis, four cases in all. "No alarming symptoms were observed," Proescher assured his readers. Effects included a reduction in blood pressure, a marked increase in red blood cells with the smaller amounts of radium and a decrease with the largest amount, a reduction in leukocytes (in one case), and pain relief. Proescher cautioned that the "marked diminution of the red blood cells in the case . . . [where the most radium was used], seems to indicate that higher doses may have an injurious effect." In his next article, however, he declared that the use of up to one milligram (1,000 micrograms) of radium element—twice the maximum dose he cautioned as dangerous before—was "perfectly harmless." During the course of his employment at Standard Chemical from 1913 to 1917, Proescher estimated that he injected radium into 1,500–2,000 patients. He, Cameron, and Viol consistently wrote that the use of radium was safe, that it involved no hazards.[16]

Joseph Flannery of the Standard Chemical Company borrowed another French idea to promote radium: in 1913 he began publishing a journal, *Radium*, which contained most of his employees' research. Physicians received a free copy. Standard Chemical reached out to physicians in other ways, too. At the 1913 meeting of the Medical Society of Pennsylvania, for instance, Cameron displayed radium salts, both as crystals and in solution for injection, ingestion, and bathing; he also gave a paper recounting the work in his company's experimental clinic. In addition, Cameron and Viol labored to establish a medical society for specialists working with radium; their efforts culminated in the formation in 1916 of the American Radium Society to promote the scientific study of radium therapeutics and the maintenance of high ethical standards. The new association invited only physicians to join but made an exception for the physicist Viol.[17]

If the radium industry tended to boost the medical use of radium, so did the spa industry, not just in Germany but in England and the United States as well. Spas were popular—in America over half a million people took a spa treatment annually—but as scientific medicine waxed, physicians at spas became increasingly uncomfortable. How could they explain why their waters worked? And more important, why did the waters work only when freshly drawn at the source itself? Waters that were bottled, shipped, and sold elsewhere cut into the spa's income. Radon was the perfect answer. Because it was a gas, it would escape into air from water transported uncapped or bottled with an airspace. Because radon's half-life was about three-and-a-half days—

versus radium-226's half-life of over a thousand years—water transported or stored for more than a week had already lost three-quarters of its radon content. Finally, radon allowed a scientific explanation for spas' effectiveness and permitted spa physicians to join the mainstream of their profession. In 1912 a doctor from the spa at Mount Clemens, Michigan, stated "that much of the humbuggery of the medical profession still lingers about mineral springs. It is indeed a humiliating acknowledgement." But, because of the discovery of radioactivity in spa waters, he predicted a "brilliant future" for "balneology"— the science of baths. "Radioactivity [is] the curative power of natural mineral waters." He was echoed, perhaps more effusively, by an English physician from Bath, who wrote:

> The discovery of radium . . . has altered our views fundamentally with respect to balneological treatment. A more scientific era has dawned. Time was when a suspicion of a man's professional integrity was raised if he called himself a balneologist; but now, instead of groping in the shades of night, our landscape is illumined by the sunshine of science and we march along with firmer footsteps and with head erect. . . . Is it, then, any wonder that we balneologists regard radio-activity as the unknown god[?].[18]

What was the radium or radon in spa waters good for? In 1913, in the British journal *Lancet*, Dr. W. Engelmann reported on research with radium emanation by physicians associated with the Kreuznach medicinal springs, also the site of Germany's only radium factory. He noted that radon had an "actuating influence" on certain "body ferments," that it increased the "coagulability" of the blood, retarded the growth of certain bacteria, lowered blood pressure, and dissolved uric acid deposits in gout. Diseases for which he recommended treatment with radon included gout, arthritis, and rheumatism; sciatica, neuritis, and neuralgia; herpes, tabes dorsalis, and myalgia; asthma, arteriosclerosis, and "diseases of the genitals in women." Observations at Kreuznach and the spa at Joachimsthal suggested emanation's effect on "the power of generation"; one researcher "succeeded in enhancing the secondary symptoms of sexual passion in water-newts by radium emanation."[19]

At least one American spa town consciously sought to exploit the new research. Saratoga Springs, a popular resort in the Gilded Age, was languishing in the twentieth century. A local doctor, Dr. Douglas Moriarata, decided in 1908 to "boom our mineral waters" and with other physicians set out to imitate at Saratoga the popular European spas. By 1909 the New York State government had become interested. It bought land to protect the springs and regulated the local bottling industry, which was beginning to deplete the gases that brought up the spring waters from the ground. In 1912 the state sent Simon Baruch, a famous hydrologist and balneologist (father of banker, Bernard Baruch), to Europe to study spa operation and treatment. Still, Moriarata considered Saratoga Springs in 1912 to be a health resort "only in a desultory manner. The local medical profession has not given these great natural therapeutic resources the consideration they deserve."[20]

New York State eventually decided to take over management of the spa at Saratoga Springs and began by building new bathhouses. A prominent physician lauded the decision because "state and municipal control . . . has placed the principal European spas upon their high plane of efficiency." He predicted better times, "now that the former crude and unscientific use of these domestic bathing waters, exploited as they were merely for temporary gain, is being replaced by intelligent management and State control."[21]

In 1914 scientists discovered that Saratoga's waters were radioactive, one of the rare instances in which spa water contained dissolved radium in addition to the more common radon. In response, Moriarata purchased some radium and began experiments; he soon wrote articles on the medical uses of radioactivity. In "Radium and Symptomatic Blood Pressure," published in 1916, he reminisced, "I have been treating patients with symptomatic blood pressure for years with our mineral waters." He defined the cause of hypertension to be "perverted metabolism" and argued that radium "corrects a perverted cell action by its influence on the enzymes." In a review of fifty-six cases treated with "radium emanation" (radon), Saratoga's mineral waters, and a "regulated regime," Moriarata claimed that every case showed a lowered blood pressure and increased blood counts. In another 1916 publication he admitted, "The physiological action of radium sounds not unlike a fairy tale." It increased the red blood count and hemoglobin, activated the digestive ferments, dilated the blood vessels, and "stimulates all cell life." "Its action in anemia is little short of marvelous." Radium was a "tonic and a stimulant." As Moriarata's practice evolved, his standard treatment came to include a 25-microgram injection of radium chloride followed by the ingestion of radon solutions over a period of days. Fortunately, he found, "No toxic or lasting ill effects have been reported." Indeed, "every radium patient would take on a feeling of well-being and contentment; many would state that they had not felt so well for months."[22]

Moriarata's fervent championship of radium might be written off as the rantings of a quack doctor, and Moriarata was hardly a respected physician: he never finished medical school. But his enthusiasm for radium was echoed in more measured tones by public officials. The New York Conservation Commission, in charge of refurbishing Saratoga's reputation as a health spa, also promoted its radioactive waters. In a 1917 report, "Saratoga Mineral Springs and Baths: Their Value as Therapeutic Agents," the commission maintained that "the medicinal value of radium is now well known and established." The report repeated many of the claims of the spa doctors: radium or its emanations cured gout, lowered blood pressure, was a powerful nerve sedative and hypnotic, and generally had "a favorable effect upon organic metabolism."[23]

Even the U.S. government accepted radium's place in internal medicine. Hot Springs, Arkansas, the biggest spa in America, was the site of an Army and Navy hospital. The spa itself was administered by the Department of the Interior. The medical corps of the army supervised spa treatments, and the spa donated to the indigent free treatments, overseen by the Public Health Service and the National Park Service. When the federal government discovered that the waters were radioactive (in 1905), it began to promote them on that basis. A 1907 government publication claimed:

> The hot waters may reasonably be expected to give relief in the following conditions: In gout or rheumatism . . . in neuralgia . . . in the early stages of chronic Bright's disease, in catarrhal conditions of the gall bladder, in certain forms of disease of the pelvic organs, and in sterility in women; in chronic malaria, alcoholism, and drug addictions . . . in many chronic skin diseases; in some forms of anemia; in syphilis; in gonorrheal rheumatism; in toxaemia and conditions of defective elimination; and in some forms of cardiovascular disease with increased tension in the blood vessels. . . . The general tonic and recuperative effects are marked in conditions of debility and neurasthenia due to strain and fatigue incident to social and business care and responsibilities.

In advertisements the U.S. Railroad Administration encouraged passengers to ride the rails to Hot Springs, asserting that "within recent years radium has become known as a powerful healing agent. . . . The waters [at Hot Springs] are radioactive, and by means of the bath every rheumatic joint, every sealed-up pore of the skin may be not only reached and cleansed of impurities, but renewed under the influence of that brain-baffling curative which we call radio-activity. . . . More than ninety per cent of those who have taken a full course of baths have been either cured or benefitted by them." Thus both New York State and the federal government added the weight of their authority to the increasingly favorable assessments of radium.[24]

Besides the radium companies and spas, one other group had a financial interest in advocating radium as an internal medicine: physicians who based entire practices on radium treatments. C. Everett Field, for example, in the years before the war could not survive on his general practice and became involved in selling proprietary drugs to other physicians. By 1914 he was working for the Radium Chemical Company, the sales arm of Standard Chemical. Standard Chemical had begun radium production in 1913 and was now ready to promote it among physicians. Field received eight weeks' training in radium therapeutics from the staff scientists Proescher and Viol, then was sent on a lecture tour that included the Albany Medical School and the Royal Medical Society in Montreal. He told this story: "To Joe Flannery [owner and president of Standard Chemical] I said, 'My Gosh, if any of those professors ask me some questions, I'm sunk!' [Flannery responded,] Dr. you have been *trained* more than anyone in America. . . . answer them anyway you choose — nobody knows enough to refut[e] you." In 1914 Field sold "the first large order of radium in America" to Joseph Bissel, founder of the Radium Institute in New York, a private hospital and clinic offering radium treatments to the public. The next year Bissel hired Field, and the two published a number of articles on radium. In "The Efficiency of Radioactive Waters for the Control of Faulty Elimination," Field made reference to "the campaign already started to acquaint the medical profession with a better understanding of hydrotherapy . . . and to promote a more liberal attitude toward the peculiar forces . . . of radioactivity." Americans, he claimed, spent between eighty and ninety million dollars annually seeking "health at the radioactive shrines of Europe." To help keep that money at home, Field was prepared to treat Americans for heart and artery diseases, kidney diseases, infections from disordered digestive ferments, and rheumatism. He assured his readers: "Radium for several years has been given internally and by injection in large doses with absolutely no disturbing symptoms. It is accepted as harmoniously by the blood stream as is sunlight by plant life." In a second article, Field declared himself ready to help those suffering from high blood pressure. He found improvement in almost 95 percent of the patients he treated. Again, he stated that radium was "accepted as harmoniously by the human system as sunlight by the plant." In both these articles, Field mistakenly claimed that injected or ingested radium was all excreted and that none was deposited permanently. His treatment consisted of one injection of 25 micrograms of radium, followed by drinking solutions containing 2 micrograms of radium each, ingested three times a week for four weeks and then twice a week for an undetermined period. "Results are prompt," Field promised. Bissel published similar articles. Radium injections relieved pain, improved the blood, caused absorption of joint deposits, lowered blood pressure, and had a hypnotic effect. Bissel treated pernicious anemia with two injections of 100 micrograms of radium eight days apart, followed by one-half a microgram every two days for two weeks and then every three to four days for six weeks. He reiterated the maxim that "radium is accepted as harmoniously in the body as is sunlight by the withering plant." Like Field, Bissel believed that radium, once injected, was completely eliminated over a period of time. "There seems to be no evidence that would point to any danger from radium when used internally," he concluded.[25]

Bissel later served as president of the American Radium Society, founded by Cameron and Viol of Standard Chemical; in 1919, at age fifty-nine, he died from leukemia, possibly caused by his work with radium. Field took over the Radium Institute, and, after recovering from an illness he assumed was caused by radium, moved the clinic to Riverside Drive, where he made a good living treating wealthy patients.[26]

By 1913 the internal use of radium had attained a modicum of legitimacy. In part, this legitimacy was based on proposed pharmacological mechanisms for radium's efficacy. In January of that year, in the American *Medical Review*, Carl von Noorden, "Director of the First Medical Clinic in Vienna," addressed the thorny question of how radium could affect so many parts of the body and improve so many diseases:

> The radioactive bodies are, as already said, carriers of energy. . . . We permeate the body with molecules which have a tendency . . . to explode. They hurl thereby, in their immediate environment, an amount of energy gigantic in proportion to their mass. . . . Other . . . molecules and cell constituents may be dissociated and it should not cause wonder that many vital expressions of the cells become thereby essentially modified. . . . The most powerful force which becomes active through the explosions of radioactive molecules is electricity. Thus, now, radioactive substances give us a method of distributing the carriers of electrical energy into the interior of the body and there engage the fluids, the protoplasm, and the cell nuclei in the immediate bombardment of electrical atom explosions.

We who know of the dangers of radioactivity cannot celebrate these events, as did von Noorden. He designated an alternate name for radioactive therapy: "internal electrotherapy." What were its effects? Increased respiratory interchange of gases signaled increased metabolism and meant that radium might be used to treat obesity. Increased uric acid output suggested that radium would alleviate gout. Stimulation of blood formation with small doses of radium, and destruction of blood cells and blood-forming cells with large doses, intimated that, with care, radium would improve pernicious anemia. Large doses destroyed white blood cells, and this might be useful in the treatment of leukemia. Radium also benefited rheumatism and arthritis, muscle pains and neuralgia, sleeplessness and nervous overexcitability, and sexual potency. To von Noorden, it was "definitely established that radioactive substances may exert an essentially stimulating influence on the generative glands and reflex apparatus." Blood pressure could be improved with radium treatment as well.[27]

Other explanations abounded for radium's efficacy in treating a number of conditions. Echoing von Noorden, C. Everett Field described radium therapy as the "administration of fluid electricity." A different idea was that radium acted catalytically, quickening chemical processes in the body, especially oxidation. "Life and oxidation are almost synonymous," wrote one researcher. This concept might be combined with the hypothesis that radium quickened the work of the "ferments," or enzymes. In the early part of the century, research on enzymes was very new and very exciting, as scientists unraveled how metabolism was regulated. Many illnesses were blamed on underactive or overactive glands. "Disturbed metabolism is nothing but disturbed ferments," was a common observation. Aging was thought to be a function of hormone imbalance: "Man is as old as his ductless glands," or "Would it not be reasonable to suppose that by taking radioactive waters in conjunction with observing the rules of health, one should live four or five times as long as it now takes to attain maturity? . . . It would be a common occurrence to live to be a hundred years old." Another claim was that radium worked on nerve endings.[28]

Scientific explanations for radium's healing powers were necessary to establish such treatment among orthodox physicians, but even more important was publication in leading medical journals. Much of the literature cited thus far was presented either in the industry-supported *Radium*, in journals of radiological societies, or in local medical journals (especially in the New York–based *Medical Record*). In 1913, the pivotal year in radium's career as an internal medicine, the *Journal of the American Medical Association* published a review of the German and Austrian literature on radium treatments. Oddly, the article ignored American contributions to the literature, as well as French and British research. In the 1,038 cases surveyed, radium was used to treat arthritis, gout, neuralgia, sciatica, lumbago, polyneuritis, neuritis, and "miscellaneous." In 837 of those cases, the various scientists had reported improvements or cures. L. G. Rowntree and W. A. Baetjer, the authors of the review, found that the "value of radium is unquestionably established in chronic and subacute arthritis of all kinds." They reported that a number of leading German researchers "speak in glowing terms of [radium's] value. The introduction of emanatoria in a large number of the German spas as well as the establishment of a radium institute in Berlin for the treatment of medical cases express confidence in [radium] on the part of the profession abroad." Their own findings, however, were disappointing: "We must state frankly that the results in our small series of cases comprising only eighteen patients have not been gratifying." Their arthritis cases had the most favorable outcomes; three of five "showed some slight but definite improvement." The investigators recommended further research.[29]

Accounts of German research appeared in other important British and American medical journals. As noted earlier, W. Engelmann of the Kreuznach spa reviewed the medical effects of water saturated with radon in the British journal *Lancet*. By 1913 Kreuznach physicians had five years of experience with radium emanation therapy; their findings were regularly published as collections of essays entitled "Radiological Communications." According to Engelmann, "one need not be too much of an optimist to be able to endorse von Noorden's declaration that radium emanation therapy may now be justly regarded as belonging to the solid stock of medical therapeutic measures."[30]

In 1913 a second review of German research was published in the *Archives of the Roentgen Ray* and the *Journal of the Roentgen Society*, both British publications (X rays were alternatively called "Roentgen rays" after their discoverer). The article had originated as an address to the American Roentgen Ray Society by Dr. S. Saubermann. Saubermann discussed the clinical work reported on hundreds of patients and the more than one thousand treatises produced on radium emanation. He asserted that this mass of work entitled radium therapy to be regarded as scientific fact: "The attitude I assume is this: I do not believe either on physiological or on biological grounds in the effect of any remedy when no information other than a statement of such action is supplied. . . . Fortunately, I do not think it impossible to render the clinical results . . . reconcilable with the general physiological results obtained for experiments with living organisms." Saubermann described the various effects of radioactivity on the human body: increased peristaltic action, increased uric acid excretion, dilation of blood vessels, diminished blood viscosity, lowered blood pressure, increased metabolism, and modification of the blood constitution. He noted radium's "nerve-soothing" qualities (with "no very assured explanation") and its ability to enhance sexual activity (due to a direct effect on the glands and to increased vitality). In addition to the usual list of diseases benefited, Saubermann promoted radium for the "degeneration of old age" — specifically, arterial calcification and high blood pressure — and for this condition "I allowed my own mother to be thus treated." He concluded by quoting another German researcher: "The continued existence and future of emanation therapy depends on the use of sufficiently strong doses."[31]

A third review, more cautious than the others, appeared in the *Maryland Medical Journal* in 1914. Ernst Zueblin, a professor of medicine at the University of Maryland who exhaustively researched the international literature on the internal use of radioactive substances, reported that the "literature on radium and radiotherapy during the last 10 years exceeds 700 articles." He covered claims of radium's and radon's healing powers (and accounts of their dangers). Zueblin predicted a "great future" for radium injections, although he cautioned that "in the future it will be necessary to watch the further development of this treatment without too much enthusiasm, but without a nihilism which opposes any new method of treatment."[32]

Reputable institutions in America were by 1913 conducting research on radium as an internal medicine. In 1909 Dr. Wickham, of French school fame, had visited Dr. Howard Kelly, professor of gynecology at Johns Hopkins Medical School. Wickham interested Kelly in using radiation for uterine cancer, but also employing radium for other uses. In 1911 Kelly and his associates acquired a small amount of radium, some for ray application and some to set up an apparatus to prepare radon-saturated water, and in 1913 they established an emanatorium. Work with radium as an internal medicine was discontinued within a few years, but it added something to radium's reputation as a medicine that Johns Hopkins, the capital of scientific medicine in America, was willing to experiment with it. Also in 1913 an emanatorium and a drinking water apparatus were introduced at the New York Post–Graduate School and Hospital.[33]

In 1905 the American Medical Association had established its Council on Pharmacy and Chemistry, charged with the study and control of patent medicines. The council published annually a report, *New and Nonofficial Remedies*, which reviewed medicines and which in 1914 included radium solutions for ingestion and injection. The standards that were set were for *minimum* strength of these solutions; the council was concerned with the fraudulent sale of solutions containing no radium at all.[34]

Thus, by 1913–14 the use of radium as an internal medicine had acquired a certain legitimacy. It had a nod from the American Medical Association, was being studied at leading research institutions, was well received in the medical literature, and was promoted by both New York State and the U.S. government.

Ernst Zueblin's 1914 review of the radium medicine literature was one of the earliest to caution users of radium's dangers. He warned that radium was stored in the bones and that depositions in human tissues could lead to ulcers and necrosis.[35] This was an unusual early synthesis of knowledge about radium's physiological route in the body and its destructive effect on tissues. Over ten years later, after the dialpainters' illnesses appeared, other scientists would "discover" a similar mechanism to explain the dialpainters' symptoms. What happened in the years between? What was the history of knowledge about the hazards of radium?

Since 1901 and the infamous Becquerel burn, scientists had known of the dangers of radium rays, of their ability to destroy tissue. Radium rays were similar to X rays—indeed, medical use of radium rays mimicked the earlier use of X rays for cancers and skin diseases—and by 1903 overexposure to X rays by patients and medical workers was a well-recognized problem. Cancer was one outcome of overexposure: between 1903 and 1911 fifty-four cases of carcinoma were attributed to X rays, mostly among physicians and technicians. Sterility, too, was known to result from overexposure to X rays.[36]

Radium experimenters also knew of the element's blood effects, but most researchers considered these therapeutic or at least not dangerous. There were some exceptions, reports linking radiation exposure to illness and death. In 1911 Viennese physicians reported the deaths of four X-ray workers and one radium worker from leukemia. Blood tests on ten healthy workers showed changes in the number and proportion of white blood cells. In 1914 an Italian physician's work with X rays was linked to his death from pernicious anemia. Also in 1914 German investigators reported a study of twelve persons working with radioactive substances; they found blood changes, skin damage, and behavioral symptoms like lassitude, sleepiness, headaches, irritability, dizziness, and fainting and recommended protective devices and techniques. The medical community remained little alarmed, for blood changes were listed among radium's beneficial effects as well.[37]

Something that might have suggested danger was the knowledge that radium accumulated in the body. Research by the French school between 1910 and 1913 demonstrated that radium salts, soluble or insoluble, remained in the body for at least a year, fixed in the skeleton. The British researcher W. S. Lazarus-Barlow in 1913 confirmed radium deposition in bone.[38]

Standard Chemical scientists knew about the deposition of radium in bone marrow. What is interesting in this instance is how the fact of radium's accumulation was interpreted. In early 1914 Frederick Proescher, then employed by Standard Chemical, described injected radium's deposition in bone but considered this a *good* thing: "the body becomes a vital emanatorium due to the deposit of radium salts." In other words, with radium permanently fixed in the body, the body was constantly exposed to the alleged healing properties of the radon and radiation continually produced. In an article appearing in 1915, fellow employees Cameron and Viol discussed radium's deposition in a deceased patient's bone marrow. Another 1915 study, published by Viol with others, focused on the elimination and deposition of radium salts; its purpose was to devise a "rational method for maintaining a certain amount of radium in the system."[39]

Less biased researchers (like Ernst Zueblin) might have considered dangerous the body's accumulation of a substance whose rays were often destructive and were suspected of producing blood changes and cancers. Indeed, ten years later, once radium was identified as the possible cause of the dial-painters' necroses, anemias, and eventual cancers, these connections were quickly made. With industry-supported scientists such as those in Paris and Pittsburgh contributing the bulk of the research, however, radium deposition was assumed to be beneficial rather than potentially harmful. Radium was promoted as a cure for anemia, whereas it turned out to be a cause. In 1915 Cameron and Viol praised their own investigations to date as "the most extensive work carried out" in the field. They had "never seen the slightest ill effect."[40]

Radium's renown as an internal medicine reached its peak in the years before America's entry into World War I, but with the war, skepticism about it surfaced. In the *Reference Handbook of the Medical Sciences*, published in 1917, an article cautioned that the value of radium had been "greatly exaggerated" and reports were "confusing." Although the author felt that there was "no question" of radium's efficacy in gout, rheumatism, and neuritis, he urged concern about its effects on the hematopoietic (blood-forming) tissues. A 1921 article complained, "When radium was first introduced here, its extraordinary physical characteristics and its undue exploitation in the public press led to expectations of its value which were undoubtedly unjustified by the actual facts." By 1922 the director of the Frank Edward Simpson Radium Institute, Simpson himself, stated, "There is so much possibility of error in estimating the value of radium when administered internally that many of the reports of benefit and cures must be accepted with the greatest caution until further experience has been accumulated." And the London Radium

Institute, involved in research on the internal administration of radium since at least 1915, admitted in 1922 that for most illnesses radium was "practically useless" or results were "doubtful," except in the relief of arthritis. For instance, the inhaling of radon in emanatoria had "obtained a great vogue on the Continent in the years immediately preceding the war. Most of the spas and bathing establishments were provided with radium emanatoria, luxuriously equipped." Although the method "had made slight headway in the British Isles," it now was "but little favored on the Continent." The report continued: "Extravagant claims were made as to its therapeutic value in the treatment of widely differing forms of diseases," but these were "poorly substantiated by subsequent clinical observation."[41]

In step with this emerging skepticism, the literature promoting the internal use of radium waned. Favorable American commentary appeared largely between 1913 and 1916; from about 1917 until 1923 such accounts were scant, even in *Radium*, Standard Chemical's own medical journal. The reasons for the increasingly doubtful and decreasingly positive reports about radium's medicinal benefits can only be speculated. One explanation takes into account continuing discoveries of the dangers posed by exposure to radium. Perhaps manufacturers and others with an interest in radium medicines were concerned or cautious enough to cease promoting their products. Between 1915 and 1921 the medical literature continued to warn of the hazards of overexposure to radiation. Radiologists were especially likely to suffer the consequences. In 1916 an article appeared—first in the *Scientific American* supplement and within a year in the *Journal of the American Medical Association*—describing changes in the fingers of radiologists in London, Vienna, and Germany. The skin grew thick, "horny," and the nails brittle; fingers became numb or painful, clumsy, and sensitive to heat and pressure. Lesions sometimes appeared. From the London Radium Institute in 1919 came similar reports, as well as warnings about systemic changes in radiation-exposed workers. Symptoms ran the gamut from fatigue to blood changes—lowered white and red blood cell counts—to amenorrhea in women. A four- to six-month vacation from radium work restored women's menstrual periods.[42]

Then a series of studies made very clear the dangers of radiation exposure. The London Radium Institute was the site of the first focused experiments on radium workers' health, instigated by three deaths among the institute's personnel between 1916 and 1920. The first three studies were published in 1920. J. C. Mottram and J. R. Clarke checked the white blood cell counts of twenty laboratory and clinical radium workers and compared these to the counts of thirty-eight normal individuals. They found the number of white blood cells and lymph cells lower in the radium workers than in "the normals," although they could not associate this with any reports of ill health. Mottram's next article revealed many instances of diminished red blood cell counts among workers exposed to radiation, including the three severe cases of anemia that led to the death of the institute's employees. The anemia was caused by damage to the "hematopoietic system," the organs that produced the blood cells, including the bone marrow, the site of red blood cell production. Laboratory experiments with rats led to a third publication, which confirmed that radium rays damaged the bone marrow. In 1921 Mottram's fourth article demonstrated that increased protection of radium workers would allow blood conditions to return to normal.[43]

These articles and the deaths that precipitated them were widely discussed, both abroad and in America. Medical centers began to take greater precautions with their employees' health. At least one physician warned the medical community that "internal administration of any radio-active substance, in view of the difficulty of its elimination and the extremely minute quantities which are necessary to produce marked results, seems to me to be a very dangerous procedure." The author also pointed to "the general relationship which, in my opinion, obtains between radiation and cancer."[44]

Another possible explanation for the decreased promotion of radium medicines focuses on other contemporaneous events. Undoubtedly World War I interfered with research and publishing, and perhaps the recession that followed the war likewise delayed radium work.

The war also cut the American medical community free from its attachment to German medicine, which had stimulated much of the interest in radium therapeutics. Despite what we now can see as a hiatus in publications about the medicinal use of radium, Charles Viol of Standard Chemical believed in 1921 that the isolation of World War I allowed the development of an "American School of Radium Therapy." If such existed, then it was marked by both the larger doses employed in this country by experimenters such as Proescher at Standard Chemical, and by an insistence on the value of empiricism in medicine, perhaps in reaction to German rigor in experimentalism. An example is Samuel Delano, a physician who prescribed radium in his Boston practice. He wrote: "To some, the glamour of laboratory tests being wanting, [reported cases] may appear not complete—perhaps not scientifically precise. But there is still a large place for empiricism in medicine. . . . No doubt many will scent a strong 'psychical' odor to them. . . . In our quest for greater so-called scientific precision in the shape of objective laboratory findings we are in danger of making 'subjective' tantamount to 'psychical.'"

In contrast to Delano's measured tone supporting empirical trials was the fervor of Ellis Fischel, a more spirited radium proponent: "There is scarcely any disease in which it has not been tried. It is such a fascinating element, such a magical element, that it has been used in practically every condition, whether indicated or not. . . . depending on the enthusiasm of the therapist, we get such different results." Although admitting the subjective biases possible in empirical work, Fischel continued to treat his Missouri clients with up to 300 micrograms of radium via radium salt injections, bolstered by radon emanatoria and drinking solutions.[45]

Another important factor, however, was that an American medical market never materialized on the scale demanded by industry: radium extractors suffered financial difficulties when the world war began due to the loss of the European market. In 1915 little radium was mined. The business recovered with the 1917 dialpainting explosion, and by 1918 an estimated 95 percent of the radium produced in America was used for luminous paint. Thus the curtailment of medical demand may have resulted in the dearth of medical research and publishing on radium.[46]

Radium production increased in America until 1921, when the announcement of the discovery of high-quality radium ore in the Belgian Congo threatened the survival of the U.S. radium industry, which could not compete given its reliance on low-concentration ores. In fact, most radium production ceased at home. The two largest producers, the Standard Chemical Company and the Radium Company of Colorado, began to cooperate with the Belgian radium manufacturers, swapping technical skills for the rights to sell Belgian radium in the United States and England. A third producer, the U.S. Radium Corporation, was able to continue manufacturing radium. U.S. Radium was the largest dialpainting firm in the United States until it gave up large-scale operations around 1920 and helped its clients, the watch and instrument manufacturers, set up their own dialpainting studios. The Orange, New Jersey, corporation continued to sell luminous paint and so, unlike other American radium extractors, U.S. Radium remained in "production" until 1926; one student of the radium business concluded that "its heavy involvement with the production of luminous paint [gave] it a position in the market that was somewhat sheltered from the Belgian competition."[47]

U.S. Radium also continued to promote radium as an internal medicine. In 1922 it offered for sale a bibliography and an abstracts collection on radium. Both included information on the internal use of radium as well as on the dangers and injurious effects of radium exposure. Clearly, the company was aware of the publications that warned of radium's effects on health. Yet when the dialpainters' illnesses became manifest in 1923, it denied that radium could be the cause.[48]

Concerns about radium continued to mount in the early 1920s. Workers exposed to radiation continued to die. Two more cases of anemia were reported in 1922, these caused by damage to blood-producing organs by X rays, similar to the effects of radium's gamma radiation. Animal experiments confirmed that radium could damage bone marrow, the site of blood cell production. In 1923 the U.S. Public Health Service reported that radium workers at the Bureau of Standards demonstrated blood changes. Medical experts were beginning to suspect that radiation caused cancer: the uranium miners whose low rate of arthritis led to theories of radon's ameliorative effect on joint diseases now were discovered to suffer a high incidence of lung cancer, attributed to radon inhalation.[49]

Despite these new fears, in 1923 Cameron and Viol of Standard Chemical's medical clinic reviewed their work on radium administration and continued to recommend radium injections. A "safe dose" was 50 to 100 micrograms of radium salts, repeated every ten days, to a maximum of 300 micrograms. The National Radium Products Company, in New York, also promoted the medicinal use of radium. In 1924 and 1925 it published five pamphlets that advertised one of its products—an emanator that made radium water by exposing water to radium, thus allowing radon to dissolve in the water—by citing the literature explored in this chapter on radium's efficacy in treating diseases. The pamphlets quoted Marie Curie, Frederick Soddy, Carl von Noorden, William Cameron, and Charles Viol; the London Radium Institute; Joseph Bissel; the surgeon general of the United States, and others not cited in this text; radium was recommended for diseases from arthritis to sexual dysfunction, leukemia to senility. In addition, radium was touted as a cure for "the greatest menace to . . . existence—civilization!" The "high tension [of] civilized living" led to increased mortality rates, and radium, "the spark of life," could supply the energy necessary to overcome this threat.[50]

J. Everett Field continued to practice out of his Radium Institute on Riverside Drive in New York City. In his words: "From 1920 on to 1930 I had among my most considerate patients a most interesting group of wealthy patients from Maine to California. . . . I had Governors from three of our eastern states and a large group of army officers or their families. . . . Boy, it was fine sleddin [*sic*] while it lasted." In 1922 Field began arguing that radioactivity might prolong life. Referring to Moriarata's work at Saratoga Springs, he insisted that "even now" life could be extended fifteen years through the use of radium, which soon would replace "monkey glands" as a rejuvenator.

"There is no fundamental reason . . . why most of us should not figure on at least a century of life, why we should not remain young twice as long as we do," Field declared in newspaper interviews.[51]

The dialpainters began working in 1917 and remained in high demand in Orange, New Jersey, until around 1920, when the export of jobs to the watch and clock companies in Waterbury, Connecticut, curtailed dialpainting in Orange. The first illnesses and deaths became known in 1923. But even before then, doubts about the internal administration of radium surfaced. To review:

First, an early detractor was Ernst Zueblin. His 1913 literature review noted the deposition of radium in bone and expressed concern about the likely ill effects of radium's destructive radiations on bone and nearby tissues.

Second, radium's effect on the composition of the blood had long been known. Because blood cells are created in the bone marrow, it seems obvious (with hindsight) that deposited radium was the cause of radium's blood effects.

Third, beginning in 1911 anemia and leukemia were discovered among X-ray and radium workers. Again, from a modern perspective this might have suggested caution to medical experimenters.

Fourth, X rays were implicated in carcinogenesis as early as 1903. Logic now indicates that researchers should have been concerned about placing radium so close to tissues, especially over the long term, as in radium plated into bone.

Enthusiastic promoters interpreted the data on radium's properties and effects quite differently. Radium's effects on blood composition were used therapeutically to alter the blood. Radium's bone deposition allowed a long-lasting effect from a powerful but expensive medicine. Researchers associated with radium businesses interpreted data in light of their hope, expectation, and commitment to radium's safety and beneficence.

What motivated researchers to such a reading of the data on radium? Four influences can be distinguished. First, radium promoters may have shared an altruistic interest in discovering new medicines to ease pain and illnesses. That is not incompatible with a second influence, their hopes for a top-selling new radium medicine. Two of the biggest dialpainter employers, U.S. Radium and Standard Chemical, profited from the radium medicine fad, and during the same years were producers and disseminators of the positive literature they later cited as their defense for endangering their employees. If radium had been deemed hazardous, that conclusion would have threatened these companies' medical markets as well as their dialpainting operations.

A third reason for radium promoters' resistance to negative interpretations of data on radium relates to their own close work with radium. Some business leaders had worked in radium production as chemists or physicists responsible for the mechanical and chemical separation of radium from ore or its transformation into various medicinal products. Some had experimented with self-dosing to determine radium medicine's effects. To admit to radium's dangers after vast personal exposure would have been to confront fears about their own mortality. And fourth, to make such an admission would mean acquisition of a heavy burden of guilt about endangering the lives of others.

For these altruistic, financial, and psychological reasons, radium entrepreneurs' reality precluded awareness that radium might be dangerous. From this perspective, it becomes difficult to answer the "Watergate question" of "what did they know and when did they know it" since "knowledge" does not accumulate reflexively from "facts." Rather, as I hope this chapter has demonstrated, "facts," "knowledge," and what we might call "perspective" are all interrelated and interdependent. But what about culpability? Does this argument forgive radium company officials for what we might label a self-interested misreading of the evidence based on profit, fear, and guilt? If self-duplicity is self-serving, can we assign blame? An exit from this quandary may be found if we consider experts' views about radium as an ideology—which I define as ideas grounded in and preserving power. Such a perspective lifts the questions out of the realm of psychology and individual culpability and places them within a sociopolitical nexus. The historian may then seek explanation for the triumph of a pro-radium ideology in social structures and relations that allowed the integrated assertions and aims of radium businesses to prevail. This is the context in which to understand why young women were set to work painting luminous dials with paint that contained radium.

SOMETHING ABOUT THAT FACTORY

The Dialpainters and the Consumers' League

The New Jersey Labor and Health Departments through March 1924 had taken no action regarding the illnesses and deaths among the dialpainters, despite three occasions on which complaints led to investigations. At the U.S. Radium Corporation, dialpainters were not waiting for official or corporate recognition of occupational poisonings. Dialpainters were voting with their feet, as this production manager report demonstrates: "In the application work considerable difficulty has been encountered in keeping our force of operators to the proper number. This has been due to rumors of sickness and disease of the mouth claimed by certain dentists to be caused by contact with our Undark material. . . . Because of this trouble we have lost a number of our older operators and work has been held up. . . . So far we have no reason to believe that there is any foundation in these reports and a thorough investigation is in process."[1]

In March 1924 Leonore Young, the Orange, New Jersey, health officer who had investigated dialpainter complaints, contacted Katherine Wiley, executive secretary of the New Jersey Consumers' League, a women's voluntary society dedicated to improving working conditions for women and children. Young asked for a conference with Wiley and Lillian Erskine, the state Labor Department official who had also looked into dialpainter complaints. Erskine had discovered yet another ailing dialpainter and was sure there was a problem at U.S. Radium, but, she reported, the Labor Department had no funds or personnel available for further study. Young added that the Health Department would undertake no further investigation of the dialpainter matter, either: "the authorities are hesitating," she told Wiley, and the Consumers' League "must keep after them to see that *something* happens."[2]

Katherine Wiley, described by one acquaintance as "intelligent, conscientious, and enterprising," accepted the challenge of the radium cases and undertook a campaign to win recognition of the disease, compensation for its victims, and prevention of future cases. Wiley began by tracking down information on the three deaths and two illnesses known to the state. She spoke to dialpainters and their medical providers and was quite moved by the plight of the sick women. After visiting one woman, whom she described as "this poor sick young thing who looked fairly transparent and whose jaw had been partly removed," Wiley resolved "to stick to this thing until there is some action somewhere."[3]

In this chapter we consider, first, why it was the Consumers' League to which the state bureaucrats Young and Erskine turned when they were frustrated by the inaction of their own agencies, and second, why the women of the Consumers' League were interested in industrial disease reform. In examining these questions, we must explore briefly the development of industrial health reform in the United States and an industrial hygiene movement of which the Consumers' League became a part. With this background, we can begin to follow the league's investigation of the dialpainters' illnesses and the continued campaign to win acknowledgment of radium poisoning from the medical community and government. This chapter details that campaign through the end of 1924, a frustrating year in which the Consumers' League seemed of little benefit to the dialpainters.

The Consumers' League was one of three organizations working for industrial disease reform in the 1920s. These groups were inheritors of the industrial hygiene movement, one of the many Progressive reform movements
of the early twentieth century. Like other Progressive concerns, industrial
health attracted interest from groups widely scattered across the social spectrum: workers and their employers, intellectuals and government bureaucrats,
medical scientists and insurance companies. George Rosen recognized the
movement in his 1930s and 1940s work, describing how "socially conscious
Americans—economists, physicians, labor leaders, lawyers, social scientists,
and others" created a "movement to better conditions in industry" by attending to the problems of occupational disease. More recently, David Rosner
and Gerald Markowitz have written about "a movement" or "a broad coalition" of "radicals, labor leaders, and even business representatives" interested
in industrial health and safety.[4]

Still, it is odd to consider the many different groups with very different
agenda a "movement." Workers sought to prevent industrial diseases and to
compensate victims. Business and insurance groups endeavored to curb government or worker interference in shop conditions and to limit their own liability for industrial diseases (but often, if only to save themselves from potential liability, they took leading roles in discovering and implementing safe and
healthy work procedures). Less polarized were three other groups—reformers, scientists, and government experts—although reformers tended to ally
themselves with workers' interests (or with what they identified as workers'
interests) to prevent and compensate industrial disease, and many scientists
and government bureaucrats hesitated to offend business. All of these groups
were active in industrial hygiene during the 1910s and 1920s, albeit with different goals and tactics. Although they often lacked a common objective, I
can think of no other word to describe the plethora of activity in industrial
health during that period than "movement."

Perhaps the best way to characterize the industrial hygiene movement is
through the life and work of Alice Hamilton. Like so many effective women
reformers, Hamilton was a product of Jane Addams's Hull-House, beginning
her residency there in 1897. And like other women who lived at the Chicago
settlement house, Hamilton sought to combine the pursuit of a career with
social reform. In her early Hull-House years, as a physician drawn to research,
she ached to find both intellectually satisfying and socially helpful medical
work. Her scientific training was excellent, including an M.D. from the University of Michigan and postgraduate work in bacteriology and pathology at

the Universities of Leipzig and Munich and at the Johns Hopkins Medical School. At Hopkins she may have imbibed the social reform philosophy of its leading medical specialists. Historian Lloyd Taylor has ascribed to her teachers, the prominent physicians William Henry Welch and William Osler, faith in the environmental causes of diseases and in government programs for their control (among them, social legislation limiting the hours that women and children could work). In addition, Welch and Osler admired the work of female reformers and appreciated feminism. From Hopkins to Hull-House was not such a leap, then. During the first ten years of her residency in Chicago Hamilton focused on public health, pathology, and bacteriology.[5]

In 1907 Hamilton read both a muckraking report of workers killed by fire at a Lake Michigan pumping station and a compendium of dangers to workers in British industries; these accounts aroused her curiosity about health and safety in U.S. industries. Hamilton was suspicious of the widely held belief that American working conditions were superior to those in Europe, and an article published in 1908 indicated her disbelief that U.S. manufacturers voluntarily took more stringent precautions than their European counterparts. If Hamilton threw doubt on one myth, that America's industries were healthier than Europe's, she originated another: that before the twentieth century American physicians had been disinterested in industrial health. Library research turned up scant information (so she reported) on American workers but plenty of data on injured Europeans. In her autobiography she complained, "When I talked to my medical friends about the strange silence on this subject in American medical magazines and textbooks, I gained the impression that here was a subject tainted with Socialism or with feminine sentimentality for the poor." Hamilton wrote in the 1940s that when she began her early research on occupational diseases, "it was all German, or British, Austrian, Dutch, Swiss, even Italian or Spanish—everything but American." Abroad, "industrial medicine was a recognized branch of the medical sciences," she continued, but "in my own country it did not exist."[6]

In fact, U.S. interest in industrial health paralleled that in Europe, although it was slowed by later industrial development and attenuated by the supposition of an American "exceptionalism" that promised immunity from the worst effects of European industrialization. In Europe, mention of occupational hazards can be traced back to the earliest physicians—for instance, Hippocrates noted lead poisoning, and Galen, illnesses from copper smelting and gypsum mining. Most accounts of the development of industrial toxicology attribute the fatherhood of that specialty to Bernardo Ramazzini, who in 1700 published an exhaustive treatise describing about one hundred occupations and the hazards inherent in each. Apparently this was an isolated publication, and further interest in such diseases generally came later in eighteenth-century Europe, where the increasing number of industrial workers spurred studies of their health, longevity, and child mortality rates. These focused attention on the conditions of labor and began a debate about the limits of individual responsibility for health and the appropriateness of state intervention.[7]

Similar studies and debates took place in the United States as industrialization intensified in the nineteenth century, paralleling events in Europe. In America by the 1830s a medical debate began over the health effects of industrialization. This debate echoed workers' concerns about industrial health during the same period. Although much of the medical discussion leaned on data from European investigations, a domestic industrial health literature did evolve. The first influential medical examination of industrial diseases in the United States was Dr. Benjamin McReady's "On the Influence of Trades, Professions and Occupations," an 1837 dissertation that won a prize from the New York State Medical Society. Other writings followed. Many of these continued to rely on European cases and data, but an 1895 text by James Henrie Lloyd utilized specifically American incidents, such as mercury poisoning among New Jersey's hatters. Phossy jaw was described in 1854, 1856, 1879, and 1898; lead poisoning was depicted in fourteen publications before 1898. In fact, at least two hundred articles on occupational health were published in the United States before 1900.[8]

The assumption that American factories were healthier than European ones limited discussion of their ill effects until the 1870s. Americans, it had been hoped, were immune to the impoverishment and radicalization of the European working classes. By the 1870s, however, the factors thought to protect America from these developments had been lost. Rural factories had given way to urban plants, the Lowell model of protected and ennobled labor had been abandoned, American and female labor was replaced in great part by immigrants and men, the sense that factory work could be a temporary stage of life was replaced with the gloomy realization that it was the lot of many for their entire lives, skilled labor was being replaced by unskilled as the division of labor brought lower incomes and deskilling, and radicalism among workers was rising. Now it was clear to observers, including physicians, that the factory was both undesirable and inevitable.[9]

Acquiescence to industrialism was only one of several factors that prompted the new interest in industrial health. A second factor was immigration. Observers' elitist and ethnocentric notions led to the premise that immigrant laborers would not protect themselves at work as well as American workers. A third factor was concern that heredity, understood to include the inheritance of acquired characteristics, would doom employees' progeny to a life—heretofore unknown in the United States—within a factory worker caste of the biologically feeble.

A new American exceptionalism took the place of the old as physicians and others, determined to use American gumption and ingenuity to rid the workplace of its worst threats, ushered in the era of factory regulation. They helped win boards of health, bureaus of labor statistics, and associated factory regulation. Late-nineteenth-century American reform is comparable with legislative overtures in Europe, argues Asger Braendgaard, if one considers actions at the state level. Protective measures for women and children, usually establishing maximum hours of work, were among the earliest laws. Factory inspection was instigated in Boston and Chicago by 1880, in fourteen states by 1894, and in all fifty states by 1920, although inspectors were often poorly trained, few in number, and short on authority. By 1890 twenty-one states had laws preventing accidents and ensuring healthy work environments; by the end of the century twenty-one mining states had safety codes, protective legislation, and/or mine inspectors. By 1910 most industrialized states regulated the workplace to some extent, although little provision was made for enforcement.[10]

Still, throughout the nineteenth and early twentieth centuries the possibility of *chronic* chemical poisoning was often ignored, reform efforts centered on workplace safety, not health, and even those health measures that were implemented often were based on suspect reasoning. For instance, health regulations frequently included ventilation requirements, such as the removal of stale air, fumes, and dusts, and the provision of fresh air. Ventilation laws, therefore, might have obviated some health hazards, but they were predicated less on the dangers of toxic fumes and dusts than on the salubrious quality of air from the outdoors.

Focused medical interest in occupational health and safety intensified in the twentieth century to such an extent that Alice Hamilton may be forgiven for thinking that she and her generation were on to something new. Reasons for the surge in medical interest include, first, physicians' enhanced social authority. As Paul Starr details in *The Social Transformation of American Medicine*, nineteenth-century doctors practiced in an overpopulated field with skills that were decreasingly respected over the century. They did not enjoy the demand, high status, or high salaries now associated with medicine. But late-nineteenth-century physicians created the American Medical Association (AMA) and instigated a campaign to increase their professional cohesiveness and to win respect for their skills. By the twentieth century, when physicians studied the effects of factories and mechanization on workers, their opinions were given credence by reformers and government officials.[11]

Medicine's prestige was bolstered by the successes of public health. A new optimism about preventive medicine beginning in the 1890s encouraged medical and lay crusaders to seek reforms in living and working conditions. Mortality and morbidity figures compiled in the United States beginning in the twentieth century led to both demand for public health work and proof of its effectiveness. Statistical methods that could make sense of such data became available at the same time.[12]

Medicine changed direction during the first decades of the twentieth century. On the one hand, there were the "medical reformers," characterized by Lloyd Taylor, who through social medicine or public health and state action sought to improve the standard of living of all members of society. Clearly, the campaign for industrial health reform drew on this strata of medical science. On the other hand, Christopher Sellers describes medical scientists with a new emphasis on experimental medicine, especially physiology. In a search for interesting research problems they turned to poisoned workers as living laboratories in which to study the physiological effects of various substances. Without obvious social agenda, such physicians rejected the Progressive link between research and policy, insisting instead that research was an inviolate realm of pure intellectual activity without ties to social or political debates. These researchers avoided partisan activity, at least overtly. Unlike Sellers, I would stress that by avoiding the moral dimensions of the problems they so dispassionately studied, the scientists tended to support the status quo of the uneven division of power between workers and their employers. Because of this, they often gained much support from business and legitimacy within government policy circles. In this sense, then, the objectivity of the medical scientists may be seen as an ideology that gained them power and widened the spheres in which they might use it.[13]

Alice Hamilton derived much of her authority as a medical reformer from her ability to bridge these two worlds. She sought, as Barbara Sicherman has written, to combine in her work both science and service, experiment and reform.

Businesses to some extent became more open to concerns about occupational health in this period. Competition among companies had eased due to the mergers and informal collusion agreements that mark the era. Managers now had financial leeway to improve working conditions in the hope of allaying worker discontent. "Welfare capitalism" involved passing on to workers some of the increased profits of the oligopolist market in the form of shortened hours, regular employment, better pay, health and life insurance, and safer and healthier jobs. American business sponsored the safety movement that began around 1908 to accept—indeed, to control—safety regulations issued by local, state, and federal governments. Workers' compensation, instigated in the 1910s, focused attention on rising insurance costs, and companies instituted safety programs to save money on their insurance. Principles of scientific management also encouraged a more efficient use of labor power. Companies began to hire physicians to weed out unhealthy workers (those most susceptible to accidents and disease) and to suggest preventive measures. "Industrial medicine," as it was called, may be seen as one of a number of welfare capitalism programs introduced by business at this time.[14]

Insurance companies also became interested in occupational diseases. One example was the Prudential Insurance Company's support of Frederick Hoffman, a valued employee who was given leeway to pursue his own research interests. Hoffman, like Alice Hamilton, moved in many social circles. While devising actuarial tables and developing statistical tools for the burgeoning life insurance industry, he began to consider the effects of various lines of work on life expectancy. This led him to cooperative research with federal agencies monitoring labor, and by the 1920s he was an established authority on industrial diseases. His reputation rested on the voluminous studies included in his 1908 and 1909 reports on "Mortality from Consumption in Dusty Trades" for the U.S. Bureau of Labor. Like other national figures, Hoffman would become involved in the dialpainters' cases.[15]

A more activist government also explains increased attention to industrial health reform. With the turn of the century, governments began to assert more control over the workplace, employing various experts, including physicians, to facilitate such intervention. One reason for this transformation was business cooperation with government regulators. A second and related factor was that citizens increasingly feared class warfare as America's industrial scene came more to resemble Europe's. A revived labor movement played an important role in intensifying the industrial health issue. Or, as Henry Sigerest put it, rather baldly, "The ruling class recognized that a sick proletariat was a menace to its own health."[16]

A final factor explaining business, medical, and governmental interest in the country's industrial health was the conservationist ideology of the era. Conservation—the planned use and scientific management of natural resources and national wealth—included a concern with the conservation of human life and health in order to maintain high industrial productivity. This dovetailed neatly with public health physicians' campaigns to clean up city streets and water and food supplies utilizing state power.

One of the earliest mergers of medicine and federal government occurred with Theodore Roosevelt's nomination of George Kober, professor of hygiene and state medicine at Georgetown's School of Medicine, to chair the Committee on Social Betterment of the President's Homes Commission. Kober was an early contributor to the founding of industrial hygiene as both a field of knowledge and a reform movement. Since joining the Georgetown University faculty in 1890, he had became involved in campaigns against typhoid and tuberculosis and for clean milk and decent housing. During his first year at Georgetown he began offering lectures in industrial hygiene, thus becoming the first university professor to do so.

In 1902 Kober recommended to the U.S. commissioner of labor, Carroll Wright, that one of Kober's students, C. F. W. Doehring, undertake a study of the effects of chemicals on workers; the findings of this inquiry were published in 1903 by the Bureau of Labor. Wright went on to sponsor studies on legal attempts to control industrial safety and health abroad and in the United States, as did the commissioners who succeeded him.[17]

As chair of the president's Committee on Social Betterment, Kober wrote a report focusing on health as "the chief asset of the working man," wealth that was threatened by diseases incident to occupations and the work environment. "It has been the aim of the committee to emphasize . . . the causes and prevention of industrial diseases and also of some of the principal preventative diseases," Kober explained. This became the subject of his book-length report entitled *Industrial and Personal Hygiene* (1908), the "first American text on industrial health," according to George Rosen. Kober addressed the connections between the dusty trades, consumption, and other respiratory diseases; tobacco work and miscarriages; work with wool and animal skins and the disease, anthrax; lead industries, including potteries, and lead poisoning; work in foundries and "brass founders' ague"; and a host of other occupational health hazards posed by various chemicals. He appealed for an activist government that would promulgate laws limiting the hours of labor and the use of child and woman labor, improve sanitation and ventilation, make employers more liable for the injuries and illnesses of their employees, and regulate manufacturing processes in dangerous industries. In sum, he believed that "workers in many industries need special protection, and . . . it must be provided for by . . . laws." Kober's work as chair of the Committee on Social Betterment and in the Georgetown Medical School marked intensifying industrial health efforts in both the government and academic sectors.[18]

Alice Hamilton, then, was not alone in her interest in industrial diseases in the first decade of the twentieth century. A number of other industrial hygiene reformers were to be found in the American Association for Labor Legislation. The AALL, a branch of the International Association for Labor Legislation, was formed in 1906; its motto was "Social Justice Is the Best Assurance against Social Unrest." The American association was closely tied in both personnel and funding sources to the National Civic Federation (NCF), the early-twentieth-century group that sought to bring labor and management together to prevent the kind of class warfare predicted by Marx and perceived by Americans looking at Europe. Labor was wooed by the NCF, but many workers were never convinced of the benefits of corporate capitalism—a capitalism that at least theoretically served both labor and capital—and others were soon disenchanted with the results of labor/capital cooperation. The National Civic Federation became anathema to many labor leaders by 1911; although Samuel Gompers, president of the American Federation of Labor, continued to support the NCF, he came to label the AALL the "association for the assassination of labor." Nonetheless, the AALL was one of the most important organizations working to improve industrial health in the early part of the century. Other concerns of the AALL were workers' compensation for work-generated injuries or illnesses, unemployment compensation, old-age pensions, and health insurance.[19]

In 1909, on beginning its study of phossy jaw, the American Association for Labor Legislation asked Alice Hamilton to take part in the phosphorous investigation but she demurred; at that early point in her industrial health research, she felt insecure about her abilities. But other opportunities presented themselves. Because of her interests and her Hull-House connections, Hamilton was appointed by the governor of Illinois to a commission formed to investigate the prevalence of industrial diseases in the state. Hamilton supervised its two-year study of industrial processes and hygiene, specializing in ascertaining the magnitude of lead poisoning in Illinois industry. Her lead work attracted the attention of the U.S. Bureau of Labor, which hired her to study lead poisoning nationally and then to investigate many different kinds of industrial poisoning.[20]

The years following 1910 were pivotal ones for the coalescing of the industrial hygiene movement. The AALL's phosphorous investigation and Hamilton's lead studies spurred the AALL to hold its First National Conference on

Industrial Diseases in Chicago in 1910; two years later a second conference would convene in New Jersey with AMA backing. The Women's Welfare Department of the New York and New Jersey Section of the National Civic Federation undertook a study of mercury poisoning that was published in 1911.[21]

Interest in industrial health reform can be seen in the establishment of a number of new medical and governmental institutions. In medical circles, the first industrial disease clinic was established at Cornell University in 1910; soon after others followed in Illinois, Ohio, and Massachusetts. The first courses in industrial hygiene were offered at the Massachusetts Institute of Technology in 1905 and at Harvard University in 1910; these two schools undertook a cooperative effort to train health officers in industrial hygiene in 1913. Departments of industrial hygiene with degree-granting status followed (Harvard's, among the earliest, is discussed below). New medical professional organizations included the Conference Board of Physicians in Industrial Practices, founded in 1914 by doctors working in industry. Those not tied to industry created a rival American Association of Industrial Physicians and Surgeons. The American Public Health Association supported a Section on Industrial Hygiene. A professional periodical, the *Journal of Industrial Hygiene*, began publication at Harvard in 1919.[22]

Within government, state and federal activities in industrial health throughout the late nineteenth century were enhanced in the twentieth by a federal Bureau of Mines, established in 1910 in part to study the health of miners. The U.S. Public Health Service created a Section on Industrial Hygiene and Sanitation in 1914; the Department of Labor, organized in 1913, continued to monitor occupational health through its Bureau of Labor Statistics. Nevertheless, federal involvement in industrial health was relatively slight, and during the 1910s and 1920s most activity was centered in state Departments of Labor (and in Connecticut in the state Health Department). Industrial hygiene sections existed in some state Labor Departments, such as New York's, founded in 1913.[23]

At least partly as a result of Alice Hamilton's lead studies in Illinois, the state in 1911 passed an occupational disease law for the purpose of preventing and reporting a few well-established industrial diseases. Few state workers' compensation laws covered industrial diseases specifically before 1917. Although a few statutes were worded in such a way that they might be interpreted to include them, by 1917 only the Massachusetts courts had accepted such an interpretation. Between 1917 and 1925 a dozen federal jurisdictions or states — including New Jersey, Connecticut, and Illinois, the states where dialpainters were to discover their illnesses — made occupational diseases compensable.

Most states adopted schedules making only listed diseases compensable; this was true of all the states adding industrial diseases to compensation laws after 1920. Still, by 1933 only a third of the fifty-one compensation jurisdictions in the nation had occupational disease compensation. The American Association for Labor Legislation was the main force behind the inclusion of industrial diseases in state compensation programs. To advance such legislation, Alice Hamilton became a member of the AALL's General Advisory Council.[24]

During World War I the federal government incorporated industrial hygiene specialists and other social experts in efforts to assure high productivity and minimize dissent. Alice Hamilton undertook studies of industrial diseases related to the manufacture of explosives for the federal Bureau of Labor Statistics. She served on the Committee on Intoxications among Munitions Workers, a branch of the National Research Council, and oversaw the health of munitions workers. Hamilton also pursued her studies in cooperation with a National Defense Council Section on Industrial Medicine. For a brief time the federal agencies supervising health and labor cooperated in industrial hygiene investigations: the Labor Department established a Working Conditions Service that employed Public Health Service physicians. These organizations, like other social engineering experiments of the war years, were dismantled when the conflict ended.[25]

Following the war, Alice Hamilton was widely acknowledged as the founder of industrial toxicology in America. Her prominence led to her appointment as the first woman faculty member of Harvard University. She joined the Department of Industrial Hygiene, part of Harvard's Medical School but later in its School of Public Health. Hamilton remained in Cambridge until her retirement, always as an assistant professor and always—at her insistence— part-time in order to pursue her investigations and reform work. Although she never wrote about any discomfort with her colleagues, the entire department did not share her reform bent, as will become clear in the history of the dialpainting cases. Pro-labor reform had its proponents at Harvard, but business interests supported academic industrial hygienists and so shaped the direction of the Industrial Hygiene Department.[26]

The 1920s brought contradictory trends within industrial health reform circles. By that decade the industrial hygiene movement could take credit for rewritten compensation and liability codes, and for new medical institutions and administrative agencies. The reform spirit in some circles began to wane: with an apparatus designed to respond to industrial diseases in place, some Progressives were satisfied as reformers, scientists, and government officials settled into the new institutions they had created. Also, workers' health and industrial health concerns tended to get lost in the celebration of declining accidents in the workplace, promoted by the industry-supported National Safety Council. The twenties, however, did not see the overall industrial hygiene movement decline, despite fading enthusiasm for occupational health reform in medical and governmental circles and in the American Association for Labor Legislation. Rather, the movement intensified during this period.

In the early 1920s two new groups joined the AALL's agitation for healthier workplaces and compensation of industrial diseases. One was the Workers' Health Bureau, an agency started by two women in 1921 to give workers access to experts and data on industrial diseases. The bureau was shaped by its founders' radical analysis that science was an adversarial activity informed by class interests and by the founders' insistence that workers control scientific research to ensure that the ensuing diagnoses and prescriptions reflected laborers' realities and needs. Recently, the Workers' Health Bureau has been the subject of two similar studies, one by Angela Nugent and the other by David Rosner and Gerald Markowitz; both studies describe the bureau's success throughout the 1920s in investigating and publicizing dangers to worker health and in winning the support of many local unions. Initially, a short-lived radical philanthropy funded the Workers' Health Bureau, and the bureau hoped to continue operations by winning support of affiliated labor unions. Local affiliated unions did offer limited funds, but the bureau needed the financial assistance of the national American Federation of Labor. A combination of factors drew hostility from conservative AFL leaders: in an era when union survival was in doubt, health research was viewed as a luxury; male union executives hesitated to support a women-led organization; the politics of the bureau's directors were more radical than those of the AFL leaders. Finally, the AFL became suspicious that the bureau aimed at dual unionism by urging support for a national labor health program from both organized and unorganized workers. In 1925 the AFL began pressuring unions to withdraw from the Workers' Health Bureau, and by 1928 the organization folded.[27]

The Consumers' League was the second organization to carry the industrial hygiene movement through the 1920s. In this work we will be concerned with both the National Consumers' League and a local branch, the Consumers' League of New Jersey. Each had roots in settlement houses. The national league coalesced from a number of local consumers' groups concerned with the plight of women and children in industry, and in 1899 Florence Kelley became its director. Kelley, whose commitment to social activism grew from a grafting of socialist principles onto Quaker scruples, had been a resident in that most influential American settlement house, Jane Addams's Hull-House in Chicago. There she had pioneered in utilizing social science methodology to document the evils of sweatshop labor for women and children. Her work led to the passage of an 1893 factory act regulating women's and children's labor, and she was selected by Illinois's governor to establish and direct an office of factory inspection to enforce that law. In the struggle against exploitation of women and children, the Consumers' League promised a new weapon—the pressure of consumers' purchasing power— and Kelley was pleased to accept appointment as general secretary. She moved from Hull-House to New York's Henry Street Settlement and began to organize leagues throughout the nation. One of the first new leagues resulted from an invitation for Kelley to speak at Whittier House, a settlement in Jersey City, New Jersey. The Consumers' League of New Jersey was formed in 1899 during Kelley's first year with the league.[28]

Originally utilizing consumer boycotts and preferential purchasing to influence manufacturers' workplace policies, the Consumers' Leagues evolved a quintessentially Progressive Era style of reform entailing social welfare investigation, community education, and protective legislation. League members studied working conditions and household budgets, compiled comparative data on legislation affecting working women and children, and wrote and lobbied for model legislation. The famous Brandeis brief, submitted to the U.S. Supreme Court in 1908 to support hours legislation for women, was researched and written by members of the National Consumers' League. The New Jersey league, in its early years, counted among its successes the establishment of the state Department of Labor (1904), a law requiring seats for women workers in commercial establishments (1909), an act limiting women to ten-hour workdays, admittedly weak on enforcement (1912), and legislation prohibiting child labor (1914).[29]

Although much of their work focused on improving working conditions for women and children through the enactment of laws that established maximum hours, minimum wages, and restrictions on employment in dangerous occupations, league members hoped to use the special needs of women and children as a wedge to open protective legislation to workers as a whole. Protective legislation for women met with much success in the 1910s and into the early 1920s; indeed, by the twenties some women began to regard protective legislation as anti-feminist because it limited women's access to the labor market. This debate would divide feminists throughout the 1920s and 1930s, with the Consumers' League always insisting that protective legislation benefited working-class women and their children.[30]

Both the local New Jersey league and the National Consumers' League remained small in size. The New Jersey branch, for instance, had only 675 members in 1922, shrinking to 500 by 1930. But the leagues' influence was out of proportion to their numbers; national and local leagues were respected and effective. Part of their impact resulted from their relatively narrow focus on helping working women and children (although often they aimed to help men as well), whereas other women's organizations, like the YWCA, the Women's Clubs, and settlement houses, called for broader reforms affecting the urban poor at home, in their neighborhoods, in politics, and at work, as well as the improvement of middle-class women's lives. Another reason for the leagues' influence was their ability to network with other reform groups. In 1920 the New Jersey Consumers' League helped form a Women's Cooperative Legislative Bureau to lobby the state government in Trenton. Five other women's groups took part, including the State Federation of Women's Clubs, the Woman's Christian Temperance Union, and the League of Women Voters. Ties between the Consumers' League and the League of Women Voters were particularly strong. The president of the Consumers' League was also the chair of the League of Women Voters' Committee on Women in Industry, and the two organizations' programs were usually identical. Katherine Wiley, executive director of the New Jersey Consumers' League, reported to the LWV's Women in Industry Committee as well as to the Consumers' League.[31]

Alice Hamilton contributed to links among the three industrial hygiene reform organizations of the 1920s. She served on both the AALL's General Advisory Council and the Workers' Health Bureau Science Advisory Board, and she helped obtain funding for the latter from the American Fund for Public Service even though she was offended by the Workers' Health Bureau's radical politics. Hamilton sought to unite these reformers' activities with similar efforts by universities, public health agencies, and, ultimately, the National Consumers' League.[32]

Hamilton was responsible for pulling the Consumers' League into the industrial hygiene movement. She had pitched industrial health reform to women's organizations since her early industrial health work in Illinois, and at least in part because of pressure from women reformers, Illinois passed the nation's first occupational disease law in 1911. The same year Hamilton described her lead research to the National Conference of Charities and Correction, a group with a large female component, and she joined its Committee on Standards of Living and Labor. Florence Kelley, general secretary of the National Consumers' League, was a friend of Hamilton's as well as a former resident of Hull-House. Kelley offered Hamilton an entrée into the National Consumers' League, and by 1923 Hamilton was its vice president. From that post she campaigned to orient the league to industrial health reform. She also created and chaired the league's Committee on Poisons in Industry. By 1923 the national league was asking state leagues to promote industrial hygiene departments within state Labor Departments and to undertake the study of dry cleaning establishments and the use of "benzol" (benzene) in various manufacturing processes. Hamilton was alarmed by the growing number of chronic benzene poisoning cases among women and girls using benzene-based glues in industries as varied as tire manufacture, shoe manufacture, and canning.[33]

Still concerned with lead poisoning, Hamilton spoke about lead-related illnesses during two reformers' meetings in 1922. At an annual National Consumers' League Conference, she reviewed literature on lead poisoning among women in potteries. In 1912 the U.S. Department of Commerce and Labor had published Hamilton's study of New Jersey potteries, which revealed that 36 percent of their employees suffered from chronic lead poisoning, a rate twice that in Britain. This report had been challenged by the manufacturers, who claimed that there was no lead poisoning among American potters. In a second study sponsored by the American Association for Labor Legislation in 1918, twenty cases of lead poisoning among potters were found within a four-day period. Hamilton also discussed the dangers of dry cleaning liquid fumes. At a second meeting, a National Social Workers' Conference, she spoke on industrial poisons during a luncheon of Consumers' League secretaries.[34]

Katherine Wiley attended both of these meetings and in April 1923 wrote to Hamilton. During the league's annual conference she had learned that her state housed the second largest pottery industry in the nation, and she wished to study the prevalence of lead poisoning among local potters.[35]

Like so many other women reformers of her period, Wiley began her career in a settlement house. Settlement houses aimed to make neighbors of the urban poor and privileged reformers so that friendly bonds between them would yield insight into the causes and solutions of poverty and its attending social problems. Settlement houses also permitted America's first generation of college-educated women to create both meaningful work and a life outside the family. Wiley, who had projected a career as a domestic science teacher, approached her first job in a Baltimore settlement house from that perspective. In trying to win over a small gang of boys, she decided to teach them to cook. "The boys were fascinated," she was pleased to discover, and while they were "learning how to boil and bake . . . I was learning more and more about life in the homes from which they came." Wiley resolved to "do more than teach cooking," and she "went in for a broader phase of social work." Her next position was in a settlement house on the upper East Side of New York City. During World War I she worked in programs for immigrants, which led her to northern New Jersey, where, after the war, she became involved in various women's political and reform organizations. By 1919 she was a member of the League of Women Voters' Women in Industry Committee, and by 1921 she was executive secretary of the New Jersey Consumers' League.[36]

Following the passage in 1923 of a league-promoted bill forbidding night work for women, Katherine Wiley began searching for a new project on which to concentrate her energies. She was the sole paid employee of New Jersey's Consumers' League, and so the direction of her work had a strong influence on the efforts of the New Jersey league at large. Health in the workplace captured her interest when she heard Alice Hamilton speak about occupational diseases.[37]

After writing to Hamilton about her industrial hygiene speeches, Wiley contacted two other experts who would figure prominently in the radium dialpainter cases. First, she wrote to John Roach, director of the Bureau of Sanitation and Hygiene in New Jersey's Department of Labor, asking for information on the pottery industry and laws controlling working conditions. Roach replied that Frederick Hoffman, a statistician and a pioneer in comparative mortality figures, had studied the pottery industry for the Prudential Insurance Company and concluded that poisonings in New Jersey were rare. A recent U.S. Public Health Service report on the industry had, Roach pointed out, mingled together statistics from Ohio, Pennsylvania, and New Jersey, and he implied that, if isolated, New Jersey's statistics demonstrated superior conditions. A final exchange of letters during this period brought Hoffman's assurance that the "risk of lead poisoning in this country is much less than that indicated by Dr. Hamilton's reports." Hoffman would play an ambiguous role in the dialpainting cases, first helping and then hindering the women's campaign for recognition of their disease.[38]

Hamilton responded to Roach's claims and Hoffman's assurances as well as to Wiley's interest. She noted that poor conditions were reported in *all* the plants studied by the health service. Part of the problem in the pottery trade, Hamilton continued, was with the workers' union. The hatters' trade, she argued, in which mercury poisoning had formerly been ubiquitous, was "revolutionized" partly because of a union "eager" to assist. In contrast, "the National Brotherhood of Operative Potters will not do much to help you," Hamilton wrote. "They have always resented their trade being regarded as dangerous because the insurance companies discriminated against them." Wiley took this argument another step. In a letter to Roach at the New Jersey Labor Department, she asserted, "Since women and girls are not organized in the pottery trade," that is, because women in the pottery industry did not have a union to look out for their interests, "we would like to go into it further."[39]

Wiley, then, was well aware of some of the problems confronting those wishing to ameliorate working conditions in the 1920s. Experts frequently disagreed with one another. It was hard to prove that a problem even existed. Unions might be uninterested or even hinder research, and women often had no union to turn to.

Despite the difficulties inherent in industrial health reform, Wiley was undaunted. During a National Consumers' League conference in Spring 1923, she met with Hamilton, Roach, and Hoffman to further discuss New Jersey's potteries. Here Roach made the relatively radical suggestion that New Jersey's compensation laws be expanded to cover all occupational diseases, a goal soon adopted by the Consumers' League.[40]

By 1924, then, the New Jersey Consumers' League, particularly its executive secretary Katherine Wiley, was primed for action. This readiness explains why it was to the Consumers' League that the frustrated state bureaucrats turned. The radium poisoning cases activated a preconceived campaign to promote industrial health.

The presence of so many women in industrial health circles has been little noted. The major reform organizations involved—the American Association for Labor Legislation, the Workers' Health Bureau, and the Consumers' League—all were shaped by women with ties to the women's reform community formed during the Progressive Era. In any list of reasons for twentieth-century attention to industrial health, the interests of such women reformers must be included.[41]

The AALL felt a strong female influence. A mainstay of its research department was Irene Chubb, and the assistant secretary of the organization was Irene Osgood. Lillian Erskine, the chief of New Jersey's Bureau of Industrial Statistics who conducted early research on the dialpainters' maladies, was earlier an industrial disease investigator for the AALL. Women from the Consumers' League, the Women's Trade Union League, settlement houses, and other reform organizations joined the AALL. Jane Addams and Lillian Wald were vice presidents; Grace Abbot, Mary Dreier, Josephine Goldmark, Alice Hamilton, Ida Tarbell, and Mary Van Kleek served on the General Administrative Council.[42]

The ties of the two founders of the Workers' Health Bureau to women's reform circles has been noted but little discussed. One of them, Grace Burnham, was swayed to choose workers' health as a career both by her father's background as a neurology professor at Yale and by her mother's work as a social reformer who resided for a time at Hull-House. Burnham gained experience through employment at the New York State Department of Labor Factory Inspection Commission and at the Joint Board of Sanitary Control of the New York City garment trades, organizations themselves shaped by women workers' health and safety concerns. The other founder, Harriet Silverman, came out of the prewar social work movement, itself a primarily woman-led enterprise until it became professionalized in the 1920s. Silverman also worked for the Joint Board of Sanitary Control. Thus, the leaders of the Workers' Health Bureau had ties to the women's culture that underlay much social reform during the Progressive Era, and both had found jobs in agencies created amid feminist-progressive-labor ferment in New York.[43]

Of the organizations promoting industrial health in the 1920s, the Consumers' League was the most long-lived and its commitment to reform outlasted all others. Why, it must be asked, were women so interested in industrial health and so prominent in industrial health reform?

In part, their interest in the workplace reflected the traditional concern of American women for the health and welfare of those around them. Nineteenth-century women expressed this largely by engaging in voluntary, charitable activities, which led in the twentieth century to nationally organized, often remunerated work as professional reformers, government officials, or both. Twentieth-century reformers did not abandon older forms of organization, however. Kathryn Kish Sklar argues that well-organized women reformers drew support and vitality from a well-maintained network of grassroots organizations, especially when compared to the American Association for Labor Legislation, which declined to nurture ties to local-level groups.[44]

Further, middle-class women's concern grew from their identification with women of other classes; gender interests allowed cross-class sympathies to develop. Working women's problems with their own health, as well as wives' and mothers' worries about the health of their families, led middle-class women reformers to adopt health issues as an important part of their program. Women reformers' health concerns, then, were never just for women's health alone, but for the health of working men and children, too. Sklar contends that gender served as a surrogate for class in these reform movements. Working-class women and men could not organize as workers or citizens because the state repressed labor organizations, the judiciary overturned labor laws, and political parties were based on coalitions that often precluded support for working-class legislation. Thus, women promoters pushed protective legislation for working women based on their gender, not their class, and hoped that this would serve as a wedge to widen political structures later to better serve all workers and their families.[45]

Finally, middle-class women reformers of this period were often drawn from the first generation of women with college educations, and they sought to build careers for themselves. Working-class health was an area open to such career and institution building, and middle-class women migrated to it partly for their own benefit. Like the women working for children's health and welfare described in Robyn Muncy's *Creating a Female Dominion in American Reform, 1890–1935,* women industrial health reformers evolved to fill a niche underpopulated by men. Sklar points out that middle-class male reformers were poorly organized both because of repression within universities of their working-class advocacy and because of traditions of limited government. The absence of men in reform circles permitted women to use gender analyses to stimulate improvements in working-class life.[46]

If self-interest as well as altruism ruled women industrial health reformers, this is no more or no less true of most figures in this history. What we must attend to as we follow Consumers' League efforts on behalf of the dialpainters is how effectively women reformers promoted the workers' agenda for recognition of their disease, compensation for their suffering, and prevention of future cases. Despite intransigent business and government leaders, the league eventually forced some, albeit limited, gains on all of these fronts. A major reason for this success was the Consumers' League's wily maneuvering to overcome the hesitancy of medical experts to take sides on a controversial issue that might affect their funding from industry.

Katherine Wiley found the dialpainters' physicians and dentists "willing and anxious to do anything possible to end this hazard to health." Hazel Vincent Kuser had consulted a New York physician and dentist, Dr. Theodor Blum. Although he did not know the exact composition of the luminous paint, he wrote to Wiley that in Kuser's case there was "practically no doubt in my mind as to the responsibility of the company" and promised his cooperation. (On the other hand, Blum apparently wrote to U.S. Radium promising to treat Kuser secretly, to not reveal to her the nature of her malady, and thus to protect the company from liability suits in exchange for its payment for Kuser's treatment.) Wiley visited Katherine Schaub to talk about her cousin

Irene's death; although Schaub was fairly well recovered from her mouth problems, her dentist told Wiley that her teeth were "radio active."[47]

To Wiley, this seemed sufficient evidence of a health problem at the Orange plant, and she tried to get the government to act. In May 1924 she presented her conclusions to Dr. Andrew McBride, commissioner of the New Jersey Department of Labor. McBride, according to Wiley, was "furious" when he learned that the Consumers' League was studying the radium cases. John Roach, who was in charge of industrial hygiene at the department, had given Wiley a list of sick women, and McBride called him into the room and "rebuked him in my presence." To this "severe calling down" of his subordinate, Wiley responded by saying to McBride "a few words about the Labor Dept. being a public agency, etc." McBride then assured Wiley that an ongoing investigation had found no connection between the women's work and their illnesses.[48]

During Wiley's confrontation with the labor commissioner, McBride told her that Dr. Harrison Martland, a physician and member of the Newark Board of Health, had studied the matter and found no evidence of an industrial health problem. It is unknown exactly what research findings Martland presented, but Wiley later conjectured that at this point his investigation consisted of no more than one meeting with several affected women in their dentist's office. Martland would later claim the "discovery" of radium poisoning.[49]

In the face of McBride's denial that the dialpainters' illnesses had an occupational origin, Wiley argued that "to the lay person" it seemed "that *something* about that factory caused those deaths and these sick girls to have identical symptoms." When McBride asked for her recommendation, she replied that she desired an inquiry by the U.S. Public Health Service. McBride asked her to put her request in writing, "which I did at once."[50]

Wiley tried other means to secure an official investigation. Working through the American Association for Labor Legislation, she suggested to the U.S. Department of Labor that it hire the noted toxicologist, Alice Hamilton, to study the radium problem. Hamilton expressed interest in the dialpainters, but the Labor Department, strapped by budget cuts during the Harding presidency, stated that it had no funds with which to proceed. The commissioner of labor statistics did promise an investigation beginning in July; however, nothing was done by the federal department during the summer or the fall.[51]

Wiley also wrote to the Public Health Service for information; officials replied that little was known about the dangers of dialpainting. An inquiry to the Massachusetts Department of Labor and Industries produced the response that the agency knew of no cases in the state. On a more hopeful note, Wiley contacted the prominent statistician and expert on industrial diseases, Frederick Hoffman. As a resident of nearby Newark, Hoffman was close at hand and, at Wiley's urging, began to look at the evidence.[52]

Meanwhile, Katherine Schaub was again bothered by problems with her jaw, and by coincidence she sought a consultation with the same New York physician and dentist who was treating Hazel Vincent Kuser. Schaub and Kuser met in the waiting room, and the two of them confronted their doctor with their similar work experience and symptoms. He was "surprised," Schaub reported. Theodor Blum labeled the condition "radium jaw" and mentioned it during a paper on jaw necroses at an American Dental Association conference. His parenthetical remarks were scarcely noted.[53]

By November 1924 Wiley disclosed with exasperation that in spite of all her networking with officials and physicians, "nothing has been done by them." In December 1924, despite her new husband's attentive care, Hazel Vincent Kuser died. Moreover, Marguerite Carlough was very ill. In January 1925 the New Jersey labor commissioner informed Frederick Hoffman that a thorough investigation had yielded no evidence that the dialpainters' illnesses were caused by their employment. The president of U.S. Radium wrote to Wiley, "If a common cause exists [in the women's illnesses], we are convinced that it lies outside of our Plant and that this is not an occupational disease."[54]

Two years had passed between the 1922 complaint by Irene Rudolph's physician about her alleged industrial disease and the 1924 denials of any connection between dialpainters' illnesses and their work by U.S. Radium and the New Jersey Labor Department. For about half that time the Consumers' League had been working for the recognition of radium poisoning. The next chapter relates how one more year's work would see radium poisoning acknowledged by important medical scientists and government bureaucrats. Radium companies, however, continued to insist that the dialpainters' illnesses were not related to their work, and they began to implement strategies to contain the threat that radium poisoning posed to their businesses.

A "HITHERTO UNRECOGNIZED"

OCCUPATIONAL HAZARD

The Discovery of Radium Poisoning

The Consumers' League campaign to prove the existence of the dialpainters' industrial disease eventually interested a number of scientific experts who would then contest for the right to claim discovery of radium poisoning. But although on the one hand radium poisoning, like the New World, was "discovered" by those who first inhabited it (in this case, the dialpainters), on the other hand "the discovery of radium poisoning" was less a discrete medical event than it was an ongoing political process. Further, questions about the existence of radium poisoning remained inseparable from other issues, such as the route radium took into and through workers' bodies. That is, whether radium poisoning existed or not was tied to debates about its etiology—how radium entered the body, how it did its damage. These etio-logical controversies, in turn, were related to what kind of radium exposure was dangerous and how much exposure was dangerous. These were political issues in that they involved negotiations among management, workers, and the state about acceptable working conditions. Disputes over radium poisoning's scientific discovery, then, were inextricable from and indeed dependent on debates regarding prevention, because how radium poisoning was defined would determine the measures required to eradicate it.

Richard Gillespie uses the phrase "accounting for illness" to describe the political process of the discovery of industrial disease and how social knowledge about "what has happened" to cause disease is inseparable from programs about "what will happen" to prevent more disease. In an account of industrial lead poisoning, Gillespie portrays scientific expertise as an ideological construct enabling businesses, scientists, and the state to collude (if unconsciously) in producing theories about industrial diseases most amenable to minimal transformations at work and in the larger political balance. This account of the discovery of radium poisoning suggests, too, that knowledge about industrial diseases is a contested site of power.[1]

Three related debates were part of the process of discovering radium poisoning. First, did radium poisoning exist nowhere else than among Orange dialpainters? Second, was any particular isotope of radium more dangerous than the rest? Could workers continue to use with abandon a "safer" radium? And third, was lippointing the source of the dialpainters' body-burden of radium, or might inhaled radium dust be a factor? Would the elimination of lippointing alone make dialpainting a healthy occupation? The business officials involved worked toward a certain set of answers to these questions. By concealing radium poisoning outside the state, the New Jersey illnesses might be attributed to local conditions, not radium exposure. If a particular radium isotope were blamed, businesses could switch to other, safe isotopes. If lippointing alone were identified as the culprit, inexpensive technological innovations could eliminate it. The Consumers' League, working for the dialpainters, continually undermined industry attempts to define the radium problem. These debates meant that radium poisoning came into existence with a political agenda at its heart.

In January 1925 Katherine Wiley of the New Jersey Consumers' League received a letter from Alice Hamilton, the eminent toxicologist, Harvard professor of industrial hygiene, and officer of the National Consumers' League.

Hamilton was writing to say that she would get herself appointed special investigator by the U.S. Bureau of Labor Statistics and travel to Orange to interview one of the dialpainters.[2]

Before she could start her investigation, however, Hamilton learned something that changed all her plans. While Wiley had been desperately trying to arrange an investigation, three of Hamilton's colleagues were working with the U.S. Radium Corporation to study the industrial hygiene of dialpainting. In March 1924, almost a year earlier and in the same month that Wiley began her involvement with the dialpainters, the radium company's president had written to Cecil Drinker, a founder of Harvard's industrial hygiene program and the school's assistant dean of public health, asking for his help.[3]

Drinker had formed a team of three physicians drawn from the Harvard program—himself, his wife Dr. Katherine Drinker, and a junior colleague, Dr. William Castle. The Drinkers had visited Orange in April 1924; Katherine Drinker and Castle made a second trip later that spring. The plant was toured and studied, the few remaining dialpainters were examined, and several ill women and various dentists and physicians were consulted.[4]

The Harvard team immediately fastened on radium as the likeliest cause of the dialpainters' difficulties, based on a review of the literature similar to the one undertaken here in Chapter 2. Radium, they knew, could linger in the body for a long time, especially in bone. Radium was chemically similar to calcium and strontium, and, like both, could deposit in bone. Radium destroyed tissue and disturbed the blood composition. Radium's gamma rays were similar to X rays, and necrotic jaws had been produced through exposure to X rays.[5]

Yet the Harvard researchers spent little time considering the lippointing route of exposure. The radium salts used in the paint were insoluble, and it was possible but not likely, they thought, that imbibed radium would be absorbed into the body. Also, Cecil Drinker was expert in the toxicology of dust and fumes, and his brother Philip, an engineer in Harvard's industrial hygiene program, was expert in the technology of removing dust and fumes, so it was on dust and fumes that the Harvard team concentrated.[6]

Dust samples taken throughout the plant—even in offices—proved to be luminous and radioactive. The workers, the Harvard study revealed, "were luminous. . . . The hair, faces, hands, arms, necks, the dresses, the underclothes, even the corsets, of the dialpainters were luminous." William Castle remembered in 1984 how the women glowed "like ghosts" in the dark. To study radium's deposition, the researchers arranged for several cats to inhale the powdered luminous paint; afterward, radium was found fixed in the cat's bones. The Harvard study concluded that radium, and likely inhaled radium, had caused the dialpainters' "jaw rot." [7]

In addition, the Drinkers and Castle considered the long-term effects of exposure to gamma radiation. At that time, a suggested safety guide for radiation exposure allowed radiation below that necessary to fog dental films in two weeks; all films left throughout the Orange plant fogged in less than two weeks and in the dialpainting studio they fogged in two or three days. Moreover, of twenty-two employees examined, "no blood was entirely normal." [8]

Cecil Drinker reported to the U.S. Radium Corporation in June 1924: "We believe the trouble . . . is due to radium." He recommended precautions that, he thought, "will not prove a serious matter financially." Arthur Roeder, president of the company, objected to Drinker's conclusions. "There is no place where Radium is not present in good amounts," he wrote, adding that radium exposure was probably even higher in other departments of his company. "If we are suffering from a new ailment caused by Radium, it should occur pretty generally throughout the plant and throughout other Radium plants." Several weeks later, in another letter to Drinker, he asserted: "Your preliminary report is rather a discussion with tentative conclusions based on evidence much of which is necessarily circumstantial. . . . Our conclusion is that there is nothing harmful anywhere in the works." [9]

During all the months when Katherine Wiley had insisted that there was "*something* about that factory," a full-fledged medical investigation had been under way. Imagine Wiley's frustration when she found out that Drinker would not release his data without the radium company's permission! And then Wiley learned that Alice Hamilton chose not to investigate the radium cases "because of her chief's position in the matter." [10]

But Wiley persevered. With Hamilton's behind-the-scenes advice and assistance, she set out in February 1925 to gain access to the Harvard data. They pursued three strategies. First, Wiley and Hamilton attempted to play on Drinker's professional jealousy. The Prudential statistician Frederick Hoffman, armed with Wiley's data, announced in March that he would deliver a paper on the radium cases at the May meeting of the American Medical Association. He sent Cecil Drinker a copy of his proposed paper revealing the existence of a new occupational disease. [11]

This ploy bore some results. Drinker immediately wrote to U.S. Radium advising immediate publication of the Harvard team's findings to forestall the forthcoming, more damaging pronouncements and to "convince the public that you have done everything humanly possible to get to the bottom of the trouble." "Dr. Drinker's attitude," Alice Hamilton noted dryly, "has changed since he finds someone else will publish if he does not." Hamilton urged Katherine Wiley to write to Roeder along similar lines, advising him that Drinker should be allowed to publish his own paper because for public relations reasons the company might want to seem to have attempted discovery of possible dangers of dialpainting.[12]

Meanwhile, Hamilton and Wiley proceeded with their second strategy. Wiley and the dialpainters had been trying for a year to find a lawyer willing to take on their cases. On consulting a judge, Wiley had learned that the women were not eligible for workers' compensation, because radium poisoning, a new disease, was not on the schedule of compensable diseases in New Jersey's workers' compensation act. If it were added to the list, compensation could not be paid retroactively. The state's legal aid office concurred with this opinion and seemed little interested in helping to file a damages suit through the civil courts. But with much effort, Wiley found a lawyer for the dialpainters, who filed a suit on March 10, 1925. The first plaintiff was Marguerite Carlough, and the plan was to subpoena Cecil Drinker and his data. Although in the long run this approach might have worked, in the short run it backfired: faced with a liability suit, the radium corporation finally informed Drinker that his study was too sensitive to release.[13]

The final strategy pursued by the Consumers' League women bore surprising fruit. Hamilton thought that the state Department of Labor might be persuaded to demand the Harvard report from U.S. Radium, in spite of her belief that the New Jersey labor commissioner was "going on the principle that capital is not getting a square deal." Much to Wiley's surprise, the Labor Department informed her that it had received a summary of the Harvard study's findings from the radium corporation almost a year earlier. The memo assured the Labor Department that U.S. Radium had been cleared of all blame in the dialpainters' illnesses.[14]

Hamilton knew this to be a lie. This time she circumvented Cecil Drinker. In yet another example of the power of women's networks, Hamilton wrote to Cecil Drinker's wife and colleague, Dr. Katherine Drinker. The radium corporation's president, she related, "tells everyone he is absolutely safe because he has a report from you exonerating him from any possible responsibility in the illnesses of the girls. . . . The Labor Department . . . has a copy of your report . . . that shows 'every girl in perfect condition.' Do you suppose Mr. Roeder [the president] could do such a thing as to issue a forged report in your name? . . . Do you not think that by misquoting you, using your names to bolster his cause, Mr. Roeder is failing in honorable dealing with you?" Katherine Drinker quickly replied, "Cecil . . . Dr. Castle and myself . . . were very indignant at the idea that Mr. Roeder might have misquoted us . . . if that were the case we should be quite justified in publishing our report without further delay. I am therefore a bit hopeful that he has proved a real villain." Subsequent checking with the Labor Department confirmed that Roeder had indeed misrepresented the Harvard study, and Cecil Drinker arranged to have it published.[15]

Once the New Jersey commissioner of labor received the full Harvard report, he finally acted, insisting that all of the recommended safety procedures be instituted. Human life, he wrote, was too important to be wasted, and U.S. Radium must now "comply or close." The company left town, although it soon reopened in New York City.[16]

Drinker's withholding of valuable evidence seems unconscionable. By refusing to release his data and conclusions, he was acquiescing in further exposures and preventing dialpainters from seeking compensation. Yet U.S. Radium fully expected his silence; its lawyer wrote Drinker that he was "astonished" that Drinker had arranged for publication. The letter is worth quoting at length:

> It is my understanding that you were retained by and at the expense of the Radium Corporation to make the investigation in question. . . . While, of course, the Radium Corporation did not desire this fact to in any way affect your investigation and conclusions, it certainly assumed that in case your conclusions were adverse to its claims you would not publish. . . . This is the first occasion in my experience covering many years with doctors and other expert witnesses of high character when any expert has ever threatened, during the pendency of litigation, to publish a report adverse to the party to such litigation who had retained him.

Furthermore, none of Drinker's colleagues — not even Alice Hamilton — ever suggested that he acted unethically. Rather, almost everyone understood that Drinker's acquiescence to U.S. Radium's suppression of his study reflected his understanding of the obligations of a compensated consultant. His colleague, William Castle, has sought to explain Drinker's actions by noting that even medical ethics have a history: basing his reply on Carleton Chapman's book, *Physicians, Laws, and Ethics*, Castle asserted that medical ethics in the early part of the century were not patient-centered but focused on relations among physicians. In this case, Drinker's ethical concern was with his relationship not to another physician but to those who paid for his study, a very different matter.[17]

Drinker's attitude can be better understood by looking at the history of Harvard's Industrial Hygiene Department and at the views of some experts in the field. The influence of business on academic industrial hygienists is obvious in the following account and explains Cecil Drinker's hesitation to publish a study that suggested the presence of radium poisoning among New Jersey's dialpainters.

The Department of Industrial Hygiene at Harvard was among the first in a wave of such departments to be established in the 1910s. In 1913 Harvard and the Massachusetts Institute of Technology offered a joint program to train health officers that included courses in industrial hygiene. By 1916 Harvard had established an Industrial Clinic to treat occupational accidents and diseases, and by 1918 the Harvard Medical School had established its own degree-granting program in industrial hygiene.[18]

As early as 1915 the Harvard faculty opened discussion on relations between industrial hygiene specialists and industry. Dr. David Edsall, a founder of the program and later a dean of both the Harvard Medical School and the School of Public Health, became involved in one such exchange. Edsall was a reformer who promoted industrial hygiene research because of a "desire to provide material that would be useful in making laws and special rules relating to occupation and its effect upon health." Nevertheless, he cautioned researchers that the university should not directly investigate individual cases nor report to government agencies because "employers are afraid of what the State [Compensation] Board may do and it might injure the hospital to be identified as being in close cooperation with them."[19]

Industrial goodwill became even more important in later years. From 1918 through 1922, industrial hygiene at Harvard was completely funded by business contributions. Dr. Frederick C. Shattuck, a professor of clinical medicine at Harvard, raised $125,000 from such giants as U.S. Steel and General Electric—leaders in promoting smoother capital-labor relations—as well as from smaller New England firms. Other businesses arranged for "consulting relations resulting in a definite service."[20]

Business financing had repercussions for the pursuit of scholarship that did not go unremarked at Harvard. For instance, Katherine Drinker, as editor of the *Journal of Industrial Hygiene,* launched at Harvard in 1919, found it necessary to exercise discretion in choosing articles for publication so as not to lose financial support for the program. Edsall reported that Alice Hamilton criticized another university's industrial hygiene program for becoming "a purely consulting and commercial affair without any special reference to teaching or real academic research. In my mind," Edsall continued, "this is the principal danger in any industrial hygiene organization in a university." During a discussion of what Harvard doctors called the "outside jobs," when a pair of physicians warned that Harvard "must not sell its services too cheaply," Edsall complained that "the matter had become very commercial . . . the University was undertaking to supply a type of service which it had no intention of giving." [21]

Also in 1919, Frederick Shattuck discussed with the Rockefeller Foundation the relation between research and business funding. After defending the scientific value of the work at Harvard, he considered the effects of business funding, writing, "Obviously, industries should pay the expenses of any particular investigation . . . but it is equally obvious to my mind that it should not be conducted on a money-making basis. . . . keep the commercial element out of it." On the other hand, he added, there was the expense of keeping equipment and personnel on hand. No single individual or corporation was willing to pay for such overhead. [22]

The Rockefeller Foundation responded that year with a conference on industrial hygiene, where other specialists commented on relations between business, labor, and industrial hygiene. Dr. J. W. Schereschewsky, assistant surgeon general of the U.S. Public Health Service, promoted the view of medicine as a mediator between capital and labor, benefiting both: "We have in Industrial Health Service a means whereby capital and labor may meet on common ground. It becomes a means of revealing the identity of their interests. It is susceptible of forming one of the most potent agencies which we can devise for welding society into a harmonious whole." Dr. Otto Geier, an industrial surgeon at the Cincinnati Milling Company, was more radical. Medicine, he argued, must be more "democratic," more available to "the average patient . . . not . . . to the rich and poor only." He continued, "Efforts must be made to democratize knowledge for the benefit of better medical service to the great producing masses of the people." Rather than fund further research, Geier hoped that the Rockefeller Foundation would help set up industrial clinics to bring already existing medical knowledge to workers. "Capital," he insisted, "must pay for the increased health facilities that are to be brought about." [23]

Others rejected both of these positions. In a 1919 Public Health Bulletin, Dr. Clarence Selby scoffed at the idea that industrial hygiene should be supported by business to help workers or to ameliorate discontent, insisting that "industrial medicine is not a gift to labor, nor a charity, but purely a function of good business." By preventing illnesses, industrial hygienists increased the efficiency and productivity of the workforce. In 1921, another industrial physician complained, "Are we in industry to do a charitable act . . . ? . . . Are we in industry to help carry out some soft, silly, social plan? Are we in industry to buy the goodwill of the employees?" and concluded, "no." Instead, "We are in industry because it is good business." [24]

At the Rockefeller conference in 1919 there were hints, too, that physicians were in industry not only to help businesses profit but also to benefit themselves. Dr. C. E. Ford of the General Chemical Corporation complained that his "great problem" was "in securing medical men with a social conscience. . . . I had only one application for the job. He demanded a salary of $10,000 a year." And Cecil Drinker reported that the Harvard postgraduate program was attracting "a dismal group of students. . . . a large proportion of them were obviously failures in general practice." These comments echo a story told on Harry Mock, a pioneer in industrial medicine. A friend once asked him why he wanted a job overseeing the health of Sears's sales staff. Mock replied, "Because I want to get married, I'm in debt, and I need to make money at once." Only on second thought did he acknowledge the "great opportunity" in studying Sears as "a great human laboratory." [25]

By 1922 direct financial support from business became less crucial to the Harvard program, as industrial hygiene was now to be funded by the Rockefeller Foundation as part of the university's new School of Public Health. Nevertheless, business still contributed money for special projects. Shattuck continued to promote the Harvard program as a way business could display interest in workers' welfare while advancing its own interests. In 1922 he reflected in a letter to his friend J. P. Morgan that Judge Gary of U.S. Steel might be willing to donate more money to Harvard. "It was thought possible," he wrote, "that Judge Gary might feel like doing something substantial to offset in part, at least, criticism to which he and the Steel Company have been subjected on the ground of inhumanity to their workers." He hoped for a $100,000 contribution. The next year Shattuck assured potential philanthropists that it was "distinctly understood that the subscriptions savored in no way of either charity or philanthropy; that they represented an investment, a sort of insurance if you like." Like some of the industrial hygienists quoted above, Shattuck sold industrial medicine not as "a soft, silly, social plan" but as good business. That it might appear to benefit labor was, however, to the good; in his promotion of the Harvard program to business, Shattuck speculated, "The employer who convinces his employees that he considers their interest equally with his own promotes an attitude of mind likely to yield a rich reward to all concerned."[26]

The chief of the Industrial Clinic in 1923 wrote Dean David Edsall that it was "desirable" that the industrial hygiene program solicit funds from industry; therefore, it was "essential that there be secured the cooperation and cordial interest of industrial organizations." He believed that the university could serve industry "without compromising itself unduly" by avoiding actual contracts in favor of "a willingness or even an eagerness" to be of assistance. "The academic freedom of the University need not be imperilled in any manner or degree," he concluded.[27]

But Cecil Drinker's unwillingness to publish data harmful to the corporation that funded his research suggests that industrial funding did compromise the university. Given his department's reliance on donations from industrialists, Drinker's reticence to offend the U.S. Radium Corporation becomes more comprehensible, if still not laudable. Few industrial hygienists ever suggested that their discipline should serve workers; at best, they promoted their field as a neutral mediator between capital and labor, and some went so far as to recommend industrial hygiene on its benefits to business alone.

Drinker's assessment of his duty to his funding source was likely shared by his colleagues. None of them indicated that he acted unethically, and Drinker eventually rose to the deanship of Harvard's School of Public Health. Only Katherine Wiley criticized his behavior; to her, he was "dishonest" and "showed a very unethical spirit."[28]

Beginning with the first two publications on radium poisoning—Frederick Hoffman's and the Harvard team's—opinions on how radium entered the body and caused damage diverged. Hoffman focused on the paintbrush route (although more on the actual application of radioactive substances to the teeth than on the swallowing of the material and subsequent consequences), whereas the Drinkers and Castle were more concerned with the dangers of inhaling radium from dust in the air. Over the next decades, attention would be paid largely to the oral route and the ingestion of radium, and the dangers of radium dust would be largely ignored, leaving dialpainters at some, if reduced, risk.[29]

The Drinkers' Harvard study was published in the *Journal of Industrial Hygiene* in August 1925, although Cecil Drinker appended "Received for publication May 25, 1925," a date that preceded Hoffman's AMA speech. The paper chronicled five cases, three of which already had resulted in death. The report emphasized the "absorption of radium through inhalation," "the literature on radium [which] indicates that . . . radium is deposited in bone," "descriptions of the effects of radium on bone already in the literature," and "over-exposure to radiation which we demonstrated in the plant we studied." [30]

Frederick Hoffman, statistician, actuarial expert, researcher for the U.S. Department of Labor, and expert on industrial diseases, published his conclusions on the radium dialpainters' maladies in the *Journal of the American Medical Association* in September 1925. Relying on Katherine Wiley's research, he discussed twelve cases: four deaths and eight illnesses. The deceased women were Irene Rudolph, Amelia Maggia, Helen Quinlan, and Hazel Vincent Kuser. In addition to the illnesses of Marguerite Carlough and Katherine Schaub, Hoffman revealed jaw problems in six other women. Both of the Smith sisters, Genevieve and Josephine, had necrosis of the teeth and jaws, according to X rays, and Josephine had suffered from an abscessed tooth and perhaps from anemia. Anita Burricelli also had been ill. She, like many of the other women, had started dialpainting as a teenager; she had worked with Katherine Schaub, and they became alarmed over blemishes on their faces they attributed to the luminous paint, though the company's owners assured them that the paint was harmless. Burricelli began suffering tooth problems in 1920 but by 1925, except for the loss of many teeth and a suspicious "swelling" in one breast, she was in good health. Another woman, Nettie Madlinger, had remained well following dental problems, but "a small growth" appeared "on the side of her neck." Grace Fryer had mostly recovered from her necrosis, which had led to the removal of teeth and bone on one side of the jaw. Hoffman did not comment on other complications that Fryer was experiencing: pains in her feet and her back. Finally, X rays of Vera Eldridge's mouth disclosed necrosis of the teeth and jaw. These cases led Hoffman to "the very obvious fact and conclusion, that the cases in question are of occupational origin." [31]

Hoffman centered on lippointing as radium's means of entry. The radium company, he wrote, "disclaimed any instructions [to lippoint] . . . but, to the contrary, . . . had warned employees against wetting the brush with their lips." (This, of course, contradicted dialpainter experience.) Hoffman tried to explain how ingested radium did harm. The Harvard study had implicated inhaled radium deposited *systemically*: radium deposited throughout the body accounted for both anemia and the decay of jaw tissue allowing subsequent infection. Hoffman ignored anemia and described how radium might act *locally* to cause jaw necrosis. Here he relied on the reasoning—indeed, quoted at length the words—of Sabin von Sochocky, the originator of the luminous paint formula and one of the founders of U.S. Radium's corporate predecessor. Von Sochocky believed that because the radium in luminous paint was insoluble, it could not be absorbed "to any large extent" into the system through the digestive track. If radium was not absorbed into the body, von Sochocky reasoned, then the small amount of paint constantly in the mouth must be the source of the dialpainters' necroses. There was too little radium in the paint, however, to have any effect: von Sochocky insisted that if such a small amount of radium were harmful, then physicists (like himself) and chemists who worked under conditions of high radium contamination would have had enough radium introduced into their mouths through smoking or dust inhalation to cause disease—but, von Sochocky mistakenly reported, workers never experienced such ill effects. Perhaps, he speculated, the jaw necroses were caused by "the joint effect of various agents." Zinc sulfide in the paint, in constant contact with mouth tissues, might irritate the jaw bones. Alpha particles from radium's radioactive decay might also be mildly "irritative," he conceded, if in constant contact with a particular spot in the mouth. This irritation might produce "local changes on the mucous membranes" sufficient to lower resistance to "pathogenic bacteria's entrance." Von Sochocky had difficulty explaining the mouth necroses because of his firm belief that small amounts of radium were harmless, and perhaps because fears about his own health led him to discount crucial evidence. Hoffman added to his report data from the U.S. Bureau of Standard's 1923 study showing that radium exposure even without actual ingestion did harm scientists. Still, Hoffman agreed with von Sochocky that the "insanitary habit of penciling the point of the brush with the lips" was at fault in introducing radium into the mouth, which then acted locally to cause necrosis.[32]

Hoffman repeated stories about the dialpainters' "luminous appearance," which might be evidence that radium was present in sufficient amounts to cause disease through inhalation. He rejected this possibility, writing, "Radium could not possibly be a factor in this luminosity." Hoffman did not explain this conclusion, but later denials of radium dust at large in the plant have pointed to radium's great expense, which logically suggests rigorous attempts to control it. According to the argument, no corporate officer would have sanctioned such expensive waste as signified by radium-laden dust. We know, however, that the president of U.S. Radium informed the Harvard team that "there is abundant Radium in other parts of the plant [besides the dialpainting studio]. . . . The entire back yard is filled in with old residues [sand left over after radium extraction and still somewhat radioactive]. There is no place where Radium is not present in good amounts." Arthur Roeder offered this as proof that radium could not have harmed the dialpainters, since, he thought, it was not harming those working elsewhere in the plant.[33]

Katherine Wiley had assured the publication of the Harvard study and supplied the data and dialpainters necessary for the Hoffman study. She could also take credit for the initiation of a federal Labor Department investigation of the dialpainting industry that began in April 1925. This was the same month in which it was established that U.S. Radium had been lying to the New Jersey Labor Department about the conclusions of the Harvard study: there may have been some connection between the New Jersey Labor Department finally concluding that there was a problem and the federal department beginning its study. Swen Kjaer of the U.S. Labor Department's Bureau of Labor Statistics was assigned the investigation. Kjaer visited Wiley, who told him that she was preparing a report on the New Jersey cases at the request of his bureau's commissioner; Kjaer decided to concentrate on cases outside of New Jersey. He also spoke to experts at the U.S. Bureau of Standards who apprised him of the medical literature concerning radium's dangers.[34]

Kjaer quickly identified U.S. Radium of Orange, New Jersey, and Standard Chemical, of Pittsburgh, Pennsylvania, as the major producers of radioactive luminous paint in America. Of approximately 120 firms nationwide using the paint, about 110 purchased it from U.S. Radium. Standard Chemical sold luminous paint to nine customers through its subsidiary, the Radium Chemical Company. Radium Chemical also undertook dialpainting itself through yet another subsidiary, the Radium Dial Company. In 1925 Radium Dial, located in Ottawa, Illinois, was the biggest dialpainting firm in the country, employing about one hundred women.[35]

The Radium Dial Company had opened its first dialpainting studio in Long Island City in 1917. Another studio, which began operations in Chicago in 1920, moved to the Illinois town of Peru. It left there because it was competing for female labor with Western Clocks, the firm for which it painted dials. Radium Dial then moved to Streator, Illinois, and in 1923 finally settled in Ottawa, where it purchased an old school building. The Ottawa dialpainters worked at school desks.[36]

Like their counterparts in New Jersey, the Illinois dialpainters enjoyed their job. They were, a journalist reported, "a happy, jolly lot. The work paid well. They had picnics and parties." They decorated their buttons and belts, and painted rings on their fingers, with the radium paint. Most were young; a few married women worked part-time.[37]

And like dialpainters elsewhere, the Illinois women were taught to lippoint their brushes. Another serious hazard, radioactive dust, may have been present. Reports about the cleanliness of the Radium Art Studio in Ottawa vary: Swen Kjaer found little dust, but one dialpainter said that the paint mixer worked with dust swirling around her; and at home, in the dark, dial-painters' hair, eyebrows, eyelashes, their fingers and mouths, would glow, perhaps from settled dust.[38]

The women working in Ottawa were assured that the luminous material was safe. Their instructor, wife of the plant manager and teacher of the lip-pointing technique, once ate the radium-laced paint from a spatula to dem-onstrate its innocuousness. The workers were told by their supervisor that radium would "put a glow in our cheeks," that "the paint would make us goodlooking."[39]

During Swen Kjaer's 1925 investigation, both the New Jersey and Illinois companies rejected theories that radium harmed the dialpainters. Arthur Roeder of U.S. Radium was more concerned about the damages suit filed by Marguerite Carlough. According to Kjaer's notes, Roeder claimed that the "ailment was not due to any cause in the factory, in fact was probably an at-tempt to palm off something on them." He also complained about the "great amount of publicity . . . given to the case of this operator through the news-papers, because women's clubs had taken it up."[40]

In Chicago, Kjaer met with a representative of the Radium Chemical Company and the Radium Dial Company (subsidiaries of Standard Chemi-cal), who discussed dialpainting in the firms' Illinois plant. Employees were being examined for possible illness, Kjaer was told, and—because the dial-painting workshop was in a small town in which it was difficult to maintain a sufficient workforce—he was asked "to handle the subject carefully, so as not to cause alarm among the workers unless circumstances justified it." Ap-parently Standard Chemical was still unsure that dialpainting held enough dangers to justify alarm among its employees. On visiting the dialpainting studio, Kjaer was assured by the woman in charge of the factory that "instead of proving detrimental to the health of the girls, she knew of several who had seemingly derived benefit from it [the radium paint] and showed decidedly physical improvement."[41]

Years later Kjaer discovered that tests of Radium Dial's workers in 1925 turned up at least one worker, Margaret Looney, with enough radium in her system to test radioactive. Another dialpainter, Catherine Wolfe, had symp-toms of radium poisoning: in 1924 pain had begun in her left ankle and spread up her leg to her hip; she also experienced fainting spells. Results from medical examinations of the Ottawa dialpainters were not made public or even released to the affected employees. In 1925 two men at Radium Dial's parent corporation, in Pittsburgh, underwent surgery for malignancies of the fingers that were attributed to their work with radium. One was Paul Hogue, who had packed radium salts since 1914. The second was Charles Viol, a national expert on radium as health tonic and health hazard, and a founder of Standard Chemical's medical clinic, in which he pioneered in the use of large amounts of radium as an internal medicine. Radium Dial successfully hid these illnesses until 1928.[42]

After a few weeks' study Kjaer concluded that radium was dangerous: he was especially concerned about, first, the research by William Cameron and Charles Viol showing that radium accumulated in bone; second, another 1923 study published in *Radium*, Standard Chemical's house journal, suggesting that overdoses of radium resulted in bone necroses and infections; and third, the London Radium Institute's reports of death and illness among radium workers. He discounted the theory that inhaled radioactive dust was a factor: the paint was very expensive, and he doubted that it was allowed to sift around the dialpainters' working quarters. Rather, he concentrated on the oral route, the tipping of the paintbrushes.[43]

Kjaer also sought to explain why cases had appeared only in U.S. Radium's New Jersey facility and not in other dialpainting centers in the country. One factor might be that U.S. Radium was the only American plant still refining radium (other plants now purchased radium from the Belgian Congo) and the only company that refined radium in the same facility in which its dialpainting was headquartered. Another factor might be U.S. Radium's use of the isotope mesothorium or radium-228 in addition to radium-226, which was used alone in most plants. Mesothorium was available in New Jersey as a by-product of the production of gas lamp mantles. Standard Chemical used pure radium-226. For some reason, Kjaer and the U.S. Labor Department curtailed their investigation at this point, after three weeks of work.[44]

Consumers' League activist Katherine Wiley might have been satisfied with the results of her work. The New Jersey commissioner of labor had closed the U.S. Radium plant in New Jersey, although Wiley's sense of victory was no doubt dulled by the plant's reopening across the river in New York. Perhaps more successful was her networking with industrial disease experts. Wiley had supplied data and dialpainters for Frederick Hoffman's study and the federal Bureau of Labor Statistics study, and she had assured the publication of the Harvard team's report.

Still, radium had not been definitely established as the cause of the dialpainters' illnesses. The radium companies were stonewalling, for one thing, and, in scientific circles, the Hoffman and Harvard conclusions were suggestive but as yet mere educated guesses. Without a sound understanding of the etiology of the disease, that is, how radium acted in the body to create the dialpainters' symptoms, radium poisoning was not demonstrated. Nor could it be prevented. Once more, Katherine Wiley helped arrange matters to bring about the discovery of radium poisoning's etiology.[45]

During the summer of 1925 she again called on the dialpainter Katherine Schaub. Schaub was recovering from a difficult winter. She had been depressed—"gloomy and morbid," in Schaub's own words. She had been ill over the winter, and Hazel Vincent Kuser, who had helped convince the physician Theodor Blum of the industrial origin of their similar symptoms, had died in December 1924. Schaub would later write of this time: "The massive cold white drifts against the windows made me think of nature in all her cruelty and destruction. . . . I was only twenty-two years old, with youth on my side, and yet no one was able to help me." By summer Schaub was feeling better, but if less depressed she was still anxious, and Wiley suggested that she consult the new county medical examiner, Dr. Harrison Martland, who had become interested in the dialpainters' cases. After Wiley left, Schaub turned to her sister and said, "It must be that I have radium poisoning." She went to see Martland the next day.[46]

Martland had known of these cases when they first arose, for as pathologist at a Newark city hospital he had been asked by the city health office to examine a few patients of the dentists Davidson and Barry. He did so but "failed to secure an autopsy," as he put it, and "lost interest in the matter." According to the commissioner of the New Jersey Labor Department, Martland had found no reason to suspect industrial poisoning. The pathologist's interest had revived by the summer of 1925, and he became intrigued by the radium cases. Following Frederick Hoffman's delivery of his paper at the AMA conference in May, Martland, newly appointed medical examiner of Essex County, autopsied a U.S. Radium Corporation chemist, Edwin Lemen, who had died from anemia and pneumonia. For help in measuring radioactivity and understanding radium's properties, Martland sought out two scientists: Sabin von Sochocky, former president of the U.S. Radium Corporation, and Howard Barker, a physicist who also served as the company's vice president. In exchange for their assistance, the scientists asked Martland to promise that he would keep his conclusions secret. Together, the three men established inhaled radium as the cause of the chemist's death and measured radioactivity in a body for the first time. Since Hoffman's study was, Martland conjectured, "based mainly on a survey, from the statistician's point of view," Martland now saw that he might win recognition for the first medical description of a new occupational disease.[47]

During the course of these investigations, Martland chanced to test von Sochocky for deposited radium. In this way, the inventor of radium paint learned of his own high body-burden of radium.[48]

The death of a fifth dialpainter—the sixth death if the chemist Lemen is counted—presented Martland his opportunity. He had been called in by a dentist, Dr. Joseph Knef, for a consultation about the illnesses of two sisters, both dialpainters. Marguerite Carlough, discussed in Chapters 1 and 3, had started dialpainting in 1919 at age eighteen. Her older sister Sarah Carlough Maillefer had begun two years earlier, when she was twenty-eight. Marguerite Carlough became ill quicker. In 1921, after two years on the job, she began suffering from fatigue, weakness, and tooth and jaw problems. She continued working until 1923. Sarah Maillefer painted dials for seven years, despite crippling pains in her left leg that began in 1923. She was discovered to be anemic as well. Maillefer developed a high temperature and was hospitalized; she died in June 1925.[49]

Tests run by Martland before and after her death showed that radioactive substances were stored in her body, especially in her spleen, liver, and bones. By this time her sister Marguerite was very ill with extensive necrosis of both jawbones: most of her teeth were missing, her lower jaw was fractured, and her palate had eroded so that it opened into her nasal passages. She was also anemic. Tests showed her to be radioactive.[50]

No promise of secrecy covered Martland's autopsy of Maillefer or his data on her sister, Carlough. Martland asked Katherine Wiley for help in finding corroborating cases, and so Wiley asked Schaub to consult with Martland. Compared to the other dialpainters, Katherine Schaub was quite healthy. She "did not appear ill," Martland noted, and her "general physique was good." Her jaw had healed, and she was only mildly anemic. But, tests proved, she was radioactive. "Who can tell when she may develop an acute fatal anemia, or a more chronic anemia, with or without local lesions and bone necrosis?," Martland wondered.[51]

Over the summer and fall Martland was beginning to understand that radium, whether ingested, inhaled, or injected, would find its way to the bones, and that deposits in the bones would act to cause the anemia common among dialpainters, technicians, and scientists. But he clung to Hoffman's view that radium introduced into the mouth by lippointing acted locally to produce jaw necroses. Lippointing, he would eventually write, produced

"local irritation, setting up in some cases gingivitis, dental infection of a peculiar intractable type and bone necrosis."[52]

Knowing (so his biographer claims) that "he had to publish his findings at
once or perhaps lose priority in the discovery" of radium poisoning, Martland
scheduled an appearance at a meeting of the New York Pathological Society
in October 1925. Both abstract and discussion were published in the *Proceedings of the New York Pathological Society*.[53]

The discussion is interesting. First Hoffman spoke, claiming that von Sochocky had explained to him that "the chief cause of trouble was mesothorium and not radium." Martland had supported this conclusion, he was
gratified to note. Actually, in his printed paper Martland noted that mesothorium emitted more alpha particles at a greater velocity than did radium
and so was more dangerous, but he did not free radium from suspicion of
harm. (The language here is confusing, as mesothorium is an isotope of
radium; recall that "radium" as commonly used referred to radium-226 and
"mesothorium" to radium-228.) Hoffman went on to conclude that the presence of mesothorium in U.S. Radium's paint formula explained why the employees of other plants had not experienced similar illnesses and death. Why,
we might ask, did no one wonder about dialpainters working elsewhere with
paint purchased from U.S. Radium?[54]

The next person to speak was Dr. Frederick Flinn, who Martland would
later learn was working for U.S. Radium. He reported on excretion experiments with guinea pigs fed mesothorium: 99.4 percent of the mesothorium
was excreted, leaving too little to cause disease, Flinn implied. Then, referring to the use of radium as an internal medicine, he asked, "Why is it that
these girls taking a small amount of radium through the alimentary tract have
this condition, and that no cases of necrosis occur by intravenous injection?"
Flinn meant to suggest that radium could not be harmful. Martland turned
again to the local explanation for jaw necrosis: the necrosis was caused not
by the radioactivity in bone but by deposits on the gum, mouth, and teeth.[55]

At this point Martland made an interesting observation. Based on literature
claiming a "stimulative" effect for radium, he noted, radium in the mouth
might stimulate the normal mouth bacteria, allowing them to aggressively
attack the oral tissues. Dr. James Ewing, a specialist on radium medicine, disagreed, saying, "The idea of the stimulation of infectivity of bacteria . . . is
to me untenable." Perhaps, he theorized, radium in the mouth irritated bone
tissue to produce an osteitis, or bone inflammation; this direct action "en-
ables the bacteria to get a foothold." It would seem a small step from this idea
to the notion that radium plated into the jaw bones and into any other bone

could damage the bone enough to make it susceptible to infection. Ewing, however, went off in a different direction. He asserted, "we are a long way from speaking of the ill-effects of radium therapy." Martland was excited by Ewing's theory about radium's action on the jaw and took it one step further: radioactivity "deep in the maxilla" produced an osteitis that, when exposed to the mouth's germs, caused necrosis. But when he published his findings, Martland still clung to the belief that jaw necrosis was caused by the action of locally deposited radium. The full paper (written with joint authors) appeared in the December issue of the *Journal of the American Medical Association*.[56]

The December publication concluded with the forthright statement that the cases described "represent practically a hitherto unrecognized form of occupational poisoning." Given that the dialpainters and the Consumers' League had been struggling for over two years for medical and legal acknowledgment of what they recognized as occupational poisoning, Martland's claim was a bit of an overreach. He did, however, clear much ground. He detected and measured radium deposited within the body, tied radium deposited in bone to anemia, and implicated various types of exposure to radioactivity as harmful, including inhalation, ingestion, penetration of external radiation, and injection. He particularly singled out the medicinal use of injected or ingested radium as dangerous and stated his opinion that "none of the known radioactive substances produce any specific or curative results." His work helped stamp out the practice of lippointing.[57]

Despite progress in compiling evidence of radium's dangers, it was still not completely established that the dialpainters suffered from an industrial disease. Business, for one, did not concede its existence. Industry leaders continued to argue that because no cases had been found outside the Orange, New Jersey, dialpainting studio, something other than the luminous paint might still be the cause of the workers' illnesses. The radium companies' defense included not only a cover-up of Illinois cases but also an attempt to hide corroborating cases of radium poisoning in Connecticut's watch factories.

At first, Katherine Schaub reacted calmly to the news that she and the other dialpainters were suffering from radium poisoning. "I was not as frightened as I thought I might be," she later wrote. "At least there was no groping in the dark now." But by the end of 1925 she was upset enough to persuade Frederick Hoffman to write to the U.S. Radium Corporation about her condition. The president of the company replied: "I have your letter . . . concerning Miss Catherine [*sic*] Schaub. I note your statement that she is in a precarious mental condition because of worry. While we regret very much the mental strain to which our former employees have been subjected because of the agitation concerning the supposed risks involved in their former employment, we are in no way responsible. . . . Miss Schaub's mental condition would probably be relieved if she would call on Dr. Frederick Flinn. . . . I am sure he will do what he can for her." [58]

Frederick Flinn had served as an associate physiologist for the U.S. Public Health Service beginning in 1919; in 1923 he earned a Ph.D. in physiology from Columbia University, where he became an assistant professor the next year. Flinn was first associated with the U.S. Radium Corporation back in 1924, when he was involved in the company's defense against the suit brought for damages from fumes emitted from the Orange plant. Now he had been hired to refute mounting evidence of radium's harmful effects. [59]

Flinn perhaps came to the notice of U.S. Radium because of his work on the tetraethyl lead cases, which attracted public attention during the same years that the radium cases were coming to light. At this point in our narrative the lead cases may serve as both a reprise and a foreshadowing of events in the radium controversy.

In 1922 scientists working for General Motors discovered that adding tetraethyl lead to gasoline made engines run better, allowing for bigger and faster cars. General Motors had an interlocking directorship with the DuPont Chemical Company, and together they affiliated with Standard Oil of New Jersey to create the Ethyl Corporation and to produce the new leaded gasoline. Illness and death struck refinery workers at Standard Oil's Elizabeth facility in October 1924. Of forty-nine employees, thirty-five became severely ill and five died of neurological disruptions caused by the organic lead compound. Because of the hallucinations produced, workers thereafter referred to leaded gas as "looney gas." Meanwhile, some scientists were concerned about the public health dangers of adding lead to the general environment. A U.S. Bureau of Mines report, paid for by the corporations and subject to their editing and approval, perhaps unsurprisingly defended leaded gas. The Harvard physicians Cecil Drinker, David Edsall, and Alice Hamilton and the staff of the Workers' Health Bureau found the study unsatisfactory and called for more research. Then it was revealed that the leaded gas corporations were hiding more deaths among workers at other manufacturing facilities in Deepwater, New Jersey, and Dayton, Ohio. The *New York Times* reported that over two years more than 300 cases of lead poisoning had developed at Deepwater, affecting about 80 percent of the workforce. [60]

As a result of revelation and disquiet, the Public Health Service called a voluntary conference on May 20, 1925. At the meeting leaded gas businesses defended their product as an important industrial advance and blamed workers' careless habits for their exposures. Detractors urged that lead as a cumulative poison should not be released into the environment and that management, not the workers, was to blame for the occupational poisonings. Until tetraethyl lead proved harmless, its detractors argued, it should be banned.[61]

In the midst of this debate stood Frederick Flinn. Flinn had been employed by the Ethyl Corporation in early 1925 to find evidence that leaded gas was safe. His position is particularly interesting given his future role in the radium cases. On the one hand, Flinn privately expressed concern "that there is some hazard." On the other hand, he insisted that his remarks be held confidential, and at the voluntary conference shortly afterward he kept silent about any possible danger. Following the conference the leaded businesses temporarily suspended production of tetraethyl lead to allow more research. Leaded gas production quickly resumed when, seven months later, an inadequate, short-term study certified the gas additive to be safe.[62]

Flinn first became interested in the dialpainting cases at U.S. Radium around March 1925, about the time that Frederick Hoffman announced he would give a paper on radium poisoning. Flinn then began the experiments in which he applied the luminous paint on the gums of guinea pigs and goats. U.S. Radium financed his work to some extent and that summer gave him a Wulf electrometer. The corporation anticipated his findings; in a letter to a colleague, a vice president wrote:

> Thanks for calling my attention to the article in the Journal of the American Medical Association. . . . Our friend Martland and some others including Sochocky are still maintaining that we are killing them [the dialpainters] off by the [dozen?]. . . . The [?] article is some of their propaganda. . . . I understand Flinn's report is to be published soon. His findings have been entirely negative and I think his report represents a very good piece of work. I am rather inclined to think that he will be given a certain amount of money to continue his work.

Flinn began to examine dialpainters at U.S. Radium and eventually workers at other companies as well.[63]

About this time Katherine Schaub received a letter directly from Frederick Flinn, echoing U.S. Radium's strong encouragement to consult with him. He claimed to be an expert on radium poisoning and to have examined dialpainters across the country and in Europe; he was anxious to examine her and any of her friends. "Let me give you an unbias[ed] opinion," he urged.[64]

Schaub accepted the offer but was likely either very glad or very confused to hear his assessment that none of her illnesses were attributable to radium. Similarly, on examining one of Schaub's coworkers who also was suffering from the effects of radium, Flinn advised her that "her health was in better condition than his, and that radioactive substances had not harmed her." [65]

Flinn bewildered not only Katherine Schaub, but also scientists watching his work. In January 1926, about the time he assured Schaub and her co-worker that radium had not harmed them, Flinn wrote Cecil Drinker, "Tho I am not saying it out loud, I cannot but feel that the . . . radium mixture in the paint is to blame in an indirect way for the girls [*sic*] conditions." The following December Flinn published the results of his studies, as anticipated by U.S. Radium. Entitled "Radioactive Material: An Industrial Hazard?," the article stated that no cases of jaw necrosis existed anywhere but in Orange, and that "these cases are by no means clear cut." Flinn suggested that syphilis, Vincent's angina (trench mouth), or bacterial infection could be responsible for the dialpainters' illnesses. He reported detecting no radioactivity in any of the women he examined, and he concluded that "an industrial hazard does not exist in the painting of luminous dials." [66]

But there were corroborating cases, and Flinn knew about them. In October 1926, two months before his first study was published, Flinn delivered a paper revealing two cases of radium poisoning among dialpainters in Waterbury, Connecticut. He claimed to have known of one case since June 1926—six months before his first published study denied the existence of cases other than in Orange. The new cases were very important. Up to their unveiling by Flinn, no cases outside of Orange had been discovered, suggesting that radium, present at all dialpainting plants, was not the cause of the Orange dialpainters' illnesses, and that some other factor existing only at the Orange plant was at fault. [67]

Dialpainting had begun in Waterbury in 1919, when the U.S. Radium Corporation began to export dialpainting work from the Orange facility to customers' factories. To facilitate the change, U.S. Radium sent instructors to train fledgling dialpainters in the appropriate techniques; lippointing was one such technique taught to employees at the Waterbury plants. [68]

Clock and watch manufacture, which had begun in Connecticut during the nineteenth century, was centered in the Naugatuck Valley, known as "Brass Valley" because of its concentration of brass industries. The husbands and fathers of female watchmakers were familiar with industrial disease: brass workers suffered from what were variously called "spelter shakes," "metal fume fever," and "brass founder's ague," fevers and chills resulting from inhalation of the fumes of metal oxide, especially zinc oxide. Their wives and daughters were hired to produce the famous Waterbury dollar watches. The Waterbury Clock Company had started as a brass mill department, became a separate entity in 1887, and by the twentieth century was producing five million watches a year. Other clock and watch companies came and went in Waterbury, but by 1925 they were consolidated within the Waterbury Clock Company (later the U.S. Time Corporation, and still later, Timex). Dial-painting also took place in firms in other Connecticut towns, such as the New Haven Clock Company, the E. Ingraham Company of Bristol, and the Seth Thomas Clock Company of Thomaston.[69]

In March 1925, when Flinn first began his work with radium and his association with U.S. Radium, publicity in New York newspapers about the New Jersey cases came to the attention of Waterbury Clock Company executives. They contacted U.S. Radium and were told of the work Flinn was doing, examining its employees. In August Flinn first visited Waterbury and with a team of medical assistants examined all twenty-six of the current dialpainters; he reported no health problems or radioactivity among them.[70]

One former dialpainter was already dead. Frances Splettstocher had been employed at the Waterbury Clock Company for four years. In early 1925, at age twenty-one, she began suffering from pain in her left face and a sore throat; she seems to have had severe tonsillitis as well as the jaw necrosis common to the early radium cases. When a tooth was pulled, part of the jaw came with it, and the surrounding tissue continued to rot and slough off, creating a hole through to her cheek; she was also anemic. She died in February 1925. The Waterbury Clock Company never admitted that Splettstocher's illness was related to her employment, but her death in February and the New Jersey publicity in March led the company to forbid further lippointing. At least one dialpainter later remembered that she "became frightened when Frances Splettstocher died, and would not work in the dial painting department again for any money." Splettstocher's father also worked at the clock company. He recalled that, "though he was sure" that radium poisoning had killed his daughter, "he did not dare make any kick about it" because he wanted to keep his job.[71]

The first case conceded by Waterbury officials as radium poisoning was that of Elizabeth Dunn, probably one of the first women hired to paint dials in Waterbury. In 1923, according to company accounts, she became "disabled," most likely from anemia, although she continued to work until 1925, when she slipped on a dance floor and broke a leg—likely already weakened from the damage to the bone by deposited radium. It is unclear whether she left the company before or after Flinn's first examinations, although he claimed not to know of her illness until June 1926. His second case, revealed in the paper he read in late 1926, may have been Helen Wall, about whom little is known. In August 1925 Flinn had found no health problems and no radioactivity among the Connecticut dialpainters. He repeated his examinations in November 1926 and again reported no problems.[72]

In May 1927 Flinn published the study he had read seven months before, including his Connecticut data. Here he concluded that "radium is partially if not the primary cause of the pathological condition described." With this admission, Harvard's Katherine Drinker was hopeful that the Connecticut cases had changed Flinn's stand on radium poisoning. "I doubt whether he will now consent to represent the United States Radium Corporation's interests," she wrote to Katherine Wiley. But Alice Hamilton learned from Katherine Schaub's lawyer that Flinn was still serving as an expert for U.S. Radium.[73]

Alice Hamilton sought a way to contain Flinn. First, she wrote to his superior at Columbia. In tentative but suggestive language, she claimed to be "most puzzled" by Flinn's erratic research conclusions. His continued connection with U.S. Radium was, she thought, "an attitude impossible to understand."[74]

Flinn was shown Hamilton's letter and replied. He had "never been employed by the United States Radium Corporation in the sense that I have received any compensation for my work," he claimed. As for his second paper, he had read it but then it was published without his knowledge. He was still studying the dialpainter situation, he wrote, and had yet to form an opinion.[75]

Hamilton took a stern tone in response, urging Flinn to "consider very seriously the stand you are taking." Mustering the authority of her preeminent professional position, she wrote: "You are a young man at the beginning of your scientific career, and it is a matter of no little importance that you should be careful not to alienate, not to cause suspicion in the minds of men older and more prominent than yourself in your chosen field. . . . If you mean to keep on in scientific work, then I think you should recognize the grave doubts and even suspicions in the minds of people whose goodwill and understanding should be of importance to you."[76]

Hamilton's warning had some effect. Flinn attempted to justify his position to her, complaining that "what you mean by 'my recent conduct' is beyond my ken" and noting that U.S. Radium had paid only a portion of his research expenses. The rest was being subsidized by unspecified "other parties." In a third publication, in 1928, Flinn tried to explain away the coincidence that his discovery of the Connecticut cases overlapped by six months his first publication asserting that no such cases existed. He admitted that "radioactive material is at the bottom of the trouble." Nevertheless, he continued to serve as a consultant for the U.S. Radium Corporation, and Hamilton found him "impossible to deal with." Katherine Wiley thought him "dishonest" and "a real villain." Even a few physicians labeled him "double dealing" and "two faced." Schaub's lawyer tried to instigate action against him through the New York Medical Board for practicing medicine without a license. All this came to naught. Flinn was able to progress in his career and eventually became chair of the Industrial Hygiene Department at Columbia University.[77]

Many people claimed to have detected radium poisoning. Frederick Hoffman wrote, "I was the discoverer and the first writer of radium necrosis as an occupational disease." Cecil Drinker maintained that he studied "an effect hitherto not observed." Harrison Martland claimed discovery of "a hitherto unrecognized form of occupational poisoning." But it was the revelation of Frederick Flinn's corroborating cases that turned the corner, once Alice Hamilton and others forced him to accept their obvious implications. Martland's etiology with Flinn's Connecticut cases put industrial radium poisoning on a firm scientific footing, accepted by most medical experts, if not by the radium companies themselves. Attention then began to shift to winning compensation for injured dialpainters and their families. Whether radium poisoning was *legally* established was a different question entirely. The issue was avoided in Connecticut because of the watch companies' reliance on private settlements, but it arose in New Jersey, where dialpainters filed civil suits.[78]

A DAVID FIGHTING THE

GOLIATH OF INDUSTRIALISM

Compensation in New Jersey and Connecticut

The dialpainters' experiences with the laws surrounding occupational health highlight the difficulties that workers generally encountered in seeking redress of industrial disease. One problem was the lack in law of a clear duty to provide a healthy workplace. Ignorance of possible dangers was an admissible defense for employers, who were little obligated to investigate the hazards of new industrial processes. Another problem was the presence in law of statutes of limitations, which stipulate a period of time after which an injured party may no longer file suit for damages. Ignorance of the harm done by a slow-acting toxin (or even delay in the toxin producing actual harm) did not prevent the statute of limitations from running and thus prevented many suits by victims of occupational diseases. That such inadequacies in the law (from the perspective of workers) were intentional is suggested by the new laws written in New Jersey and Connecticut at the time dialpainters sought financial redress for their injuries. Statutes of limitations were shortened in both states.

In New Jersey, the dialpainters' attorneys attempted to circumvent the statute of limitations through pleas to special, "equity" courts for relief. Employers avoided precedent-setting equity decisions: facing possible loss in court, they chose to settle out of court. Out-of-court settlements helped the few dialpainters thus rewarded but left untouched the barriers preventing other dialpainters' redress in court.

Dialpainters filed claims for compensation in both New Jersey and Connecticut soon after scientists began to support their belief that occupational exposure to radium had caused their illnesses. The course that settlements took in the two states was quite different. In Connecticut, workers and their employers settled quietly, without publicity or state intervention, whereas in New Jersey, dialpainters appealed to state agencies and to courts, receiving wide publicity. Two factors explain the difference. First, the watch manufacturers in Connecticut enjoyed greater influence than did the U.S. Radium Corporation in New Jersey, and business interests controlled Connecticut's state institutions to a greater extent than they did New Jersey's. Second, the New Jersey Consumers' League arranged publicity to help the dialpainters win settlements and, together with other women's organizations, worked to make radium poisoning compensable by law.

Publicity about the dialpainters began when the first suits seeking redress for radium poisoning were filed in New Jersey in 1925. By then, five U.S. Radium employees—the dialpainters Amelia Maggia, Helen Quinlan, Irene Rudolph, Hazel Vincent Kuser, and the chemist Edwin Lemen—were known to have died from radium poisoning, and Rudolph's cousin Katherine Schaub and the sisters Sarah Maillefer and Marguerite Carlough were known to be ill with related maladies. The Consumers' League, as part of its strategy to win access to the Harvard study, and, of course, to aid the dialpainters directly, had helped Marguerite Carlough find an attorney who was willing to take her case. In March 1925, claiming that she was ill because her employers "compelled [her] to wet the tip of her brush with her lips," Carlough filed suit for $75,000. Her parents planned to initiate a similar suit, for $15,000, for injuries to her sister Sarah Maillefer. Soon Hazel Kuser's family announced their suit for $10,000. In June Maillefer and a second U.S. Radium chemist, Frederick Starke, both died, the sixth and seventh work-related deaths at U.S. Radium.[1]

Injured workers had recourse to the common law covering relations between "master and servant" but had several barriers to surmount. First, under tort law, which ruled the dialpainters' civil damages suits, plaintiffs had to prove that they had been harmed. With a straightforward accident, this was not difficult, but chemical exposure was not necessarily harmful unless and until it caused a disease. The dialpainters, for instance, could not bring suit against their employer until they became sick from the radium deposited in their bodies. Even if the dialpainters knew that they had radium plated in their bones as a result of their employment, they had no cause of action until they were actually injured by the radium, that is, until they became sick. Second, intent mattered. Liability did not generally attend an accidental injury unless negligence was proved. This raised difficult questions regarding what the dialpainters' employer knew about the dangers of dialpainting, or, given their employer's expert knowledge about radium's properties, what the company ought to have known about dialpainting's dangers. A third problem was the statute of limitations—the cap placed on the time allowed between an injury and the filing of a suit for damages arising from that injury—which in New Jersey at the time was two years for damages suits. The statute of limitations in conjunction with the necessity to prove harm disqualified most dialpainters from filing a sound suit. Many of the workers in New Jersey stopped dialpainting when U.S. Radium shifted that job to the watch and clock companies around 1920. In the way the law was then interpreted, dialpainters were barred from bringing suit against their former employer two years later—even though in 1922 the harm from radium poisoning had yet to manifest itself and so qualify them to file suit. That is, if the dialpainters became sick more than two years after their exposure to radium ceased, as was usually the case, the statute of limitations protected U.S. Radium from liability.[2]

Legal authorities considering occupational poisoning suits still argue about when the statute of limitations begins to "toll." Should the starting point be the first exposure to toxic chemicals or perhaps the last? Should the statute-of-limitations period start when the injured employee leaves employment? Should it start when a cause of action is first discovered—when the poisoning is found, or when symptoms of the poisoning begin, or when the employee learns that the poisoning is the result of negligence or that a suit may be brought against the employer? What is a fair period in which to limit the opportunity to file a suit? Questions such as these, dealing with the fairness of law, might qualify a case to be heard in an equity court, which might, for instance, adjust the period of the statute of limitations to preserve a case for an injured party.[3]

All of these issues were present in the first dialpainters' cases filed. To illustrate, Hazel Kuser left U.S. Radium in 1920; she would have had to enter her suit against the company by 1922—two years later—even though her symptoms of radium poisoning did not appear until 1923, and even though radium poisoning was not even suspected to exist until 1923. Of the three plaintiffs, only Sarah Maillefer had filed within the period of the statute of limitations, as she worked for U.S. Radium through March 1925, three months before she died.[4]

In these first three suits, the legal intricacies were avoided. U.S. Radium tried to win dismissal on the grounds that they were workers' compensation actions. When this failed, in May 1926 the corporation agreed to out-of-court settlements with the families of the three dialpainters: $9,000 for Marguerite Carlough, $3,000 for Sarah Maillefer, and $1,000 for Hazel Kuser. Kuser's family had spent almost $9,000 on her medical care; her husband had mortgaged their home and lost it to foreclosure, and her father had spent all the money he had saved for his retirement. We do not know why U.S. Radium agreed to these settlements or why Carlough's estate was awarded so much more than the others; seemingly, Sarah Maillefer, whose suit was filed within the time frame of the statute of limitations—and who was the only one of the three with a dependent child—had the better case, and her family might have demanded more in settlement.[5]

Meanwhile, in December 1925, at age twenty-four, Marguerite Carlough had died of anemia, pneumonia, and, her niece recalled, "a horrible jaw problem." Thus, by 1926 six New Jersey dialpainters and two New Jersey chemists were known to have died from radium poisoning.[6]

In 1926 the Consumers' League of New Jersey sought to make industrial radium poisoning compensable under the state workers' compensation act. The law presently granted compensation for only those industrial diseases attached in a "schedule." The schedule of compensable diseases included poisoning from lead, mercury, arsenic, phosphorous, benzene and related compounds, methanol, and chromium compounds. As reformers of the time were quick to note, the common weakness of such schedules was that they could not include an occupational disease until it became known, and thus victims of newly discovered diseases would have no recourse under the compensation law. This was the case with radium poisoning: it was not listed in the law's schedule and thus was not compensable.[7]

Many women's groups had set up committees to lobby in the state capital for desired legislation. Some had cooperated since 1920 with the Women's Cooperative Legislative Bureau; in 1924 ten women's organizations coalesced into the New Jersey Council of State Organizations for Information and Legislation. The council was formed by personnel shared between the Consumers' League and the League of Women Voters, including Katherine Wiley. When the Consumers' League sought to amend the workers' compensation act, it was to this council that it turned for support.[8]

Early in 1926 Alice Jones reported to the Women's Club of Orange and to the New Jersey League of Women Voters on the introduction of an act declaring "radium Necrosis" compensable. Women reformers had instigated this act, which easily passed both houses and was signed into law by March. Katherine Wiley wrote a colleague, "I had a nice note from the Governor when he signed the necrosis bill—I feel like its grandmother."[9]

Women's support, however, may have been superfluous. First, 1926 was not a banner year for women in politics. "Women have never counted so little at Trenton as this year," complained Lillian Feickert at a special meeting called in 1926 to discuss women's "failures" in the state capital. As a former suffragist and a founder of the New Jersey League of Women Voters, Feickert had first-hand knowledge of the problems women faced in the political arena. When women won the right to vote, party strategists appointed her vice chair of the Republican State Committee, with special responsibility for bringing women into the GOP. Shortly afterward she became dissatisfied with the League of Women Voters's nonpartisanship, which she read as a refusal to employ the power now available to women. She quit the league and organized the New Jersey Women's Republican Club, serving as president. Though supported by Republican funds, her new group maintained independence from the party, a situation sure to pique male Republicans. Male resentment surfaced when Feickert used her club as a podium from which to chastise Republicans who hesitated to back legislation that she desired. When Feickert then campaigned against those Republicans who would not toe her line, she not surprisingly lost her seat on the Republican State Committee; by 1925 the party excluded planks backed by the Women's Republican Club from its platform. By 1926 Feickert was disillusioned with party politics and was ready to return to the nonpartisanship she had earlier disdained. Despite renewed efforts to reunite women around women's issues within women's organizations, most politically active women relished the rewards of party loyalty, a preference, of course, encouraged by the political parties. Women's role as an independent political force declined in New Jersey after 1925.[10]

A second reason for questioning the importance of women's support for the radium necrosis bill was that it apparently was "not unpopular" with the state Manufacturers' Association, or so it was reported to the League of Women Voters' legislative committee. Several aspects of the measure left it of little threat to business. The necrosis bill could not make liability retroactive, and so no one exposed to radium before 1926 could claim compensation. The bill contained a five-month statute of limitations—the maximum time allowed between injury and claims for damages—considerably shorter than the two-year limit under the existing employers' liability law. Finally, the new regulation required compensation only for radium necrosis, the jaw infection believed at the time to be caused by local irritation of oral tissues in lippointing. Once radium companies eliminated lippointing, it seemed likely that no one would contract jaw necrosis. The law did not make compensable the other harmful conditions common to radium-exposed employees, such as anemia, fragile bones, and the cancers soon to be discovered among them. In short, the bill was designed so that no one would ever collect compensation. The Consumers' League in sponsoring the bill seemed not to understand this. In contrast to the business-sanctioned radium necrosis law, a companion measure to make compensable a much more common disease, silicosis, failed because of opposition from the Manufacturers' Association.[11]

The radium necrosis bill was written for the Consumers' League by Frederick Hoffman, the Prudential Insurance Company statistician who had earlier helped the league substantiate dialpainter claims of occupational illness. That the statute absolutely precluded claims ever being paid suggests that Hoffman's insurance connections may have led him to purposely write an ineffectual bill. Against this interpretation, however, stands his later support for a blanket workers' compensation law to cover all industrial diseases. An alternate explanation for the weaknesses of his bill sees Hoffman as so fixated on his "discovery" of "radium necrosis" that he shaped the legislation around his accomplishment. The Consumers' League, on learning of its mistake, initiated a campaign to change "radium necrosis" to "radium poisoning" (but would not succeed until 1931).[12]

Harrison Martland led researchers to new understandings about the perils of radium. Martland followed his 1925 AMA paper on radium poisoning with several more. In 1926 he published with G. S. Reitter an account of a 1925 autopsy of the U.S. Radium chemist Edwin Lemen. With help from a corporation vice president, Howard Barker, Martland refined his radium detection techniques and began attempts to quantify deposited radium. Lemen, he determined, had accumulated in his skeleton about 14 micrograms of radium (or related radioactive compounds calculated in radium equivalents) and in his lungs a single microgram, for a total of 15 micrograms. Lemen had never ingested radium but had accumulated all his body-burden from inhalation of radium dust and radon. The Drinker team's concerns about inhalation of radium were validated. Next, Martland published an account of his autopsy of Marguerite Carlough. She had accumulated 150 micrograms of radium in her bones. Although differing by an order of ten, both amounts—15 and 150 micrograms—were fatal.[13]

Why did the U.S. Radium vice president cooperate with Martland's research? Apparently Barker thought that the findings would not be damning, for his own paper published in 1929 contained virtually the same data as those appearing in the 1926 essay by Reitter and Martland. Barker, however, concluded that the "quantity of radioactive material found in the body examined was hardly sufficient to produce the systemic changes which resulted in death."[14]

In a review of his radium work, also published in 1929, Martland reckoned that around 10 micrograms of deposited radium might be a "lethal dose." Because deaths among radioactive dialpainters had slowed down, he theorized that there might be two kinds of dialpainter cases, early ones and late ones. The early cases were marked by severe anemia and jaw necrosis. Martland now had arrived at the modern understanding of jaw necrosis. Radium was deposited in jaw bone just as in any other bone, but in the jaw, dead bone could be contaminated with bacteria via tooth decay or oral surgery, thus allowing severe bone infection and necrosis to set in. The late cases lacked (or had recovered from) the anemia and jaw infections of the early cases. Martland thought that perhaps the natural decay of mesothorium might account for the difference. Mesothorium or radium-228 had a half-life of about six years, whereas radium-226 had a half-life of about 1,600 years. Mesothorium's decay also produced more of the dangerous alpha particles than did radium-226's. Martland thought it possible that for those workers who had accumulated mesothorium as a large percentage of their total radium body-burden, the danger would be high at first but would decrease over time as the mesothorium in their bones decayed. If such workers survived the early maladies, they had a fair chance of surviving radium poisoning altogether. As "late cases," however, they tended to suffer from severe "radiation osteitis," or damage to bones from radioactivity. Osteitis weakened the bone, allowing it to crumble or snap under little pressure. The late cases, then, tended to be crippled but might not die.[15]

In 1925 Martland had known the names of fifty dialpainters, and he listed them in a notebook. As the years passed and dialpainters died, he made a ritual of crossing their names off his list one by one. By 1931 Martland had attributed to radium some earlier deaths—like dialpainter Amelia Maggia's in 1922 and chemical worker Michael Whalen's in 1922 of pernicious anemia. He had added some new deaths, including three more chemists or physicists and several more dialpainters, for a total of eighteen deaths attributed to work at U.S. Radium.[16]

In 1924 Jennie Stocker, who had worked at U.S. Radium from 1918 to 1922, had died of cancer originating in her knee; she had also suffered from jaw necrosis. In 1927 Eleanor Eckert, employed by the company from 1917 to 1919, died of an osteogenic sarcoma of the scapula (cancer of the shoulder bone). Eckert was the thirteenth U.S. Radium worker to die, the tenth dialpainter. Two bone cancers among that small number seemed like more than a coincidence to Martland, and he wrote a paper suggesting very tentatively that radium might be implicated in the dialpainters' cancers. By 1931 three more dialpainters—all "late cases"—had died of osteogenic sarcoma, and Martland was convinced that radium was the cause. In a major paper published in the *American Journal of Cancer*, he observed that as little as 2 micrograms of deposited radium were known to have injured a dialpainter. "Some have thought that 10 micrograms . . . is probably just within the limits of tolerance of the average person," but, he stated, "From my experience . . . less than one half of a microgram is dangerous." Martland concluded his essay by suggesting that cancers in general might be caused by "minute amounts of radio-active substances to which the human body, in its normal environment, is exposed," and that therefore *any* exposure to radiation might be dangerous. Martland seemed to be suggesting what is now referred to as a "zero-tolerance" policy, that is, a policy permitting absolutely no exposure to a dangerous physical or chemical agent.[17]

In May 1927 Raymond Berry, a young attorney "not long out of Harvard," approached the president of the New Jersey Consumers' League; he had heard about the "possibilities" of the radium cases and wanted to take them on. In the same month Berry filed suit for the dialpainter Grace Fryer in the New Jersey Supreme Court.[18]

Fryer was one of the women appearing in Frederick Hoffman's 1925 "Radium Necrosis" paper. When examined by Hoffman, she had recovered from a jaw necrosis, after at least fifteen operations to remove teeth and bone, but was suffering from pains in her feet and back that had begun in 1922. Harrison Martland diagnosed her radium poisoning in July 1925, attributing her pains to a crushed bone in her foot and two crushed vertebrae; radium had weakened her bones. To compensate, Fryer wore braces on her foot and back. She had started working at U.S. Radium in 1917 at age eighteen and left in January 1920.[19]

Fryer was soon joined in her suit by four other dialpainters. One of them was Katherine Schaub. Two other women in the suit, Quinta McDonald and Albina Larice, were sisters of Amelia Maggia, the first known victim of chronic industrial radium poisoning, who had died in 1922. Born in 1900, McDonald joined U.S. Radium in 1917, married in 1918, and left her job in 1919 to give birth to the first of her two children (the second was born in 1923). In 1924 her teeth began falling out and her hips caused problems that necessitated a cast from her chest to her knees for half a year; Martland diagnosed her condition in 1925 as radium poisoning. McDonald's hips, though better for a while, by 1925 had locked so that she could not move her legs at the hip joint, preventing much normal movement, like sitting. Larice, born in 1895, was five years older; she worked at U.S. Radium from 1917 to 1921, when she married. Of her two pregnancies, one ended in miscarriage and the second child died in infancy; Larice blamed radium for the loss of her children. Her illness began with pains in one knee in 1925 and then spread to a hip; within several years the one leg was four inches shorter than the other, and she was incapacitated. She suffered from jaw problems as well and eventually lost all her teeth. In 1928, because of her illnesses and poor prognosis, she elected to have a therapeutic abortion of her third pregnancy. Edna Hussman was the fifth dialpainter in the lawsuit. She had worked at U.S. Radium from 1917 to 1922 and then at a different dialpainting studio from 1922 to 1925. She had married in 1922, at age twenty-one, but never had children. Her symptoms began in 1925 with pains in a knee and a hip; these spread to an elbow and a shoulder, and soon she could not raise one arm at all. She also suffered minor tooth problems, which cleared up after several extractions.[20]

Raymond Berry devised two possible routes to circumvent the statute of limitations, both of which entailed rulings on the law's fairness by an equity court—called the Chancery Court in New Jersey. First, he prepared to attack the statute of limitations directly with explicit arguments that it was unfair to begin "tolling" the statute (to begin counting the prescribed period) until the dialpainters knew that they had a cause of action; that is, until they were aware that radium from work had harmed them. Not until Martland used physics equipment to determine that the women were radioactive and diagnosed their illnesses as caused by radium did the women know that they had been harmed by U.S. Radium. By this reckoning, the two-year statute of limitations would have begun to toll in July 1925 for Grace Fryer, Katherine Schaub, and Quinta McDonald, October 1925 for Albina Larice, and May 1927 for Edna Hussman. With the women's suits filed in May (Fryer), June (Schaub and McDonald), and July (Hussman and Larice) of 1927, this line of argument preserved cases for all of them. If this direct challenge to the statute of limitations failed, Berry would try to prove that U.S. Radium fraudulently concealed from the dialpainters the source of their illnesses, that is, that the company had prevented them from bringing suit within the time frame of the statute by suppressing the evidence that would have permitted them to do so. An equity court might rule that the statute of limitations had begun to "toll" at the time the dialpainters discovered a cause of action, that is, when the dialpainters learned that radium had made them ill despite the attempts of U.S. Radium to hide the truth from them. Its denial that radium was dangerous would be fraudulent concealment if it could be shown that U.S. Radium officials knew of radium's dangers while they were denying them.[21]

If the Chancery ruled in the dialpainters' favor, their case would precede to the New Jersey Supreme Court. There, the dialpainters would have to prove not only that their jobs at U.S. Radium had caused their illnesses, but also that the company was *negligent* in its management practices. In what lawyers call a "tort," a harm is done to someone by someone. At the heart of tort law are two principles: the plaintiff must prove first that he or she has been harmed (the dialpainters had to prove that radium poisoning had made them ill), and the plaintiff must prove that the defendant caused the harm (the dialpainters had to prove that U.S. Radium was responsible for their accumulation of radium). The dialpainters did not have to show that U.S. Radium's managers intentionally harmed them, but they did have to demonstrate that their harm was more than accidental on their employers' part. The issue was thus one of negligence, which in turn depended on what was known by managers about the dangers of radium, for if they *knowingly* created a dangerous situation and did not warn anyone about it or take steps to improve safety, then they were negligent and liable for subsequent harm. But if they unknowingly created a dangerous situation, they were not negligent and not liable.

The question in Berry's mind was whether he must prove only that company officials knew that radium was a dangerous substance, or whether he had to prove they knew that painting dials with paint containing radium was dangerous. Berry hoped to employ principles embodied within a controversial British negligence case, *Rylands v. Fletcher* (1868), to avoid having to prove that U.S. Radium officials knew specifically about the danger in using radium in dialpainting or that they neglected precautions contemporarily deemed sufficient. *Rylands* was concerned with the law of negligence involving dangerous conditions, or the "doctrine of extrahazardous danger." It held that if a dangerous substance is brought onto one's land and then gets out of control, then negligence is assumed, that is, negligent acts or omittances need not be proved to give an injured person a cause of action, but the cause of action is grounded in the existence of something likely to do damage and proof that damage has been done.[22]

Unfortunately for Berry, in a related decision, *Marshall v. Wellwood* (1876), the New Jersey Supreme Court had limited negligence in such cases to those in which "a want of care or skill" could be proved. Later cases, Berry thought, returned the rule to the *Rylands* formulation: specifically, he cited the *Ballantine* cases (1908, 1914) to show that a dangerous condition ("proof of something likely to do damage") plus "proof of damage" could be considered proof of negligence without demonstrating an actual negligent act or omission. He cited New Jersey cases concerning explosives (*Cuff's v. Newark and New York Railroads*, 1870, and *McAndrews v. Collard*, 1880) to confirm that U.S. Radium might be liable even without foreknowledge of danger and want of care. Or, as Berry summarized in a "Memorandum of Law," "negligence" in court could be inferred by "the exceptional nature of matter collected upon the land," and radium was exceptionally dangerous.[23]

The dialpainters' attorney was prepared for the possibility that U.S. Radium might use the "assumption of risk" defense. This doctrine held that if a worker accepted a dangerous job, the danger was inherent in the contract between worker and employer, and so by accepting the job, the worker accepted the risk and assumed liability for any injuries to himself or herself. Berry cited a number of cases in support of another New Jersey decision concerning explosives, *Smith v. Oxford Iron Co.* (1880), which stipulated that employers had an obligation to explain the cautions necessary with highly dangerous substances. In addition, employers were liable for harm done by a dangerous substance "whether or not the company was aware of its dangerous quality, or furnished it for use without having taken steps to obtain such knowledge." Again, Berry sought to circumvent the necessity of proving that company officials knew of radium's dangers.[24]

Further, Berry planned to argue that U.S. Radium had a *duty* to know the dangers inherent in work with radium and a *duty* to warn their employees of the dangers, because their employees were so often minors. Under law, employers had a special duty to ensure that minors worked in safe surroundings and to see that minors understood how to work safely.[25]

Berry, then, had a complicated case to construct. To show in the equity case in Chancery Court that U.S. Radium had fraudulently concealed from the dialpainters the cause of their illnesses, he had to demonstrate, first, that company officials knew that radium had harmed their employees during the period of the statute of limitations, and second, that the officials took actions to prevent the dialpainters from learning that radium had harmed them (alternately, he could argue that tolling the statute of limitations from the last radium exposure was unfair and that, instead, it should toll from the time when the dialpainters first knew they had been harmed in the workplace). Third, and finally, Berry had to prove that U.S. Radium officials knew of the dangers of radium while the dialpainters were working. His case would be more solid if he could show that the dangers inherent specifically in dialpainting with radium were known.

To prove the first and second points, that managers at U.S. Radium knew that radium had harmed their employees during the period of the statute of limitations and that they took actions to prevent the dialpainters from learning that radium had harmed them, Berry was relying in part on Katherine Wiley of the Consumers' League and on Katherine Schaub and Grace Fryer, former dialpainters. As evidence of concealment, Wiley would attest to letters and statements by the president of U.S. Radium claiming that radium was not harmful and that all his employees were well. Schaub would testify that the physiologist Frederick Flinn, a representative of U.S. Radium, had denied that radium was a cause of her illnesses. Grace Fryer had also been examined by Flinn and pronounced healthy.[26]

In addition, Berry was counting on testimony by the Harvard team—Cecil Drinker, Katherine Drinker and William Castle—to help prove that the radium company both knew about radium's effects and concealed them. Cecil Drinker, for example, was in possession of letters that Berry thought demonstrated that U.S. Radium knew dialpainting was causing harm but refused to act on that knowledge. Drinker's testimony also was necessary to show that Frederick Flinn had acted to camouflage the source of the dialpainters' illnesses. Flinn had written to Drinker in early 1926, "Tho I am not saying it out loud, I cannot but feel that the radium mixture in the paint is to blame in an indirect way for the girls' conditions," whereas at the end of that year he published a paper denying that radium was an industrial poison. Moreover, Drinker could attest to lies that U.S. Radium had told the New

Jersey Department of Labor about the Harvard findings and to its attempt to quash publication of the Harvard paper.[27]

The third point to prove, as Berry put it, was that U.S. Radium officials "knew of the dangerous and destructive nature of radium . . . and knew that [it] . . . would cause the very bodily injury which it did cause . . . but gave complainants no protection or warning of the same." There was but little direct evidence of this. One argument took the use of lead screens by some U.S. Radium laboratory workers as evidence that the corporation knew of the dangers of radioactivity. Another involved stories that Sabin von Sochocky, as one of the founders of U.S. Radium's predecessor firm, knew of radium's threat to the dialpainters' health "and endeavored to restrain certain members of the corporation without avail." Berry had heard about an instance where von Sochocky warned corporation executives about the dangers of radium but was ignored; in his files he noted that Sochocky had "admitted" to Frederick Hoffman, the Prudential statistician, "that he knew at the time girls were working there they would be injured." Apparently von Sochocky made similar comments to Harrison Martland. Several dialpainters also claimed that von Sochocky had warned them of dangers. Berry had reason to think that von Sochocky might testify against U.S. Radium: an informant wrote Berry that von Sochocky was "not friendly towards the controlling elements" of the company, and U.S. Radium considered it likely that von Sochocky would testify against the firm. While Berry was struggling to make his case, investigators for U.S. Radium found evidence that von Sochocky, who had been fired following a corporate reorganization in 1921, was "in back of all these cases." Certainly Berry was relying on von Sochocky's testimony to prove that the company knew about the dangers of radium and was negligent in not acting on that knowledge.[28]

Berry hoped he could demonstrate that corporate officials had scientific backgrounds sufficient to make them aware of literature detailing radium's destructive effects. U.S. Radium had employed several leading authorities on radium. First, there was the luminous paint's inventor, Sabin von Sochocky, who had studied the medicinal use of radium. Second, there was the physician with whom Sochocky had founded the radium firm, who employed radium in his practice, and who therefore likely also knew of its dangers (although penciled in next to this point in Berry's notes is "Get more evidence"). The third expert was Dr. Victor Hess, who would later win a Nobel Prize in physics for work on radiation. Further, Berry reasoned, U.S. Radium sold radium medicinal products, many of them based on the changes radium wrought on body tissues, and therefore most officials were apprised of radium's effects. To demonstrate such knowledge, Berry consulted a bibliography and a volume of abstracts published by U.S. Radium in 1922, both of which outlined the literature on the medicinal use of radium. Each, Berry wrote, "contains articles setting forth and describing the very injuries from which the complainants suffer from the year 1906 on" (he must mean that articles from as early as 1906 were included).[29]

To describe the literature detailing radium's dangers, Berry was again relying on the Harvard team, especially its senior member, Cecil Drinker. In preparing their report for U.S. Radium, the Harvard physicians had searched the literature, found instances of radium doing harm or causing effects, and linked this evidence with the condition of the women to argue that radium was causing the women's illnesses. U.S. Radium could have done the same, was the implication.[30]

Berry, then, was relying on Cecil Drinker's testimony to substantiate many points. As we have seen, Drinker was crucial, first, to show Flinn's alleged cover-up of the dangers of radium; second, to prove that by 1924 at the latest U.S. Radium had known of radium's dangers because the Harvard team reported its findings to the company; third, to demonstrate U.S. Radium's attempts to keep the Harvard study under wraps; and fourth, to show that the medical literature existing by 1925 was sufficient to indicate the dangers of radium.

Unfortunately, Drinker refused to get involved. Excerpts from his correspondence confirm his sympathy for the plight of U.S. Radium. As early as 1924, when he was still haggling with Arthur Roeder over his acceptance of the Harvard team's conclusions, Drinker had written another industrialist for whom he had worked that "the unfortunate economic situation in which he [Roeder] finds himself makes it very hard for him to take any stand save one in regard to radium, namely, that it is a harmless, beneficient [sic] substance which we all ought to have around as much as possible." In a more serious tone, Drinker explained that "though their conduct may have been questionable following our report, it does not seem to me that they could be blamed for events occurring well ahead of that time." Drinker told Berry that it was "his policy" not to testify "in cases of this sort." Berry then appealed to William Castle, Drinker's junior colleague on the Harvard study. Castle wrote to Drinker that Berry was "a very decent sort of fellow" who "appreciated the attitude that you and I take about testifying" and asked Drinker to consider talking to the lawyer. Drinker responded at length:

I rather imagine that the sort of information Berry wants from me is exactly the kind of thing I do not care to be involved in giving. It is to be a description of the attitude of the Radium Corporation and does not have anything to do with our real opinion as to the condition of the employees. . . . While I think that the Radium Corporation behaved rather outrageously toward us as well as toward their employees, I see no reason why I should make this public. . . . I don't want to see this man and I think that if I did see him I would not tell him anything that would be of help.[31]

Under subpoena, however, Cecil Drinker together with Katherine Drinker, William Castle, and Alice Hamilton submitted to questioning by both Berry and U.S. Radium attorneys in a special hearing held in Boston in November 1927. Cecil brought copies of letters from Roeder rejecting his conclusions in the Harvard study, as well as of Drinker's correspondence with Frederick Flinn. Berry asked Cecil Drinker, "Did you receive any remuneration for the work you did . . . ?" Drinker replied, "It depends on what you mean by remuneration. We received some money. I should not call it remuneration myself."[32]

Meanwhile, U.S. Radium planned a defense. Radium experts, many of them employed by the industry, agreed to testify that no one could have known of the dangers inherent in dialpainting before 1925 studies revealed the slow accumulation of radium in the body due to constant exposure at low levels. Preparation for the defense also included investigating the dialpainters bringing the suit. Katherine Wiley discovered this: "I was told by the New York World reporter that he knew that the company employed a detective to follow these five women who were suing to try to get something on them that would detract from public sympathy which is favoring them." This tactic seems to have failed, as nothing inimical to the dialpainters' characters ever became public knowledge.[33]

By September 1927 the five dialpainters' cases were transferred from the New Jersey Supreme Court to Chancery Court and consolidated. During two days of hearings in early January 1928, Grace Fryer testified that von Sochocky had warned her about ingesting the radium paint, and that her supervisor had denied that radium could harm her.[34]

When hearings resumed in April 1928, Berry's hopes that von Sochocky would testify against U.S. Radium were dashed. On the stand, von Sochocky denied that his warning to Grace Fryer about lippointing brushes involved fears about radium; rather, he said, he was worried about bacteria that might grow in the radium mixture used to paint the watches. He himself, he insisted, had never suspected that radium could be harmful. Berry brought up a conversation reported by Fryer and coplaintiff McDonald in which they had asked von Sochocky why he had never warned them about radium, to which he had answered that it was not his jurisdiction but the responsibility of the plant manager, Arthur Roeder, later president of the company. Von Sochocky denied that the conversation ever took place. Similarly, he denied telling Harrison Martland that he had known that dialpainting was dangerous but could do nothing about it.[35]

It is conceivable that Berry had been misinformed about von Sochocky's claims of early knowledge of dialpainting's dangers and U.S. Radium's refusal to change production methods, but given the number of reports from varied sources, the physicist probably had made at least some of the charges. Another possibility is that von Sochocky, bitter about losing his company through the maneuvers of people with whom he had worked, invented stories of corporate negligence either to seek revenge or some kind of leverage over his former colleagues. Or perhaps the stories were true, but U.S. Radium officials, who believed that von Sochocky was behind the suits, somehow persuaded him not to testify against them. Perhaps von Sochocky's own failing health played a role in his testimony: diagnosed by Martland as radioactive in 1925, he had become ill by 1928 and would die the following November. Von Sochocky's beliefs and attitudes are not clear, but when in court he denied all reports of his charges against the corporation, it was an unpleasant surprise for Berry and the dialpainters. Berry had called him as a witness for the plaintiffs, but during his questioning of von Sochocky Berry began referring to him as a "hostile witness."[36]

Berry also called as a witness Dr. Elizabeth Hughes, who had worked as a physicist for U.S. Radium before moving to the radium laboratory of the U.S. Bureau of Standards. She testified that, using an instrument called a Lind electroscope, she had determined that the dialpainters were indeed radioactive.[37]

Perhaps more important, Hughes testified that radium had been known to be dangerous for many years. She discussed articles that as early as 1914 suggested that workers with radium should use caution. She also testified that two U.S. Radium scientists, von Sochocky and Victor Hess, were leaders in their field and should have been aware of radium's dangers to employees. A physician, Armin St. George, similarly cited early literature warning of radium's dangers, submitting as evidence articles dating from 1911. Counsel for U.S. Radium objected: "If this evidence is to go in in this form, we can bring along books to show that radium is a wonderful thing for human ills." In fact, the corporation attorneys did just that. On cross-examination of Hughes, the lawyers introduced articles touting radium's healing effects, most notably those published in Standard Chemical's journal, *Radium*, and written by its inhouse researchers. During St. George's cross-examination, the conclusions of Dr. Findley John, a proponent of the medical use of radium, were added: "I have used radium chloride in a variety of aliments during the past four years, with surprising and gratifying results . . . I have not observed a single bad effect in any of them . . . I am convinced that this radiant energy is one of the most valuable therapeutic agents at our command."[38]

The dialpainters testified, too. Katherine Schaub noted Frederick Flinn's diagnosis that radium had not made her sick. She swore that lippointing was company policy and that as an "instructress" she taught others to point the paintbrush with their lips. She and her colleagues talked about their illnesses and how they learned from Harrison Martland that radium had caused them.[39]

In questioning the dialpainters, U.S. Radium attorneys tried to hang them on the horns of a dilemma: either the women knew about the cause of their illnesses long before Martland diagnosed them, and therefore the statute of limitations even liberally interpreted forbade their suits, or they could not have known about the cause of their illnesses until Martland told them, because until then no one knew about radium poisoning, and therefore the radium corporation could not have known about the dangers of dialpainting, either, and so were not culpable. The dialpainters tried to dodge both sides of this issue: they had suspected that their diseases were work related but had no definite knowledge that this was so until 1925, when Martland gave them a diagnosis.[40]

The most dramatic testimony was that of Harrison Martland himself, a "star witness," according to newspaper accounts. The press was fascinated by his prediction that all the dialpainter-plaintiffs would be dead within a year. Perhaps equally intriguing were the "photographs" he had made of radium in the bones of living dialpainters by wrapping X-ray film around their legs; radiation from their bones marked the film. Martland reviewed his own work on radium poisoning: autopsies of the dialpainters Sarah Maillefer, Marguerite Carlough, and Eleanor Eckert; tests on the five plaintiffs in court that demonstrated "beyond doubt" that the women suffered from radium poisoning; and medical problems facing the dialpainters. "His forthright, uncompromising testimony stood out conspicuously," one newspaper extolled.[41]

At this point U.S. Radium won an adjournment until September 1928.

Raymond Berry, members of the Consumers' League, and the dialpainters were horrified. To them it seemed that the company was banking on waiting out the deaths of the dialpainters in the suits. (If the dialpainters died, their suits would be void, although family members could file new suits for harm to themselves. Of course, the statute of limitations might preclude these new suits.) Undoubtedly, U.S. Radium also hoped to wait until publicity about the cases had died down.

Newspaper coverage of the five "Radium Girls" was extensive, especially in the tabloids, and often maudlin. Grace Fryer became a favorite of the press, as in this example: "Grace is a pretty girl with dark brown curly hair, a lovely forehead, brown eyes and clear-cut features. She does not look as old as she is. She is the kind of girl who loves dancing and laughing and children. She likes pretty clothes, too, and whoever likes pretty clothes likes a healthy lithesome body." Naturally, there were details of her physical ailments—the ravaged mouth, the crushed bones. But "Pretty Grace Fryer" bore up well under her adversity in this portrayal: " 'And it hurts to smile, but still I smile, anyway,' she said, smiling in spite of the tears in her hazel eyes." With such coverage of their suits, the Radium Girls won the sympathy of newspaper readers.[42]

These stories reveal as much about their authors and intended audience as about the dialpainters. For instance, the largely sensationalist accounts excluded actions that the dialpainters had taken on their own behalf. The papers presented the dialpainters solely as victims. So although the publicity won them public sympathy, it confirmed prevailing social sentiments about women's passivity and proper social sphere.

The press accounts reflect early-twentieth-century expectations for women. As the number and proportion of white women working outside the home increased, a debate raged about the effects of such employment on American families and society. Motherhood and housewifery were the approved roles for women, and outside work was acceptable only insofar as it was necessary to support the family and permitted a woman to fulfill her familial roles. Although the first condition was most often met by daughters and wives who contributed to the household financially, the second was more difficult for wives and mothers to meet if they labored ten hour days in a factory and then faced the additional heavy load of cleaning, cooking, child care, and other chores on returning home. For this reason, in the 1920s and before, white women workers tended to be unmarried or newly married without children, and mothers tended to stay home. The majority of dialpainters were young women in their teens or early twenties, mostly unmarried or newly married; on marrying or becoming pregnant, they quit their jobs. Radium poisoning was the more horrible because it struck down young victims—young women victims. But even more horrifying to readers of that period, radium poisoning rendered dialpainters unfit to marry or to bear children. Thus, it seemed to destroy their very womanhood.

The dialpainters, as young women, were seen as particularly innocent and undeserving victims. Quinta McDonald and Albina Larice were two of the many Maggia sisters who worked at the Orange dialpainting studio. Amelia Maggia had died in 1922, the first known dialpainter death from radium poisoning. By 1928 McDonald and Larice were both suffering from various manifestations of radium poisoning. McDonald was concerned about Larice: "Albina is so unhappy. She lives with her husband and sister and feels she's a burden to everybody with all this terrible trouble. . . . Like me, she thinks it's fate. Like me, she wonders why it should be us girls. We never hurt anybody in our lives." Quinta McDonald shared these feelings. "It's fate I guess," she said. "I can't see why we girls should be chosen to suffer, though." "Fate" implied that no one was at fault. With fate the culprit, there was no one and nothing at which to be angry.[43]

This raised another issue. The "sob stories" in the press presented the women as innocent victims of a horrible fate, but this emphasis on fate avoided placing blame at anyone's door—for instance, linking the dialpainters' suffering with decisions made by their employer. Radium was a stalking monster. No person or institution was villainous, evil—just the inanimate element radium. This narrative strategy obscured both possible U.S. Radium culpability for its employees' illnesses and the dialpainters' anger, resentment, and consequent actions.

Anger was considered an unacceptable response for women. The dialpainters were expected to endure their pain stoically, lending an aura of high tragedy to their suffering. As Quinta McDonald put it: "My husband tries to be brave. He calls me a good sport . . . I'm happier when I keep busy. . . . Of course I can't do much. I can't bend over now! . . . I better stop—you'll think I'm complaining. But somehow I can't be a Pollyanna all the time." McDonald did occasionally lapse into bitterness, even while insisting on proper behavior for herself: "Don't write all this stuff in the papers about our bearing up wonderfully. . . . I am neither a martyr nor a saint. But if you know you haven't very long to live, you might as well make the last of it as pleasant for yourself and those around you as possible." Albina Larice was more assertive, saying, "Well—I always did my best for that company, and I don't see why they don't do something for us now to make things easier." Grace Fryer, though, "meets her trouble like a Spartan," one newspaper reported. "I have no resentment," she claimed, but thought that the company should have warned the dialpainters about radium's dangers.[44]

Religion was seen as an appropriate coping device. If she had money, Katherine Schaub said, "I'd go to the shrines in Canada with the faith and hope that something can be done for me by prayer." Edna Hussman complained that "I cannot even keep my little home, our bungalow." But the *New York World* called her "peaceful and resigned," a "gentle, little blond" who was coping with her illness in an appropriately female way: "I'm religious. Perhaps that is why I'm not angry at anyone for what has occurred. Though we all feel that some one should have warned us. None of us knew that paint paste was dangerous. . . . We were only girls, 15, 17, and 19 years old."[45]

In the newspapers, the dialpainters' youth and sex made them innocent victims, and their refusal to be angry deepened the tragedy of their victimization. The publicity gives us little real understanding of the dialpainters' attitudes about their lawsuits, their illnesses, and their former employer. The press reports were filtered through, first, what reformers and journalists thought of as appropriate sympathy-gathering statements, and second, what tabloids believed would be sensational enough to sell papers.

Still, the portrait of women denying anger but seeking justice is confirmed by an autobiographical account written by Katherine Schaub and published in a magazine in 1932. As a more reliable source on the dialpainters' feelings, it bears quoting at length:

I could hardly sleep the night before the January day set for the court hearing, for I had been waiting for ages, it seemed, to see this very day. . . . At the end of the day everything was going along splendidly, I thought. Tomorrow there would be another court [hearing], and the day after, still another, and so on until the entire case was heard. And then — the court would give its verdict. Then perhaps I could get away from everything and forget. I was awakened from my dreaming by the sound of the vice-chancellor's gavel hitting the desk. The vice-chancellor was speaking. The next court day, he said, would be April 26. I could have given way to tears, but tears would not do any good, I knew. I must summon all the courage I had, and fight against sickness and worry and financial difficulties. . . .

April 26 we were again in court . . . that afternoon and the next day the medical witnesses were heard. I thought it would never end, this excruciating, horrible testimony. It had to be done though, had to be told, or else how would we be able to fight for the justice that was due us?

After all our witnesses had been heard the lawyers for the defense said that one of their important witnesses was obliged to leave town, making it impossible to complete the case at that time. The vice-chancellor said that the first day open on the court calendar was September 24. This delay was heartless and inhuman, I felt.

Again, the prevailing tone was one of bearing up but anticipating justice. Anger, an emotion that might be expected, was almost absent: perhaps nice, working-class girls of the 1920s did not get angry. This assumption was perhaps best expressed in a caption to a photograph of Grace Fryer appearing in May 1928: "Doomed, She Smiles."[46]

More than just the purported innocence of their sex and age made the dialpainters' diseases so horrible. Marriage and motherhood were often tragic for the dialpainters. Some women gave birth to healthy children only to become too ill to care for them. Other women first suffered the symptoms of radium poisoning with the onset of pregnancy. Perhaps the calcium mobilization of pregnancy left their bones more susceptible to the effects of radium. A few workers, such as Albina Larice, were unable to bear healthy children and blamed radium. Overall, the dialpainters had a lower fertility rate than comparable women, but whether because of radium's effects on reproductive physiology or because a significant number of women chose to forgo or terminate pregnancy due to their illnesses, we do not know.[47]

To Americans in the 1920s, women who could not bear children, or who could not care for their children, husbands, and homes, or who could not even marry because of illness were particularly pathetic figures. Grace Fryer, a newspaper reported, wanted "love and a home of her own," but this was unlikely. "Would any girl want to live—" she was quoted, "if she knew she could never bear a child—if she could never get married—if she could never lead a natural, normal life[?]." Some women, out of fear of dying or of being rejected socially, denied or hid their illnesses. Katherine Schaub, herself quite ill with radium necrosis and anemia, spoke scornfully of this: "So many of the girls I know won't own up. . . . they say they are alright. . . . they're afraid of losing their boyfriends and the good times. They know it isn't rheumatism they've got . . . that makes them limp. . . . God—what fools—pathetic fools! . . . Afraid of being ostracized!"[48]

The dialpainters' chronicle, then, can be likened to a horror story, which is how it was portrayed in the yellow journalism of the twenties and how it is remembered in workers' mythology today. Like a horror story, the dialpainters' narrative began with innocent young women who were unknowingly stalked by an invisible, heartless enemy. At moments of their greatest happiness, newly married, newly with child, the women were struck down. Their disease had no known cure and meant certain death.

This story is compelling, but it presents several difficulties. The first is that the villain of the piece tends to be radium itself, rather than flaws in human behavior and human institutions that might be studied in order to make changes promoting safer and healthier workplaces. A second problem is with the portrayal of the dialpainters solely as victims. An examination of their actions shows that this was not so. Early on the dialpainters realized that they faced an industrial disease, even when government and medical personnel disagreed. They struggled to win recognition of their disease, compensation from their employer, and measures to prevent future cases. In such a revision of the dialpainters' story, they are not just the victims but the heroes of the piece.

The Consumers' League helped organize much of the publicity outlined here in order to pressure U.S. Radium into resuming the hearings. In addition, Florence Kelley, general secretary of the National Consumers' League, used contacts at the *New York World* to instigate an editorial campaign castigating U.S. Radium for heartlessness and the courts for allowing the hearings to be stalled. "This is one of the most damnable travesties of justice that has ever come to our attention," the *World* declared, labeling it an example of "Jersey justice" at its worst. An editorial cartoon illustrated the doors to justice barred to five "doomed radium victims." Another critic was Norman Thomas, the Socialist candidate for president. Speaking at a Workers' Circle Lyceum in Newark, he observed that the radium poisonings were a "vivid example of the ways of an unutterably selfish capitalist system."[49]

U.S. Radium fought back against this calculated publicity. Once again the corporation turned to the Columbia University physiologist, Frederick Flinn. He called a press conference and denied that the dialpainters were radioactive and would soon die. Frederick Hoffman responded with his own press conference: the women would die. Hoffman denounced U.S. Radium's refusal to pay the medical bills of those employees who had resisted filing suits. He castigated the corporation for its "thoroughly cynical attitude," which it exhibited by using the statute of limitations as a bar to dialpainters seeking

legal redress. A newspaper editorial accused Flinn of timing his announcement about the dialpainters' health to help U.S. Radium in court; the radium
corporation and its insurance company were "heartless . . . unmanly, unjust,
and cruel." U.S. Radium's attempt at crisis management had backfired. In the
public eye, the dialpainters were proper "womanly" victims and their employers "unmanly" blackguards.[50]

Despite the outcry against U.S. Radium, there was little suggestion in the
press that the company was responsible for the dialpainters' illnesses. Responsibility tended to get foisted onto inanimate "radium" or "fate." Workers were
innocent victims, and management was an innocent bystander. U.S. Radium
was censured not for shamelessly endangering its workers but for heartlessly
refusing to succor faithful employees.

During the summer hiatus in the equity hearings, the women's reform
community resumed efforts to help the dialpainters. The state Assembly
member, clubwoman, and League of Women Voters supporter Agnes Jones,
who had introduced the radium necrosis amendment to the workers' compensation act, initiated legislation to lengthen the period of the statute of
limitations and to begin tolling the statute when an illness was discovered.
She also promised to fight for a state commission to study occupational diseases. The New Jersey League of Women Voters announced a "message of
sympathy" for the dialpainters, who suffered "as gruesome a tragedy as any
in the world's history," and promised to work for the regulation of chemicals used in New Jersey's industrial plants. The Consumers' League naturally
intervened, too. Executive secretary Katherine Wiley explained to journalists that the league was fighting for "human beings who serve as experiments
for chemicals in industry." The statute of limitations bill, however, never became law, and although the Assembly passed Jones's motion to establish a
commission to study occupational diseases, that bill disappeared forever into
a committee in the state senate.[51]

Publicity took its toll on the judiciary as well as on U.S. Radium. Several
justices helped smooth the way to an out-of-court settlement. First, the Chancery Court judge went public with his opinion that the dialpainter cases never
belonged in his court to begin with but could have been heard in the original
trial court. He suggested a novel interpretation of statute of limitations law:
because the dialpainters' bones contained radium and the radium was still
harming them, they were still being injured; therefore, the statute began tolling anew each moment of that injury. His reasoning probably would not have
withstood a legal test, but a trial court judge picked up this ball and ran with

it, scheduling hearings for June 1928. Then a federal judge, officially unconnected with the cases, took an interest and decided to help arrange a private settlement. The settlement was finalized on June 4, 1928.[52]

The judiciary may have been moved to broker the settlement because of press criticism of the legal system. Another explanation is that the federal judge who negotiated the agreement was friendly to business interests wishing to silence the public outcry and thus may not have been a totally disinterested party. Apparently, he and the dialpainter attorney Raymond Berry had worked together earlier in their careers for a law firm that counseled U.S. Radium. Berry later informed the Consumers' League activist Josephine Goldmark that the federal judge was friendly with the directors of a firm that had a controlling interest in U.S. Radium (although I have found no record that U.S. Radium was controlled by an outside firm). Nevertheless, the judge acted in this matter as a private citizen, so there is no suggestion of impropriety.[53]

What is clear were the advantages to business of a timely out-of-court settlement. Why was it "timely"? First, the settlement forestalled further publicity and protest over the radium industry's refusal to acknowledge radium poisoning. Second, concurrent with newspaper accounts of the radium cases were stories that New Jersey workers were being poisoned by the new gasoline additive, tetraethyl lead. DuPont and Standard Oil, the firms involved, were threatened by a public outcry, and they, together with other firms whose workers were endangered by toxic chemicals, may have wished to minimize publicity about industrial diseases in general. Third, although a few dialpainters did win generous terms out of court, legal issues associated with their lawsuits were never settled; resolution of these issues might have made it easier for subsequent victims of industrial disease to sue their employers. Certainly, most New Jersey dialpainters failed to obtain awards, and none ever won them in court. Indeed, by 1935 U.S. Radium would be exonerated of any fault in their employees' illnesses.[54]

Under the 1928 settlement, U.S. Radium did not have to admit guilt. The dialpainters each received a $15,000 lump sum payment, a $600 annual pension while incapacitated from radium's effects, and payment of all future medical expenses related to injuries from radium. Berry collected $3,500 in legal fees. The dialpainters' compensation was more liberal than the $1,000 to $9,000 paid in the first three New Jersey cases settled out of court. Katherine Wiley thought that Raymond Berry had accomplished a "very brilliant piece of work." No doubt the publicity manipulated by the Consumers' League made a difference. A U.S. Radium official blamed the settlement on the "cleverly designed campaign of publicity" through which "the public was appealed to and the appeal met a responsive chord. The human aspect of live women doomed to die was played up in an appealing manner." [55]

The one potential flaw in the settlement was its stipulation that a board of physicians would control disbursements. The board was to be comprised of one physician selected by the dialpainters, one chosen by U.S. Radium, and a third acceptable to both parties. If two of the three physicians agreed that a woman no longer suffered from radium illness, the corporation would not be responsible for her medical care. Finally, payment of the women's annuities and medical expenses was made contingent on a finding by the board that the dialpainters' ailments were the result of radium poisoning. "The obvious intention," Berry wrote, is for "the corporation to clear itself in the future. . . . The corporation's lawyers have not, in any way, concealed from me that this is what they are driving at." Berry's fears were not unrealistic. At the time of the settlement, the vice president of U.S. Radium informed New York City's commissioner of public health that "there is a very serious question as to whether any of the complainants is radio-active, . . . there is a suspicion that other things may have caused the condition complained of." [56]

Choosing the physicians to manage the women's medical care would be a delicate task, and attention now shifted to the board's composition. U.S. Radium selected for its representative Dr. L. F. Craver, of Memorial Hospital, who researched radium's medical applications. It proposed as the "neutral" party Dr. James Ewing, a colleague of Craver's at Memorial Hospital. Ewing, like Craver, studied the therapeutic uses of radium. A colleague of Harrison Martland was concerned about these choices and warned him that Ewing might not be appropriate for the board's swing vote because he had once attacked Martland's conclusions about the dangers of radium. Martland replied that he questioned the selection of two board members from Memorial Hospital, an institution "so closely allied with the use of radium and [the] radium industry," but that "we must trust in the integrity of Dr. Ewing." Martland then took a hand in finding a representative for the dialpainters, encouraging his fellow pathologist E. B. Krumbhaar of the University of Pennsylvania to accept the appointment. Martland wrote Krumbhaar that his great scientific ability made him "unapproachable from the commercial standpoint" and thus protected Martland's work. "I would hate to have my work . . . be discredited by a committee of inexperienced men," Martland declared. Krumbhaar accepted the position. [57]

The question of the existence of radium poisoning was immediately taken up by the three-physician panel stipulated in the settlement. Craver, the doctor selected by U.S. Radium, attempted to discredit evidence of radium poisoning but was stymied by the other two board members. Krumbhaar, the dialpainters' choice, attempted to have the panel issue a public statement verifying the existence of radium poisoning, with no more luck. The board authorized payments to the dialpainters but never offered any public, official opinion about the existence of radium poisoning. Still, in order to continue the payments, which were contingent on the board finding that the dialpainters did in fact suffer from radium poisoning, the three physicians agreed privately on a diagnosis of radium poisoning.[58]

Berry later wrote that the medical community "in reality took the place of court and jury" in establishing the existence of radium poisoning. The radium cases are a good example of an evolving strategy of granting authority over occupational diseases to scientific bodies. David Rosner and Gerald Markowitz, in a study of silicosis, identify this transition as a 1940s phenomenon involving the deflection of a class issue to a technical issue, whereby the right to assess industrial diseases passed from workers to scientific elites. But the trend started in the 1920s with the tetraethyl lead and radium cases.[59]

Upon settlement of his first five radium cases, Raymond Berry immediately filed one more. Mae Cubberly Canfield had been employed by U.S. Radium from 1915 to 1918. She married a laboratory worker there in 1917 and subsequently left her job to have their first child. In 1928 she had all her teeth extracted following a diagnosis of "radium jaw." She had suffered since 1926 from pains in her hips, knees, and back; she thought she had rheumatism. Berry again began to construct a case.[60]

This time it was harder. The statute of limitations issue was more difficult to surmount in Canfield's case because she was filing even later than had the other women, although Berry would claim that she did not know she was ill from radium until 1928. Canfield was not as sick as the women in the earlier suit, so sympathy was harder to play on. And, whereas U.S. Radium was able to line up a number of experts willing to testify that it could not have known of the dangers of radium before 1925, Berry struggled to find anyone to argue that the company should have known radium was dangerous.[61]

Locating medical experts to attest that Canfield suffered from radium poisoning, or that radium poisoning even existed, proved even more challenging. The Harvard professionals again resisted testifying; although eventually William Castle agreed to give a statement, the Drinkers were obstinate. Swen Kjaer of the Labor Department refused to become involved, claiming that as a federal investigator he was barred from testifying. Berry hoped to introduce in court the conclusions of the Craver-Ewing-Krumbhaar board of physicians that radium had harmed the dialpainters. In a preliminary hearing, he called on James Ewing, the neutral member of the board. But Ewing refused to answer questions, claiming that his relationship with the dialpainters was confidential. Berry noted that the five dialpainters to whom Ewing referred had waived their right to confidentiality in support of Canfield's suit. Nevertheless, Ewing insisted that the women were incompetent to judge their rights, and he continued to refuse to answer questions. Berry then called on the dialpainters' representative, Edward Krumbhaar. Krumbhaar consented to answer Berry's questions, but when he began to speak he was interrupted by corporate attorneys, who accused him of divulging information gleaned while in a confidential relationship with U.S. Radium. Berry argued that Krumbhaar had no legal relationship with the radium company and was responsible only to the dialpainters. The opposing attorneys then threatened to sue Krumbhaar for breach of a confidential relationship. That warning silenced Krumbhaar.[62]

Harrison Martland also declined to testify. Annoyed by publicity and angered by what he termed "misstatements" about his earlier testimony in the dialpainters' hearings, he refused to undertake any further work on behalf of workers exposed to radium. He would no longer study living patients, even to confirm the presence of radium in their bodies (he eventually relented and measured the radioactivity in two women with whom he had long been in contact). He would no longer be available for testimony concerning the source of dialpainters' illnesses. As county medical examiner, Martland conceded, he would, of course, still autopsy and determine the cause of death of deceased dialpainters. And he agreed to share information with other physicians working with dialpainters but would not work with them directly or participate in consultations.[63]

This left dialpainters and their lawyers in an untenable position. Few physicians had access to the equipment needed to diagnose radium poisoning, and without such a diagnosis, the women's lawsuits were without substance. Berry returned to the physicist Elizabeth Hughes, who agreed to measure Canfield's radioactivity and to testify in court in return for 10 percent of any settlement.[64]

Several days of testimony were heard in the Canfield case in 1929, and then in 1930, her suit, too, was settled out of court. This time the terms were less generous. Of a total settlement figure of $8,000, Mae Canfield received $3,500 and her husband, $1,000. No annuities and no continuing payment of medical bills were included.[65]

Of the rest of the settlement, Hughes presumably received her 10 percent, or $800, and, after subtracting expenses, Berry took the balance, perhaps $2,000. This amount compares rather well to the $3,500 Berry collected for the five cases settled in 1928. However, as part of the Canfield settlement he agreed to bring no more lawsuits himself and to refuse help, information, and advice to anyone else suing U.S. Radium. The loss of Berry as advocate together with Martland as physician further undermined the attempts of New Jersey dialpainters to win legal redress for their illnesses.[66]

In contrast to New Jersey, the Connecticut dialpainters had recourse to a workers' compensation act that covered their illnesses and a fairly reasonable statute-of-limitations period of five years (subsequently shortened to three). But both the state agents and the watch company representatives encouraged private settlements that included medical care, supervised by Frederick Flinn, and small monthly payments or small total cash settlements.

The Connecticut cases were successfully managed by the officers of the companies in which they occurred. Management of these cases was "successful" in that they were largely kept from public notice and from the kind of publicity that attended the cases in Orange, New Jersey, and eventually in Ottawa, Illinois; they remained out of the courts and outside the workers' compensation apparatus; and the financial settlements, made privately, were low and were awarded only if the employee or her family absolved the company of all liability. Because of this, our knowledge about the Waterbury, Connecticut, cases is limited; we cannot rely on a surfeit of newspaper articles or on legal proceedings, as in the other instances.

The political climate in Connecticut favored business. It would seem that, in the minds of most historians of the state, "Conservative" and "Connecticut" are practically synonymous. For example, the guide to Connecticut produced by the Federal Writers' Project stresses "conservatism" as characteristic of the commonwealth and refers to "aversion to change" as "almost a second religion" of political and social leaders. Connecticut government, the account continues, was "aristocratic" and "paternalistic" and "popular only in the sense that elections were held." Connecticut called itself "The Land of Steady Habits."[67]

In response to nineteenth-century changes—a business boom, urbanization, and an increasing immigrant population—Connecticut Yankees searched for means to stay in control. To the insurance companies moving into the state, Connecticut granted reduced taxes and the election or appointment of their representatives to all branches of government. Nativism was strong, and the American Protective Association successfully denied political office to Catholics in the 1890s and controlled the Republican Party in urban areas. Immigrants did capture the Democratic Party in cities, but Yankee Democrats switched sides and voted for Republicans to limit the immigrants' power. The state retained an Assembly structure that gave rural, Yankee sectors inordinate representation at the expense of the urban areas. By World War I two-thirds of Connecticut's population was first- or second-generation foreign born, but in 1924 one-sixth of the people, from rural areas, elected two-thirds of the house of representatives. In such a climate, labor and progressive reform was limited. Yet labor did have periods of influence: in 1867 workers contributed to the election of a pro-labor governor who helped pass an eight-hour law, although it was unenforceable; in the 1880s the Knights of Labor were strong, won General Assembly seats, and helped enact numerous labor and factory laws—many of them ineffectual or unenforceable as well. By 1896 state Republicans began a reign that persisted to the 1930s. In the twentieth century, nativism was evident in local manifestations of the World War I Americanization campaigns, the postwar Red Scare complete with illegal deportations of suspected radicals, and the Ku Klux Klan. Employer resistance by 1905 had destroyed labor's influence: whereas union membership in the United States doubled from 1905 to 1917, it declined in Connecticut.[68]

Connecticut politics also handicapped women's causes. Business control of legislation impeded social feminists' advancement of social welfare programs. Connecticut was one of only three states that rejected federal funds from the Sheppard-Towner Maternity and Infancy program, passed in Washington, D.C., in response to woman suffrage. Similarly, despite the participation of Connecticut women in the early twentieth-century mass suffrage movement (membership in Connecticut suffrage organizations grew dramatically in 1910 and peaked at 32,000 in 1917), political leaders stalled consideration of the suffrage amendment until its passage nationwide was already assured—Connecticut jumped in as the thirty-seventh state to ratify when only thirty-six approvals were necessary. In response, the state Woman Suffrage Association ignored pressure from its national board and opposed the Republican Party in the 1920 elections, though without much impact. Connecticut did have its share of women's organizations, which in the twenties included various clubs and local branches of the League of Women Voters, the Women's Party, the Business and Professional Women's Association, and the Consumers' League. As elsewhere, the reforming organizations were linked by shared personnel, like Mary Cromwell Welles, a Goucher College professor, who was also a paid lobbyist for the Consumers' League, a chair of the League of Women Voters' Women in Industry Committee, and a supporter of the American Association for Labor Legislation. Still, what emerges from study of the few available sources on women's organizations in the state is that a small number of Connecticut women fought insurmountable odds. The Republican Party monopolized power and was unresponsive to feminists or feminist-backed social reform.[69]

In the 1920s Connecticut government was controlled by a political machine interlocked with a tightly knit business leadership. The Manufacturers' Association and the state Chamber of Commerce worked with the J. Henry Roraback machine to win Republicans a six-to-one advantage in the house of representatives and a four-to-one advantage in the senate. During the decade governors were elected from the state's leading businesses. Taxes from corporations declined from 20 to 8 percent of total tax receipts. All labor legislation, including regulation of the workplace, minimum-wage laws, old-age pensions, and health insurance, was blocked in committee with the exception of one amendment.[70]

Workers' compensation had been passed in Connecticut in 1913. Whether the "personal injury" covered in the act included industrial diseases was debated in the courts and generally answered in the negative. In 1919 and again in 1927 the legislature amended the compensation act to include industrial diseases. Given business's control of the legislature, historians have asked why workers' compensation was amended to include a relatively liberal occupational health clause. The conclusion has been that business sought to remove "an emotional issue from the jurisdiction of the courts."[71]

Management of the dialpainters' occupational illnesses demonstrated the effectiveness of the business-controlled state government. Cases never reached the courts. The Waterbury Clock Company's private settlements with the dialpainters were given cursory approval by workers' compensation officials. Part of the reason for this was that the Connecticut watch company managers found in Frederick Flinn a compliant resource who was willing to play a two-faced role: to the dialpainters, he presented himself as a concerned medical expert, whereas for the company he persuaded dialpainters to accept financial settlements that explicitly freed the company from further liability.

Frederick Flinn was not a physician—he held a doctorate in physiology— yet he took control of the dialpainters' medical management. Recall that in August 1925 and November 1926 Flinn had studied all the currently employed dialpainters and pronounced them healthy and not radioactive, though he did know about two former Connecticut dialpainters who were both radioactive and symptomatic. Flinn repeated his examinations of the dialpainting workforce in June 1927 and February 1928, with similar results. Then, in

October 1928 he discovered that five then-employed Connecticut workers were radioactive. What accounted for the change? A few months earlier Flinn had been hired by the New Jersey radium company to measure radium in the dialpainters who were bringing suit against it, and at a crucial moment in court deliberations he had announced to the press that none of the plaintiffs was radioactive at all! Already threatened by Alice Hamilton and others with charges that he practiced medicine without a license in New Jersey, Flinn was then publicly denounced by Frederick Hoffman. Perhaps he deemed it a good time to recalibrate his electrometer. In Connecticut, Flinn diagnosed dialpainters' ailments, prescribed their treatment, even arranged for their care by physicians and dentists at the firm's expense while "holding off unnecessary work to avoid extortionate bills." He treated "all the . . . Waterbury victims," a newspaper reported in 1930.[72]

In addition to his medical management, Flinn also arranged financial settlements with the dialpainters or their families: he informed Alice Hamilton that the Connecticut cases were settled out of court under his auspices. In 1927 Elizabeth Dunn and Helen Wall, the two Connecticut workers whose radium poisoning Flinn had revealed in 1926, both died from jaw necroses. After Dunn's death, Flinn helped the company settle with her family "without the compensation question arising," a vice president of the firm wrote in 1936. Dunn's family received $1,500. Wall's family settled for "quite a considerable sum," according to the same official, who did not specify the amount. Around 1928 illness struck another former dialpainter, Marjorie Dumschott. Following a tooth extraction, her jaw became painful, swollen, and, eventually, abscessed; two sinuses formed in her lower jaw and remained open and draining until her death in 1931. In 1928 she agreed to settle for an unknown amount with the Waterbury Clock Company, where she had worked in 1924 and 1925.[73]

Details of some of these events became known to the federal Labor Department investigator, Swen Kjaer, when he revived his agency's inquiry into the radium poisonings in December 1928 and February 1929. On visiting Waterbury's health officer, he learned little except that none of the Waterbury cases had been referred to the state Board of Health. "All proceedings . . . [are] kept practically secret by the company," Kjaer was told. He next visited the Waterbury workers' compensation commissioner. No claims had been brought to him, the commissioner reported, and besides, the Waterbury Clock Company was doing "far more" than required by law. The firm's vice president then arrived at the compensation commission offices, at which point the commissioner "began to eulogize the Waterbury Clock Company for its philanthropy and good treatment of its employees." The vice president said that "adjustments had been made by the firm" and there was no need to provide compensation for the two dead and one ill workers. On visiting the vice president a few days later, Kjaer learned that "it was customary to place the weakest and frailest girls at dial painting." Perhaps dialpainting was easy work compared to other jobs assigned to women employees, but Kjaer was to conclude that they were likely to get sick and die anyway. Finally, Kjaer visited a dentist who had treated Elizabeth Dunn and other dialpainters as directed by Frederick Flinn. The dentist was paid by the clock company. "The discussion was interspersed with heavy praise of the firm," Kjaer reported. These details suggest that Waterbury was a small city where government officials and medical professionals served the interests of the clock companies.[74]

The clock manufacturers' influence operated at the state level as well. Faced with potential liability suits, the companies helped rewrite the workers' compensation act in 1927. Occupational diseases were still explicitly covered, but the time period for filing for compensation was shortened from five years to three. A Waterbury Clock Company executive discussed this with Swen Kjaer. The company still protested consideration of the dialpainter Frances Splettstocher's death as a radium poisoning case, but, the executive gloated, "it did not matter anyhow as the case is outlawed under the law passed by the last State legislature. Under it, the compensation is outlawed, and no other recourse can be had."[75]

The next dialpainter to die in Connecticut was Mildred Williams Cardow, in 1929. She was twenty-two years old when she succumbed to a cancer arising from her sinuses, a cancer soon to be attributed to radium deposition in the bones. Newspapers reported that initially her husband William, a car mechanic, refused to settle with the company. Apparently, while his wife lay dying, he was given a form to sign absolving the company of all responsibility in Mildred's illness and death in exchange for half of her $13 wage, or $6.50, for a retroactive thirteen-week period, plus reimbursement for medical expenses. He demurred and planned to sue for damages. Florence Kelley, general secretary of the National Consumers' League, briefed the league's board of directors on the case:

> In the radium industry, this year's horror has been the death of an [employee] of the Waterbury Clock Works in Waterbury, Connecticut. . . . When, in 1927, two of its women employees died of the poison, the corporation promptly procured a weakening of the statute of limitation which formerly ran five years. This was shortened to three years so that the hus-

band of Mrs. Cardow, this year's victim, could not bring suit. . . . He was offered . . . the insulting sum of $43.75, which he indignantly refused.

The *New York World* referred to William Cardow as "a David fighting the Goliath of Connecticut industrialism." This time, David lost. William Cardow learned that the statute of limitations did prevent his bringing suit: because his wife had left dialpainting more than three years before her illness began, her employer was no longer liable for damages to her health. Within three months of his wife's death, Cardow accepted a settlement from the company; we do not know what he received.[76]

After that, settlements seem to have proceeded smoothly under Flinn's management. Some of the financial deals are known. Marion Demolis accepted compensation in 1929 and died three months later. She was an archetypical dialpainter, starting her job at fifteen and working for about a year. Five years later she began suffering from a bad tooth. She had it pulled, but the affected area would not heal; instead, her jaw swelled, hurt, and oozed a foul discharge. Demolis died of infection and anemia, the fifth Connecticut dialpainter to lose her life. Louise Pine died in 1931, and a year and a half later her family settled with the Waterbury firm for $250. (Her mother tried to sue a New York–based dialpainting factory for which her daughter had also worked but was blocked by New York's statute of limitations.) Edith Lapiana settled in 1931 and died ten months later; Mabel Adkins settled in 1934 a few months before her death; Anna Mullenite settled in 1932 and died in 1935. Florence Koss made an agreement with the company in 1929; after her death in 1935, her husband then signed a full release.[77]

By 1936 the Waterbury Clock Company acknowledged ten deaths from radium poisoning (it still did not count Frances Splettstocher, the first to die). The company had made final settlements with two more women who were ill. One was Katherine Moore, who had been told by Flinn eight times that she was not carrying radium in her body. In 1935 she wrote to an expert on radium poisoning: "I have had the advantage of the examination given the girls of the radium department by a doctor who came from New York about once a year or more. After each examination I was dismissed as being negative." In 1934 her jaw had spontaneously dislocated: "Gradually my jaw seemed to work out of place and now it has such a deformed appearance it has practically made a recluse of me, and it is in such a condition that I cannot go about in public without pulling my coat collar around my jaw to hide it." By 1936 a grapefruit-sized tumor had grown from her chin; she died that year. Her settlement amounted to $1,000 from one watch company for which she had worked and $1,500 from the Waterbury Clock Company. The second ill woman was Ethel Daniels, who settled with the Waterbury company in 1930; seven years later she would also die of cancer of the jaw. Four other women were being carried on the company payroll because of radium-related illnesses, at rates ranging from a low of $8.71 a week to a high of $23.00 a week. Another case was under negotiation. The vice president expected that five more "cases" would become disabled over the next five years, and settlements would then have to be made.[78]

From 1926 to early 1936 the Waterbury Clock Company spent almost $90,000 in settlements, support, and medical costs for the dialpainters, or, so the vice president explained, about $9,000 a year. A reserve of $10,000 a year was set aside to cover future costs. Listed in the vice president's report were the names of ten deceased women who had accepted settlements, two ill women who had been compensated, and four ill women receiving weekly stipends—sixteen women in all. On average, the clock company had spent for medical bills and damages about $5,600 per affected employee. If the cases had gone to court or to the compensation commission, the costs would have been far higher, the vice president observed: "I feel that our manner of handling these cases by voluntary agreements has saved the Waterbury Clock Company very many thousands of dollars. In addition to this we have the Compensation Commissioner one hundred percent with us and if we had fought these cases I believe that the sums expanded under awards would in many cases have been doubled."[79]

It is difficult to judge whether the Connecticut firm made the wiser choices from an economic perspective. Compared to some of the New Jersey settlements, these expenditures were indeed low. The five dialpainters whose suits garnered the most publicity cost U.S. Radium over $15,000 apiece, but the other four settlements to date had cost $1,000, $3,000, $8,000, and $9,000 respectively. One may roughly calculate an average cost per settlement of over $10,000, which, compared to Waterbury's $5,600, is high. On the other hand, despite the bad publicity experienced by U.S. Radium, it had taken steps that made it unlikely that any other dialpainters could file, let alone win, negligence cases against the company. Still, its legal expenses must have been high, probably far higher than the cost of Frederick Flinn to the Waterbury Clock Company.

The Orange, New Jersey, dentist Joseph Knef was so impressed with Frederick Flinn's position with the radium companies that he tried to duplicate it and even improve on it. Between 1921 and 1925 Knef claimed to have treated thirty-seven dialpainters, but he seems to have had trouble getting paid for his services. In 1924 he tried to arrange a suit against U.S. Radium over Amelia Maggia's death to recover what was owed him. Then, in September 1925 Knef informed the Waterbury Clock Company that he would be willing to examine "a few girls" for $100 a day. He declared himself to be the only expert knowledgeable about treating dialpainters' illnesses. He suggested that if "the Orange people" had "heeded my advise . . . I could have saved them many thousands of dollars." His offer was ignored, as Flinn already had begun his Connecticut examinations in August.[80]

Knef was at the forefront of events in Orange. He had treated some of the earliest dialpainters who had become ill, including Amelia Maggia, the first dialpainter known to have died from radium poisoning, as well as Sarah Maillefer, Marguerite Carlough, and Hazel Kuser. Knef was one of the dentists Katherine Wiley spoke to early in her investigation, and she arranged for him to share his information with both Frederick Hoffman and Harrison Martland. Knef contributed to Martland's first paper on radium poisoning to such an extent that Martland listed him as a coauthor, although Knef later suggested that Martland stole the "discovery" of radium poisoning from him.[81]

In May 1926 Knef approached the U.S. Radium Corporation with a proposal. If the firm would engage him to treat all of its injured dialpainters, he would keep quiet about the causes of the women's illnesses and use his influence to prevent the dialpainters from filing suits. If legal proceedings should arise, he would testify that jaw bone changes were due to pyorrhea, not radium necrosis. "I will keep my mouth shut. . . . I will use my influence. . . . I can hold these girls off for four or five years," he promised. On the other hand, if the company refused his offer, Knef would sue the dialpainters for his fees. The dialpainters would then be forced to sue U.S. Radium, and in those suits he would testify that injured dialpainters showed radium necrosis in their oral X rays. A U.S. Radium attorney later remembered, "he said that an expert always testifies on the side that is paying him." "It is a question," as Knef posed it, of "Do you want me as a friend or do you want me as an enemy." "I ask $10,000," he said. "I am going to get this one way or the other."[82]

Suspicious of Knef, the radium company had taped this conversation, and the president now rejected his proposal. "Your proposition is immoral; we will have nothing whatever to do with it," Arthur Roeder asserted. At some point the company hired women detectives to visit Knef, posing as dialpainters with sore mouths. Reportedly, he asked them to strip to the waist (without a nurse present), x-rayed their mouths, and told them that although the X rays showed no necrosis, they should sue U.S. Radium anyway. He would take care of the proof. Knef next turned to the children of dialpainters: whereas other clinicians found them free of radium, Knef declared those he tested to be radioactive.[83]

I tell this story because its very tawdriness puts in perspective the activities of scientific experts. Knef was nefarious, but his sins differed more in degree than in kind from the actions of other scientists. Knef promised to mislead the dialpainters about the cause of their illnesses if U.S. Radium would pay their dental bills; so had Theodor Blum. Knef promised that as a paid consultant of U.S. Radium he would keep silent about radium poisoning; so had Cecil Drinker. Knef promised to testify as a paid consultant for the radium company; so had Frederick Flinn. Knef took these attitudes beyond then-acceptable limits. He baldly demanded money for his silence and for his outright invention of testimony supporting the company's positions. He threatened U.S. Radium with lawsuits and adverse testimony if they rejected his offer. Knef was a blackmailer. But his assumption that "an expert always testifies on the side that is paying them" is a frank formulation of expectations motivating more circumspect scientists.

The intervention of the Consumers' League in the New Jersey cases was crucial to their outcome. Using what Alice Hamilton called its "weapon of publicity,"[84] the league broadcast the plight of the dialpainters. It was assisted by newspapers that printed often lurid feature articles bemoaning with necrophilic intensity the impending deaths of the doomed "Radium Girls." The Consumers' League also helped the dialpainters find lawyers, contributed to the planning of lawyers' offensives, and publicized legal maneuvers of U.S. Radium to stall the cases on their way to court. For workers made ill in the future, the league, together with other middle-class women's reform organizations, sought to add radium poisoning to the schedule of compensable industrial diseases under New Jersey's workers' compensation law.

The Consumers' League had its successes, but they were blunted. Although a few New Jersey dialpainters had won settlements, most had not. After 1925 radium poisoning was generally accepted medically, but it was not recognized in court or by the legislature until 1931. Medical experts recruited by the Consumers' League to diagnose radium poisoning were little interested in rendering further aid to the dialpainters. Two "victories" offered the dialpainters as a group no help at all. First, out-of-court settlements avoided the issue of establishing legally the chronic toxicity of radium as well as the liability of the radium companies. Second, the radium necrosis bill disregarded other manifestations of radium poisoning and brought it under the jurisdiction of the workers' compensation bill with a statute of limitations far too short to permit claims ever to be filed.

Besides the support of the Consumers' League for the New Jersey dialpainters, a relevant variable explaining events in that state was the decidedly different status of the small and relatively unimportant U.S. Radium Corporation in Orange versus the position and authority of the watch companies in Connecticut. Certainly many New Jersey politicians were pro-business — Commissioner of Labor Andrew McBride, who had sought to limit investigation of the radium cases, was motivated, according to Alice Hamilton, by the premise that "capital was not getting a fair deal" [85] — but such biases favored business in a generic sense and did not call forth a spirited defense of the radium industry's right to conduct itself in any way it saw fit.

In New Jersey, the Consumers' League continued efforts on behalf of the dialpainters. The ground of the struggle shifted to federal courts and agencies. The pattern of Pyrrhic victories would be repeated at the federal level.

IS THAT WATCH FAD WORTH THE PRICE?

Industrial Radium Poisoning and

Federal Courts and Agencies

In late 1928, following the publicity attending the New Jersey settlements and the Connecticut revelations, the Consumers' League and the dialpainters again turned to federal agencies for the regulation of dialpainters' work practices. Compensation suits also moved to the national level as dialpainters' cases entered federal court. Industrial health issues raised by the dialpainters were much clarified at the national level; federal courts and agencies continually confronted issues that may be distilled into three sets of questions about radium poisoning specifically and industrial disease generally.

The first set of questions revolved around what should take priority, worker health or industrial development. Should a new industrial process be held in abeyance until its safety was assured? If injury followed a new process, were negligence and liability to be assumed, or was ignorance a defense? The concerns about industrial development were part of a larger issue of social utility: could it be said that "society as a whole" benefited from products if they were made with workers at risk? What constituted "society" if workers were excluded?

The second set of questions involved the validity of establishing a "tolerance level" specifying a safe level of exposure to a toxic chemical. Was some amount of worker exposure to a poison safe or an acceptable risk? Was this true for cumulative poisons as well as acute poisons? In the case of the dialpainters, discussion though the 1920s and 1930s focused on whether it would be sufficient to eliminate lippointing while permitting radioactive dust and radon to remain in the air. A related issue was individual susceptibility. In identical situations, why did some workers accumulate radium and some not? Of workers with the same body-burdens, why did some get sick whereas others did not? Were affected workers somehow especially susceptible, whereas other workers found exposure safe?

Finally, debates centered on whether industrial diseases should be considered primarily a labor problem (within the jurisdiction of labor authorities) or a medical problem (subject to the control of medical authorities). Were industrial diseases better handled through dialogue between workers and managers or through discussions among medical specialists? A related issue was how best to study industrial diseases—through retrospective studies of workers previously exposed, prospective studies of those currently working, or animal studies.

These issues arose in four settings: a 1928 Public Health Service "voluntary conference" on radium poisoning, a 1929 Bureau of Labor Statistics investigation and report, a 1929–30 Public Health Service study, and a 1935 federal court hearing on New Jersey dialpainter compensation cases. In all instances, business perspectives prevailed, especially in regard to valuing industrial development and social utility over worker health.

Medical specialists were the other beneficiaries of political developments regarding industrial health, as they were empowered to define and control industrial diseases, in part through their determination of acceptable "tolerance levels" for body accumulation of radium. Tolerance levels touched on the three parts of industrial disease politics—recognition of a new disease, compensation of injured parties, and prevention of future cases — by naming a level of toxicity separating safe from dangerous exposure.

Since at least 1914 the Public Health Service and the federal Labor Department had been jockeying for jurisdiction over industrial health, a battle that would continue through the 1930s. A major difference in orientation divided the two agencies: the Labor Department was concerned with workers, the Public Health Service with diseases. As a working-class issue, industrial diseases loomed as a large problem, but as a medical issue, they were considered of less importance than other diseases. The Public Health Service, then, seemed less interested in investigating industrial diseases than in preventing the Labor Department from gaining control of them. The two departments faced off during a 1928 conference on industrial radium poisoning.[1]

Initially, the Public Health Service had little interest in industrial radium poisoning. Following a June 1928 Consumers' League request for its involvement, the agency responded that it was willing to help, that it did not require letters and publicity in order to act, but that money and personnel to undertake a study were lacking "at present." A few weeks later the surgeon general turned down an appeal from the New York City health commissioner for a radium investigation, explaining that the health service was concerned with communicable diseases, not industrial hazards.[2]

At this point the National Consumers' League's general secretary, Florence Kelley, intervened. Kelley with Alice Hamilton persuaded eminent physicians to sign an open letter requesting that the Public Health Service sponsor a conference for those concerned about dialpainting. The two reformers wanted this conference modeled on the Tetraethyl Lead Conference held in 1925. The sympathetic *New York World* publicized the physicians' petition and supported the meeting editorially. The surgeon general then agreed to hold a conference. After waiting several months, the reformers arranged for a physician who had autopsied several deceased dialpainters to speak at the Consumers' League annual meeting, with Eleanor Roosevelt as "toastmistress." Publicity supporting this event finally brought the Public Health Service to heel, and the long-awaited conference was scheduled for December 1928.[3]

The subsequent Radium Conference had marked similarities to the Tetraethyl Lead Conference: both were voluntary; both involved representatives of business, government, science and medicine, and reform; and both neglected to include affected workers. These gatherings could be considered the federal government's 1920s solution to industrial health controversies.

They fit interpretations of government during the second and third decades of the twentieth century that found operative an ideology of corporatism, that is, an ideology that both independence and cooperation among all sectors of society were possible if each sector acted responsibly for the good of the whole. This vision, variously labeled "corporate liberalism," "associated individualism," or "the associative state," emphasized voluntarist methods of managing conflict. As Joan Hoff Wilson put it, voluntary cooperation between public agencies and private organizations was seen as a solution to (or a compromise between?) disorderly laissez-faire capitalism and dangerous state control. Embedded in this voluntarist solution was the assignment of control to allegedly neutral scientists whose "neutrality" was most marked by lack of sympathy for labor.[4]

Attending the Radium Conference were executives of the dialpainting and watch companies in New Jersey, Connecticut, and Illinois; representatives of the federal Labor Department, Public Health Service, Bureau of Mines, and Bureau of Standards; officials of the New Jersey Department of Labor and the Connecticut Department of Health; scientists and medical specialists interested in radium, including Frederick Flinn, Harrison Martland, and Alice Hamilton; and reformers Florence Kelley and Katherine Wiley. No dialpainters were there. The American Federation of Labor sent an observer.[5]

Early in the conference, Ethelbert Stewart, head of the U.S. Bureau of Labor Statistics, raised the issue of the social utility of luminous watches compared to the cost in workers' health. He reckoned that in the industry there were now thirteen women and seven or eight men known dead and eighteen women and five men alive but ailing. "And what is it all about?," he asked, "The luminous watch . . . purely a fad." "My question is this," he continued, "Is there enough real utility in the manufacture of luminous watch dials . . . to pay for what is happening? . . . Is that watch fad worth the price?" Professor R. B. Moore, of Purdue University, a man with close ties to the radium industry, argued for valuing industrial development over worker health. "There is not any industry which has not any industrial hazard," he said. "If we shut down because of a hazard, we would shut down every industry." Arguing for the social utility of radium, he claimed that for every death in the radium industry, "hundreds of lives had been saved by its use." Dudley Ingraham, representing his Connecticut watch company, also defended radium's usefulness, citing letters received from invalids telling of the comfort given by luminous dials. He added that his own mother, mere hours before her death, told him that her alarm clock with its radium dial "meant more to her than anything else."[6]

These larger questions were dropped from consideration when, after lunch, the manufacturers' representatives returned armed with an agenda. They had caucused over lunch and now proposed that the Public Health Service appoint two committees. C. H. Granger, vice president of the Waterbury Clock Company, served as spokesperson.

One committee would address regulation of the workplace by establishing codes embodying the best protections and safeguards known. There were no objections to this, but it was noted that safety and health codes would depend on how much deposited radium was deemed dangerous and on whether lippointing alone or also inhalation of radium dust were identified as hazardous. To answer these questions would seem to have required a retrospective study comparing dialpainters formerly employed before lippointing was banished to dialpainters currently employed under hopefully safer conditions. A representative from the New Jersey Labor Department broached this issue, asking if currently employed dialpainters would be studied. No, spokesperson Granger replied.[7]

The second committee proposed would consider regulation of the workforce. Earlier in the conference, Granger had noted that in Connecticut the "weakest girls" were assigned to dialpainting, suggesting that they were somehow predisposed to radium poisoning by their own feeble constitutions. He now explained that some people were "unfit" for radium work because of such conditions as anemia, while others with good general health and blood were "best suited for [the] work." The second committee would determine health standards for radium workers and would devise tests to detect radium accumulation and blood changes indicative of radium poisoning. Katherine Wiley of the New Jersey Consumers' League asked a sharp question. Were the companies proposing "to ascertain what people are immune" to radium? The radium executive's remarks were typical of twenties' business leaders who sought to shift responsibility for industrial poisoning from unsafe workplaces to susceptible employees.[8]

Three agreements emerged from the voluntary conference. The two industry-proposed committees were approved, and the Public Health Service would appoint industrial and public health workers to staff them. There is no record that the committees ever were formed or met. Under the third agreement, the Public Health Service would conduct a review of current dialpainters and their working conditions, a study delayed for several years but finally published in 1933. Dialpainters hired after 1926, the findings revealed, were accumulating radium at an average of one-half microgram each, when

body-burdens as low as one microgram were known by then to be dangerous. Yet no state or federal regulations had been promulgated to protect these
workers.[9]

There were several interesting aspects to the Radium Conference. First,
although workers suspected the source of their illnesses earlier than anyone else and were certainly the most interested party, they were not invited.
Workers were sure to have knowledge about how dialpainting might be per-
formed both safely and efficiently, but in the 1920s they were not consulted
about their work organization. By 1923, as David Montgomery has written,
American management had largely won a prolonged battle with labor over
the control of production and conditions of labor inside the factory and were
little inclined to solicit worker expertise.[10]

Second, the conference was voluntary and unofficial. No one was required
to attend, and its resolutions had no force of law. No participants suggested
preventive measures to industry. Although seemingly the two committees
proposed by the conference never formed, a group of advisers did meet to
consider the protocol for the study that was undertaken by the Public Health
Service. Apparently those attending (including Harrison Martland) chiefly
discussed methods of radium detection.[11]

Third, the conference generated no consensus that radium poisoning
existed or that radium was dangerous to workers; it was merely agreed to study
these matters. Nonetheless, although radium poisoning as a disease and its
dangers for dialpainters were not recognized officially, the meeting marked
tacit acceptance of radium poisoning's existence by government and business.

Prominent students of the radium illnesses were dissatisfied. The radium
poisoning expert Harrison Martland thought that the conference resolutions
were useless. John Roach of the New Jersey Labor Department considered
the meeting a whitewash of U.S. Radium. Katherine Wiley of the New Jersey Consumers' League could not understand why Alice Hamilton had been
pleased with the outcome of the meeting.[12]

In part, Hamilton's position was a practical one. She was convinced that
radium work would not be shut down "so long as well managed watch works
can show that they have had no cases of radium poisoning." She believed
that the Radium Conference offered a good opportunity for publicity, and
that "the weapon of publicity which we hold up our sleeves" would cow the
manufacturers into becoming meek and ready do-gooders. Further, she was
confident that the Public Health Service would conduct a sound investigation and that unrecognized cases of radium poisoning would come to light,
perhaps leading to plant closings where necessary.[13]

Hamilton praised such conferences as efficient. For one thing, she thought that they were prompt responses to new industrial diseases, although emphasizing this aspect would suggest that the meetings were automatic reactions to public outcry rather than, as was the case with the Radium Conference, concessions to organized lobbying. Second, the conferences gathered together all interested parties (overlooking the absence of workers) for discussion and debate that could translate smoothly into policy. Hamilton drew on two personal sources for her faith that data reflexively yielded policy. First, as a scientist, she felt that open scholarly discussion of disparate perspectives would lead to ratification of those facts and theories most closely reflecting the truth. Second, she drew on her Progressive faith in an almost spontaneous evolution from investigation to education to legislation. Hamilton combined scientific edification and moral suasion in her own professional life. She seldom had governmental authority but often educated and reasoned with managers to win a change in working conditions.[14]

Angela Nugent has probed Hamilton's attitudes about the conference method, revealing that in part her support derived from her professional interests. From this perspective, Hamilton favored the conference because ultimately it benefited industrial hygiene professionals. As at the Radium Conference, so in general did 1920s debates over industrial diseases tend to move away from the larger, macropolitical questions of the relative value of commodities' social utility, industries' development, and workers' health, toward the more limited, technical problems of scientific technique, clinical evaluation, and epidemiological interpretation. During and after the Radium Conference, scientists argued about the best methods to detect and quantify radium body-burdens, about the definition of normal versus abnormal blood and bone profiles, and about statistically significant levels of disease. Yet despite these narrowed concerns (or perhaps because of them), other interested parties ceded to scientists authority over identification, compensation, and prevention of industrial diseases.[15]

Industrial hygienists gained new authority less because of their technical expertise than because of the relative political positions of labor, management, and government in the 1920s. Labor perhaps had little choice but to cede authority to scientists, as scientific control of workplace health was by now the only option to complete management control. Labor had lost oversight of work procedures and, in an era of weak labor, was not likely to *regain* it through direct action in the workplace. For suppression of employer-mandated dangerous working conditions, employees could but ask for government intervention. Political and legal barriers stood between workers and effective state reform, however. Politically, employees had little influence, business much more. Legally, on workers lay the burden of proof to demonstrate the danger of industrial chemicals; industry did not have to prove the chemicals' safety. Labor organizations could hardly afford to finance scientific studies, especially in an era when union survival was at issue, and workers thus were dependent on scientists' goodwill and concern for their welfare. Of course, scientists would benefit little from activities generous to workers but stood to gain a great deal from business acquiescence to their professional expertise and support of their research.

Business in the 1920s had little reason to fear technical control by government experts, given its generally positive experience with the "technocrats" of World War I bureaucracies and Herbert Hoover's Department of Commerce. In the radium cases, industry demonstrated savvy manipulation of scientific debate, shaping research agendas, data, and interpretation through its support of scientific research.

Finally, government officials avoided overt political decisions but gained control over labor-management disputes by referring industrial disease matters to technical bureaucrats. Thus, industrial hygiene professionals acquired authority because they held crucial territory between the terrain of labor, business, and government. Like those in any other "neutral territory" separating powerful adversaries, individual scientists might be more or less sympathetic to various antagonists. But on the whole, they benefited from a reputation for neutrality.

The political map, then, more than technological breakthroughs, empowered industrial hygiene professionals. Nugent has written that industrial disease experts won recognition for a unique ability to evaluate occupational hazards, but their increased power did not develop necessarily from their scientific competence. Causality could flow the other way, as increased power enhanced scientific exploration. For instance, Martland was not particularly interested in the dialpainters' illnesses until authorized, as Essex County medical examiner, to investigate the causes of their deaths; once so empowered, he embarked on his creative research into the etiology of radium poisoning, inventing new techniques for the detection and measurement of radium. Although experts like Martland did develop methods and instruments that were useful in defining and controlling industrial diseases, these inventions and discoveries were more encouraged by than the cause of the widened scope of industrial disease investigation in the 1920s.[16]

Alice Hamilton celebrated the ascendancy of scientific expertise in the realm of occupational health. Of course, she believed that she and others of her time and persuasion had identified a new social need and created a new science to meet it. Hamilton was recognized as a founder, in America at least, of industrial toxicology. To see power granted to those in a field she had helped create must have been a heady experience, satisfying to both her scientist's ambition and her reformer's mission.

In praising scientists and open scholarly debate, Hamilton at one time both acknowledged professional differences and obscured them; scientific discussion involved airing differences but finally rectifying them. Both the Tetraethyl Lead and Radium Conferences focused on scientific debate and referred questions to committees of scientists for resolution; thus the conference method admitted scientific dissent yet pushed for its closure.

Privately, Hamilton criticized biased business-financed studies undertaken both within and outside the government, but publicly, in support of the conference method, she praised the objectivity of industrial health scientists, especially compared to partisan labor and management. In this she betrayed her own partisanship. Blinded perhaps by her own wedding of science with service, she assumed that scientists generally worked for the public good. This justified her pursuit of greater authority for the scientific community. Hamilton promoted the welfare of industrial hygienists as avidly as she promoted the welfare of workers.

The U.S. Bureau of Labor Statistics, perhaps because of financial constraints, was slow to adopt the dialpainters' agenda for recognition, compensation, and prevention of radium poisoning. But by the time of the Radium Conference the bureau and especially its commissioner, Ethelbert Stewart, emerged as the dialpainters' federal champion.[17]

Bureau action on behalf of the dialpainters began in 1925 and proceeded intermittently through 1929. Examiner Swen Kjaer undertook his first investigation of dialpainting companies during three weeks in April 1925 (see Chapter 4). In the November 1925 issue of its *Monthly Labor Review*, the bureau discussed its "superficial" research and tentative conclusions. The report asserted that, despite confusion about the action of mesothorium versus radium-226, both kinds of radium likely had caused the dialpainters' necroses. "There is a health hazard in working with radium," it concluded. The text included a summary of research by Theodor Blum, Frederick Hoffman, and the Cecil Drinker team. In January 1926 the bureau reviewed Harrison Martland's work, but in May it noted that the lack of resources had led the agency to discontinue its investigation. At that time it provided a "brief recapitulation" of the New Jersey radium cases, again reviewing Kjaer's 1925 research and published work. The report, although appearing before the Connecticut poisonings had surfaced, concluded that illness outside New Jersey could not be ruled out until all former workers were located and examined. Kjaer conducted a second probe—this time, of the Connecticut cases over a six-day period—immediately preceding the Radium Conference.[18]

After the conference Ethelbert Stewart kept his promise to continue his bureau's investigation of radium poisoning. Swen Kjaer now undertook a six-week investigation in which he inspected the large dialpainting firms in New Jersey and Connecticut as well as companies in New York City; Philadelphia; Waltham, Massachusetts; and Elgin, Springfield, and Ottawa, Illinois. He discovered more deaths and illnesses due to work with radium. Kjaer looked into the deaths of three men employed in Standard Chemical's radium refining facility in Pittsburgh. Glenn Kammer, a chemist, had extracted radium from ore between 1912 and 1925 and after that manufactured luminous paint. He died in 1927 of leukemia. Paul Hogue, who had packed radium salts between 1914 and 1923, developed cancer in his fingers, suppressed by surgery. His "blood condition" then deteriorated, and he died in 1927 of anemia and pneumonia. Another chemist, Charles Viol, had handled enough radium to cause cancer in his fingers; unchecked by multiple surgeries, the disease spread up his arm and metastasized to his lungs. He died in 1928.[19]

In Ottawa, Kjaer learned that the Standard Chemical subsidiary, Radium Dial, had adopted the position that mesothorium caused the dialpainters' deaths in New Jersey. The company insisted to Illinois dialpainters that radium-226 was safe. For instance, in April and May 1928, national attention was given to the court testimony and out-of-court settlement in New Jersey. Two days after the settlement was announced, Radium Dial published an advertisement in the local paper in response to "the wide circulation given reports of poisoning caused by paint used for luminous watch dials." The ad maintained that whereas the New Jersey women's "condition is always called 'radium' poisoning," no cases occurred in plants where radium alone was used; rather, "all the cases . . . occurred in establishments that . . . have used mesothorium." It then cited Frederick Hoffman's conjecture that mesothorium was responsible, not the isotope most often referred to as "radium" (thus ignoring Harrison Martland's work that, beginning in 1925, substantiated the damage done by internally deposited radium). The Radium Dial Corporation had employed over a thousand "girls" during eleven years of business, had arranged for their medical examination "at frequent intervals," and had found "nothing even approaching such symptoms and conditions" as those revealed in the New Jersey dialpainters. "If we at any time had reason to believe that any conditions of the work endangered the health of our employees we would at once have suspended operations," the ad concluded. "The health of the employees . . . is always in the minds of [Radium Dial] officials." A copy of this notice was hung in the dialpainting studio, and all the workers were advised to read it.[20]

Some Ottawa dialpainters were ill, Kjaer discovered. Radium Dial had begun testing the women for radioactivity in 1925 but kept its findings from both the dialpainters and the government. Margaret Looney, Kjaer learned, had tested radioactive in 1925 and 1928; in between those years she had begun to experience tooth and jaw trouble. Ella Cruse also demonstrated the classic oral symptoms of radium poisoning. Since at least 1925 Catherine Wolfe had suffered from pain in her left ankle that spread up the leg into the hip; by 1929 her bad leg and hip had worsened and she had a pronounced limp. Wolfe recalled that at one point, everyone at work but her was tested. Further, "Marie Rossiter [another dialpainter] and I, we inquired of Mr. Reed [their supervisor] why the company didn't put up the medical report that the company had gotten from the physical examination of the girls after that terrible thing in New Jersey, and he says, 'My dear girls, if we were to give a medical report to you girls there would be a riot in the place.'" The day after the company published its ad, Ella Cruse filed a petition with the Illinois Industrial Commission for workers' compensation for a facial infection that had commenced in 1927.[21]

Ottawa was as protective of Radium Dial as Waterbury had been of its watch and clock companies. When Kjaer sought information about dialpainter illnesses, local physicians, dentists, and health officers reported no symptoms of radium poisoning. Cruse's lawyer could find no physician or dentist who could or would help him prosecute her case. The Illinois Department of Factory Inspection claimed that, following publicity about the New Jersey cases, it had inspected every dialpainting establishment in the state and found all possible safeguards in place; at some point, most likely during 1926 or 1927, Radium Dial had switched from brushes to glass styluses for dialpainting.[22]

In 1929 Kjaer's work was discussed in a major report in the *Monthly Labor Review*. The article condemned the radium industry for its "appalling" disease rate and for its continued disregard of workers' health. The bureau had studied thirty-one establishments that had employed a total of 2,000 workers (253 workers currently) handling radioactive substances. Twenty-three radium-related fatalities, nineteen "living cases," and nine "possible" cases had been identified; of these fifty-one employees, thirty-three were dialpainters. Of the cases discussed in detail, twenty-nine were from the U.S. Radium plant in New Jersey, six from Waterbury, Connecticut, and two from Ottawa, Illinois. In particularly scathing language, the report criticized industry officials who had heard of the New Jersey cases, examined their own workers and found some of them radioactive, and then "carefully concealed" their discoveries because they "feared disruption of their business if the facts became known." "Even the victims had not been informed of their condition, nor the cause, through fear of panic among the workers."[23]

The Bureau of Labor Statistics recommended safety measures but seemed to acquiesce to dialpainters facing some radium exposure. First, it advised, potential workers should be examined and those in ill health, especially those with tooth decay or anemia, excluded from the workforce. Employees should be examined every two months for deposited radium and blood changes. If found affected by her job, a dialpainter should be switched to other work. Second, the women should paint luminous dials for no more than two years. They should work no more than eight hours a day, or preferably seven, and their workweek should be shorter than six days. Third, dialpainters should be educated about the dangers of their work and trained to take precautions. Fourth, such precautions should include avoiding contact with the radio-active paint. Lippointing, of course, would be banned but also pointing of brushes with the fingers. Food should not be consumed in the workroom. Personal hygiene, such as hand washing, was important, and companies must provide proper facilities. Paper to cover work surfaces should be provided and changed daily. Long-handled receptacles should be offered to decrease manual contact and proximity to the paint. Ready-mixed paint instead of powdered paint should be used to minimize inhalation of radium dust. Fifth, well-ventilated work spaces also would reduce inhalation of radium dust and radon. Sixth, exposure to radiation should be minimized by providing lead paint containers, lead work surfaces, and well-spaced work stations.[24]

In contrast to these recommendations, Commissioner Ethelbert Stewart reiterated his position that luminous watches and clocks were "a fad without economic value." "The only method of safety . . . is to abolish the industry entirely," he insisted. If it were to continue, he stressed that eliminating lippointing would not prevent radium accumulation, citing the deaths of chemists and physicists from inhaled radium and radon. Beyond inhalation, Stewart was concerned about exposure to external gamma radiation, as "present safe-guards are not adequate."[25]

Following the publication of the Bureau of Labor Statistics report, the Public Health Service began a detailed investigation of radium poisoning in the dialpainting industry. An advisory committee composed of five university scientists, a Bureau of Standards representative, and Harrison Martland made technical suggestions. Data were collected between June 1929 and March 1930.[26]

The study's protocol was somewhat suspect, the interpretation and presentation of data were even more so. It was decided to compare workers hired before January 1, 1927, with those hired afterward in order to determine whether changes in work processes, especially the elimination of lippointing, had made dialpainting safer. Manufacturers had also abandoned the use of mesothorium. Oddly, only dialpainters then currently employed but whose service had started before 1927 were included in the "early group," whereas women who had worked in the early years of the industry but since left were excluded; the latter were likely the majority of early workers. Those included in the early group, then, were long-term employees and would be compared to much shorter-term employees in the post–1926 group. The report conceded that the employees in the later group could be expected to have accumulated less radium in their short period of service relative to the longer-term employees of the early group. Built into the protocol, then, was a bias making the later group's accumulation seem relatively slighter. On the other hand, excluded from the early group were "workers who had been found by the companies' examinations to be definitely radioactive." The report acknowledged that the early group "therefore gave a more favorable picture of the old conditions than if all the former employees had been subjected to examination."[27]

The study revealed that dialpainters were still accumulating radium despite the new precautions. James Leake, a physician associated with the Public Health Service's Office of Industrial Hygiene and Sanitation, delivered a preliminary report on the study at the June 1931 AMA conference. Those in the post–1926 group who had worked for at least two years showed an *average* accumulation of half a microgram of radium, from which we may infer that some women had taken in more than that amount.[28]

Leake's interpretation of this accumulation was a conflicting mix of concern and optimism. He noted that environmental studies had found that dust in the dialpainting studios was radioactive, and that there was enough radioactive dust "to account, in part at least, for radioactivity of some of the workers."

Jaw X rays revealed changes in the bones of a significant number of those with but one microgram of radium in their system, and Leake conceded that exposure over the long term might be dangerous. More blithely, however, he referred to the radium currently being accumulated as a "small" amount relative to the radium stored by those with "serious or fatal cases of radium poisoning." Moreover, "there is no indication that the accumulation since [1926] has in any individual case been sufficient to injure the worker." Although Leake admitted that the "evidence does . . . show the necessity for a still further and marked reduction of . . . exposure," he concluded that "it appears it should be possible for the industry to be conducted with entire safety."[29]

The sanguine tone may have been adopted in response to industry criticism of the report; drafts had been made available to radium business leaders before the AMA conference. U.S. Radium's vice president, Howard Barker, responding for his own firm and several others, wrote a lengthy critique. Barker first criticized the study's protocol and data. The health service had "over-emphasized" the study of the early population (which, recall, did not include women who had left dialpainting or those known to be radioactive). Data on the post–1926 cohort were suspect. Some of the workers, Barker claimed, were lying about how much time they had spent in the industry, or at least they were mistaken about it. The implication was either that dialpainters had accumulated radium over a longer period than they claimed, and thus had amassed annual amounts of radium that were lower than those calculated, or that dialpainters in the post–1926 group might have worked during 1926 or before, when annual radium accumulations were higher, and thus their current annual build-up was lower than calculated. Barker thought that many of the radioactive dialpainters were "borderline cases"—their measured body-burdens were low—and he wanted them excluded from those counted as radioactive. Others' radioactivity he sought to explain through exposure outside of work: "we must remember there is radon in normal air," he wrote, suggesting that dialpainters accumulated radium from breathing ambient air.[30]

Next, Barker disparaged the document's conclusions. The report was "written in a rather drastic form." "Some of the clock companies," he wrote, found its judgments "rather damaging," and officials were "quite upset." The study findings, he continued, "might lead to the assumption that anyone engaged in this type of industry would accumulate one microgram of radium element for every five years of service. I don't believe this is the meaning you desire to convey." Employees and interested citizens, if allowed access to the report, might "become unduly alarmed, and the more they read . . . the less solace they will obtain." Barker explained that the dialpainting industry had believed that by abolishing lippointing and by taking "other obvious precautions," the hazard to workers would be largely removed. Only because they thought the problem solved had business leaders at the Radium Conference approved a study of post–1926 workers' safety and health, he revealed. Of course, he added, "none of us would under any circumstances care to carry on the work if we felt that by so doing we were unduly jeopardizing the health of those employed in the industry." But he believed the current working conditions to be safe.[31]

Discussion following the paper's delivery at the AMA conference centered on possible tolerance levels for accumulated radium. The industrial hygienist Emery Hayhurst cited University of Missouri research (the Missouri team was headed by a longtime associate of Howard Barker). The Missouri investigators favored a tolerance limit set at 10 micrograms. Harrison Martland, who had once floated the 10-microgram limit, now disagreed. Referring to his cancer studies, he insisted that "the normal radioactivity of the human body should not be increased in any way." Frederick Flinn then rose and claimed that "Dr. Martland and I have agreed that 10 micrograms seems to be the tolerance limit." No injury had been found in a dialpainter with less than 10 micrograms of deposited radium, Flinn added. He now thought, however, that the "limit should be cut down to less than that."[32]

The full Public Health Service report, published in 1933, seems to have been written with industry objections to the preliminary report in mind. On the one hand, it reiterated most of the findings of the preliminary report. Workers were still exposed to radioactive substances: the dust in the air was radioactive; radium was found accumulated on objects, on the floor, and in cracks of the floor; radon was present even in undusty air, in amounts averaging two thousand times that of normal air; workers were exposed to beta and gamma radiation in generally radioactive work environments. Post–1926 workers still accumulated radium.[33]

On the other hand, the report accentuated the positive and glossed over the negative. It emphasized that no bone changes were found in the jaws of post–1926 employees "which could not be attributed to dental causes." For the post–1926 group, red cell counts and hemoglobin levels, when averaged, were only slightly lower than that of a control group, and the difference was not statistically significant. The late group's blood picture was better than the early group's. Most tellingly, the full report did not repeat from the preliminary report the average figure of one-half microgram absorbed.[34]

Reading carefully, however, one can determine that red cell counts and hemoglobin levels were inversely correlated with the amounts of radium in the body, indicating that workers' health might be threatened by radium exposure. Further, since "there was [a] definite indication that the accumulation of radium in the body of the workers was associated with the degree of radioactivity of the atmospheric dust to which they were exposed," women working in the less hygienic plants accumulated radium faster than others. Radium dust posed a clear danger to dialpainters.[35]

Despite shaping some aspects of the report in consideration of industry views, the Public Health Service did not shy away from suggesting the need for improved dust control or from criticizing other unhealthy conditions. Safety precautions such as those recommended by the Bureau of Labor Statistics were not being followed in the seven plants studied, the service reported. Most employees ate in their workrooms. Few plants demanded or provided for personal hygiene procedures, such as hand washing. But the main problem was the inhaled radioactive dust and radon. Together with personal cleanliness on the part of workers and good plant housekeeping on the part of the employer, the Public Health Service recommended adequate ventilation.[36]

Despite government agencies' recognition of radium poisoning, dialpainters throughout the 1920s had never proved in court that their jobs made them sick or that they were entitled to compensation because of their employers' negligence. After the Canfield settlement in 1930, records about litigation and settlements get spotty. Apparently, after Canfield eleven more suits were filed. One of these entered into settlement—the dialpainter Helen Tuck was offered $10,000—but U.S. Radium's insurance companies refused to pay. Their refusal was held legal in the New Jersey Supreme Court in 1935; the policies covered only injuries to employees through accidents, not through occupational diseases. It is unknown whether Tuck received her money. Another case, in which a suit seems never to have been filed, was settled for $3,000 in 1931, although the dialpainter, Anna Stasi, had died the year before. Eleven suits, including Tuck's, were transferred from state courts to federal courts at the request of the dialpainters' attorneys, for reasons that are unclear but involved procedural advantages in the federal courts. At least eight of these suits survived until 1935, when the U.S. District Court rendered a decision in one of them.[37]

The suit of Irene Corby LaPorte became the lead case in federal court. LaPorte had worked as a dialpainter at U.S. Radium from 1917 to 1918 and for a few months in 1920. After her marriage in 1921, three pregnancies ended in miscarriage. In 1927 she experienced some tooth problems and pains in her jaw and face; she began to fear that, like some of her friends, she suffered from radium poisoning. Both her dentist and her physician assured her that she was fine. In 1930 LaPorte noticed pains in her legs and joints. Later that year she found a lump in her vagina that led to the discovery of an underlying tumor in a pelvic bone. Medical examiner Harrison Martland was then persuaded to study a jaw X ray taken in 1925 and on that basis diagnosed her ailment as radium poisoning. In 1931, a few months before she died, LaPorte notified U.S. Radium that she would claim damages; her husband filed a suit for her injuries and death in 1932. When the company raised the statute of limitations as a bar to the suit, Mr. LaPorte turned to the U.S. District Court for an equity hearing, hoping the court would enjoin the radium company from setting up the statute of limitations as a defense.[38]

The district court judge found for U.S. Radium in 1935, upholding its raising of the statute of limitations. His reasoning was, first, that under tort law, the statute begins to toll when a plaintiff's "cause of action" arises, that is, when a plaintiff is harmed. He expressed some confusion as to just when Irene LaPorte's "cause of action" arose but conservatively assumed that it must have arisen at some point during employment, in 1920 at the latest. Therefore, the statute of limitations would have barred her suit some time in 1922. The judge could have stipulated that in fairness LaPorte's cause of action arose when she was diagnosed with radium poisoning in 1930, thus preserving her case.[39]

Second, the judge reasoned that if U.S. Radium were to be enjoined from raising the statute of limitations as a bar to liability because it fraudulently concealed the plaintiff's cause of action, then it would have to be shown that the company knew of the dangers of dialpainting and concealed them during the period when the plaintiff could legally have brought charges. The question of concealment was laid aside as the court considered what U.S. Radium knew about the dangers of small quantities of radium in 1920, 1921, and 1922.[40]

In essence, the court took up what one might call the "Watergate question": what did U.S. Radium officials know and when did they know it? This included consideration of what they should have known or could have known. The court widened its focus to include the years between 1917, when dialpainting began, and 1925, when, with Martland's etiology, the existence of radium poisoning was generally acknowledged. In effect, the court ruled on whether U.S. Radium had been negligent in providing an unsafe workplace during those years. The company was exonerated. The court decision as rendered reads quite logically. There are, however, grounds on which to question it.

The court reached four conclusions about the state of knowledge of U.S. Radium during the years in question. First, the judge concluded that U.S. Radium officials could not have known definitely of radium's dangers until after Martland's 1925 study. Evidence available before then was suggestive, but, though it may have led some experts to worry about radium, it did not prove that all experts shared concern over radioactivity. Even after 1925, nationally renowned radium scientists (at least those who testified for U.S. Radium) remained unconvinced of those dangers. Basically, the judge argued that evidence of danger did not automatically bring knowledge of danger. Second, he interpreted the literature on radium's beneficial effects as evidence that U.S. Radium reasonably denied that the radium in its luminous paint was harmful. Third, he reasoned that the firm had responded diligently to the dialpainters' concerns about their health by hiring the Life Extension Institute and Cecil Drinker and his Harvard colleagues to study the workers, their studio, and the dialpainting process. And fourth, because LaPorte herself was unsure of her condition until 1930, U.S. Radium could not then be held negligent for being unaware of the dangers of dialpainting.[41]

We do not have a record of the arguments of LaPorte's attorney to contrast his reasoning with the judge's. We may consider, however, Raymond Berry's rationale in 1928. Berry contended that U.S. Radium officials, knowing that radium was dangerous under some conditions, were then *bound* to investigate the dangers inherent in using it in radium paint. That is, allowing themselves to linger in a restricted state of knowledge constituted negligence in itself. As for the literature on radium's beneficial effects, I have argued that it was often produced by persons with a financial stake in radium; it was hardly based on data from disinterested scientific studies. Because companies including U.S. Radium produced and promoted radium for both medicine and luminous paint, they interpreted evidence of radium's possible dangers as evidence of its medicinal efficacy. Berry contended that because of their medical research, company officials knew that radium accumulated in the body and that gross quantities of the element were dangerous; this ought to have suggested careful use of even minute quantities of radium by patients and workers. Berry had also considered U.S. Radium's resistance to the conclusions of the Harvard team and its misrepresentation of that report to the New Jersey Labor Department as evidence of the company's negligence and concealment of the industrial origin of dialpainters' illnesses. Berry also might have argued that dialpainters, unlike their employer, were not experts on radium, and that the latter's ignorance should not be excused because of the former's.[42]

The district court exonerated U.S. Radium. All of the other existing lawsuits were dismissed, no further New Jersey dialpainter suits were allowed, and no more out-of-court settlements were made.[43] The judge appended a personal comment to his decision. "Naturally," he wrote, "there is no question as to where the sympathies of any human being would lie in a case of this sort." The court, he explained, denied LaPorte's suit not because it lacked compassion but because it could not adjust the law to meet unforeseen circumstances. In this vein, the judge called for "forward looking, intelligent legislation" to meet changing social conditions.[44]

LaPorte's death should not be blamed on her employer, according to the judge; instead, "responsibility . . . [should] be laid to the tremendous progress made in science." Uneven scientific progress then served the judge as both villain and savior. Advances in science had caused LaPorte's death. Missing pieces of scientific knowledge excused U.S. Radium from culpability in its employees' deaths and illnesses. Scientific progress eventually revealed the cause of the dialpainters' disease and would prevent more suffering. In this interpretation, "scientific progress" was an independent social force that blindly rolled over unwary and helpless people, rather than an intellectual tool employed in various ways to myriad ends. The judge denied the agency of both dialpainters, who fought for recognition of industrial radium poisoning, and radium manufacturers, who chose what studies to pursue and how to interpret their data.[45]

To call for an interpretation that recognizes individual agency does not mean we should ignore limitations on agency. Workers and employers were restrained by their positions in a hierarchical profit-dependent enterprise. In such a situation, who can be held responsible for workers' health? Workers in the 1920s were refused the authority or knowledge necessary to protect their health but nonetheless tried to protect themselves. Employers had the authority and could have pursued the knowledge had not entrepreneurial optimism and bottom-line austerity taken precedence over caution. The best defense of the radium entrepreneurs is that many of them died along with the workers in their factories and the consumers of their medicines. Still, there is a difference between choosing risks for oneself and choosing them for others. Officials of the U.S. Radium Corporation, in pursuit of profit, were willing to risk not only their own health but that of workers and consumers. It was not ephemeral scientific progress but calculated human choice (if among limited alternatives) that brought the dialpainters to their deaths. Even if by law the judge could not condemn that choice, he might have censured it.

In 1935 (but before the LaPorte decision) U.S. Radium sought permission from the New Jersey Labor Department to resume dialpainting at its Orange plant. Corporate officers were told that they must win Harrison Martland's approval. Martland, in a 1932 paper demonstrating cancer among the dialpainters, had advanced the opinion that the body's radioactivity should not be increased at all. He had noted that, according to the U.S. Public Health Service research that followed the Radium Conference, dialpainters were accumulating an average of one-half microgram of radium in their bones while working in plants where lippointing had ceased and where the workers were strictly supervised and safety regulations enforced. "It is a question whether under these conditions we should be satisfied to regard the industry as a safe one," he suggested. It is difficult to explain, then, why Martland acceded to U.S. Radium resuming its dialpainting operations in New Jersey in 1935. In a reversal of his earlier reasoning, he stressed that no deaths from radium had been found among dialpainters hired since the abandonment of lippointing. Ignoring evidence of radium accumulation under the best conditions, Martland consented to the reopening of the Orange studio provided the company agreed to follow Public Health Service guidelines. Although (for unknown reasons) U.S. Radium did not resume dialpainting at this time, in 1939 it began selling luminous paint to another company that opened a dialpainting facility in the state, despite a protest by the labor commissioner. Martland tested the new dialpainters for accumulated radium every six months and over the next two years found build-ups in some on the order of 0.1 to 0.6 micrograms; one woman was measured with 1.0 microgram of deposited radium. In 1941 John Roach of the New Jersey Labor Department reported on conditions at the dialpainting facility: he was concerned that a long day (ten hours) and a bonus system (rewarding high production) fostered fatigue and carelessness and caused the two spills of radium-laced paint he had witnessed. The dialpainters, he thought, were not worried enough about the risks of their job.[46]

Between 1928 and 1935, in debates over radium poisoning—as reflected in the voluntary Radium Conference, the Bureau of Labor Statistics studies, the Public Health Service studies, and the U.S. District Court decision in the LaPorte case—the federal government verified the existence of radium poisoning in all venues. U.S. Radium of New Jersey and Radium Dial in Illinois still denied the element's chronic toxicity; the Connecticut watch companies, less dependent on radium for their profits, conceded its toxicity. Neither the New Jersey nor the Connecticut firms were held liable for industrial radium poisoning, and neither were regulated except by informal voluntary agreement.

Bureau of Labor Statistics commissioner Ethelbert Stewart had asked that worker health be valued higher than commodity production. He went unheeded. At the Radium Conference industry representatives refocused discussion on employee susceptibility—blaming radium poisoning on workers—

and on limited schemes to improve safety; they had discarded lippointing by 1928 but seemed to ignore evidence that dialpainters were accumulating inhaled radium. Stewart himself tacitly conceded that radium poisoning would continue when he including safety suggestions in the bureau's 1929 "Radium Poisoning" report. The Public Health Service overshadowed alarming accumulations of radium among workers with pronouncements that dialpainting could be conducted safely. The federal court pronounced the dialpainters unintended martyrs to scientific-industrial progress. Through the 1930s safeguards for dialpainters' health remained voluntary.

The issues of establishing radium poisoning, assigning liability, and preventing more cases arose again. Efforts to create a domestic market for radium as an internal medicine succeeded with horrifying consequences. Radium poisoning, not as an industrial disease but as an iatrogenic disease—a disease caused by medical treatment—would call forth different responses from medical experts and government agencies.

GIMME A GAMMA

Iatrogenic Radium Poisoning

Much has been made in this study about the failure of state and
federal agencies to halt the endangerment of dialpainters. It might be answered that the anticipation of government intervention in reaction to industrial radium poisoning is anachronistic: that government in the 1920s was not
expected nor was it designed to investigate and regulate threats to the health
of its citizenry. During the decade, however, the Food and Drug Administration and the Federal Trade Commission did explore the effect of radium
medicines on health and were empowered to regulate the sale of radium
products to the public. The implicit contrast is between the relative inaction

of government in response to industrial radium poisoning compared to its stand against iatrogenic radium poisoning.

Similarly, this study has suggested that health professionals offered little aid to dialpainters in their pursuit of recognition, compensation, and prevention of industrial disease. In the campaign against iatrogenic radium poisoning, on the other hand, two of these experts, Frederick Flinn and Harrison Martland, were enthusiastic participants. Martland's interlocked progressivism and professionalism explain his more fervent championing of medical over industrial radium poisoning victims.

Doubts about radium's efficacy and safety as an internal medicine began during World War I. Martland's 1925 paper established the etiology of industrial radium poisoning and the dangers of deposited radium. In that essay he criticized the medical use of radium. Radioactivity had not proven therapeutic, he wrote. Intravenous injections and the internal administration of radioactive compounds were "not warranted in any medical condition" and were "highly dangerous." Martland assumed that solutions made by dissolving radon in water were too weak and the radioactivity too short-lived to be harmful but also unlikely to be of any benefit.[1]

Immediately, radium promoters mounted a defense of their products. A 1926 article in the *American Journal of Roentgenology* argued that there was "a great difference" between radium poisoning in dialpainters and the medical injection of radium chloride. In dialpainters, other ingredients in the luminous paint "probably" had sensitized tissues to the actions of radium. In addition, dialpainters would absorb more radium than would patients treated by injection (no foundation was laid for this conclusion). Thus, the author rejected the conclusions of Martland's 1925 paper and reported on cases where injected radium chloride improved patients with leukemia or lymph system tumors.[2]

Martland's next paper on radium poisoning appeared in the same issue of the *American Journal of Roentgenology*. In this article he presented evidence that radium had accumulated in the body of the U.S. Radium chemist, Edwin Lemen, through inhalation of radioactive dust alone. Radium's effects on bone marrow were considered the cause of the anemia that killed Lemen. Again, Martland warned of the dangers of introducing radioactivity into the body. He stepped up his admonition to include the inhalation of radon, which, with its short half-life, might break down in sufficient quantity in the lungs to deposit radioactive daughter compounds; these, in turn, could be absorbed into the body and deposited in the bones with damaging results.[3]

Although radium still had its defenders after 1925, the internal use of radium medicine was abandoned by most mainstream practitioners and became the province of quacks. One of the more colorful and successful of the remaining radium proponents was William J. A. Bailey, a self-styled "consultant to the medical profession on radioactivity and endocrinology." The American Medical Association maintained a file on Bailey, a practiced fraud in their eyes: for one scheme, involving mail-order automobiles that never materialized, Bailey served a short jail term. In 1925, using radium and the radium isotope, mesothorium, both purchased from U.S. Radium, Bailey began producing and promoting Radithor, a radium tonic packaged in handy, one-dose vials that he sold by the case.[4]

Most of his competitors marketed "radium" solutions that contained no radium at all. Bailey sold the real thing: this time he was not a fraud. In the Bailey Radium Laboratories in Orange, New Jersey, he carefully manufactured twenty-five milliliter bottles of solution, each containing one microgram of radium and one microgram of mesothorium. For a case of thirty bottles, containing $3.60 of radioactive compounds, physicians paid $25 and the general public $30.[5]

Bailey cleverly promoted Radithor as a "Fountain of Youth." At least one newspaper extolled its virtues: " 'Gimme a gamma' is the cry of prematurely old humanity in search of rejuvenation," it reported in 1925. The next year a Dr. Charles Morris published *Modern Rejuvenation Methods*, which offered this challenge: "Can modern science make men and women young again? Can the abundant life, health and spirit of youth be renewed? Foremost scientists claim that this is possible and point with pride to the fact that already withered, ailing, wrinkled, forgetful 'has-beens' have been turned into alert, healthy, vigorous persons fired with the energy of romantic youth." Morris credited the "eminent scientist," Dr. William J. A. Bailey, director of the Bailey Radium Laboratories of East Orange, New Jersey, with the research into radioactivity and endocrinology that made rejuvenescence possible.[6]

In Morris's book, discoveries in the field of endocrinology fed extrapolated claims that "the endocrines are the indomitable and autocratic controllers of human life. . . . Life is what your endocrines make it. . . . Old age is depleted endocrines." Not just health, but all aspects of personality were at the whim of the endocrine glands. Bailey was cited: "We are creatures of these glands. . . . This is the modern conception of humanity and human nature. . . . The endocrine glands . . . absolutely dominate and decide what you are." The author attributed to a Dr. Louis Berman this quotation: "Life, body and soul emerge from the activities of the magic ooze of [the glands'] silent chemistry." Examples were given. Geniuses and perverts, morons and imbeciles, were all created by their glandular conditions: "Get clearly the idea that the thing we call consciousness is a . . . chemistry reaction." The text continued: "Men and women of great natural ability have something physi- cal. . . . With a good pituitary . . . they start off with the proper capacity. With a good thyroid they have energy, zeal or ambition. With good gonads and adrenals they have the courage and power to carry through." Sexuality was, of course, affected, not only sexual performance but also gender and sexual identity, although environment—particularly the femininity of the mother— was conceded to affect the endocrine balance. Most problems, though, could be overcome with proper treatment of the endocrines; love itself was a func- tion of the glands. Unhappy couples no longer need seek recourse to divorce courts because their failing glands could now be corrected. The "nagging wife," the "effeminate man" could be cured.[7]

Glands could be corrected through the introduction of radioactivity into the body. "The ionizing process of Alpha rays sets up revitalizing forces in [the] glands, pouring renewed streams of hormones into the blood," went one explanation. According to a metaphorical description, "The effect of radium water on the cells of the aged is similar to that observed when a plant that has been in partial darkness for some time is placed in the sunlight." This line of reasoning was extended: "Few germs can survive where there is oxy- gen and sunshine. Radithor furnishes ample oxygen to the cells and is like concentrated sunshine in the body." Alternately, radioactivity could clean out clogged glands, restoring their function. Radithor was the recommended compound, "a solution of the genuine radium and mesothorium elements themselves."[8]

Rejuvenation was not the only benefit promised by Radithor's sponsors. *Modern Rejuvenation Methods* included a list of seventy-six conditions re- mediable by drinking Radithor regularly. The alphabetized list ran from Acidosis to Wrinkles, including Anemia, Arthritis, High Blood Pressure, Dia- betes, Heart Trouble, Melancholia, Nephritis, Obesity, and Sinusitis—some- thing for everybody! William J. A. Bailey wrote a series of pamphlets on some of these diseases, all chronic conditions that medicine of the day—and even of today—had difficulty alleviating. Indeed, Morris's *Modern Rejuvenation Methods* actually had been written by Bailey. Bailey had proposed that Morris produce the book, but Morris only contributed several pages of notes taken during a conversation with Bailey. Bailey, the author and publisher of the volume, borrowed Morris's name for the cover.[9]

Radithor was completely safe, Morris (or Bailey) insisted. Referring to Martland's concerns about radium's use as an internal medicine, the author declared: "If any doctor or other person states that radium water is injurious he is not telling the truth. Recently some doctors, who I learn have never had the slightest experience with radium or mesothorium, and know no more about it than a school boy, have been trying to garner some publicity by claiming harmful effects from these products. Their statements are perfectly ridiculous and in opposition to twenty years of experience in hospitals and clinics." Of course, these denials echoed those of U.S. Radium and the Radium Dial Company, both of which continued to deny throughout the 1920s that radium deposited internally could be harmful.[10]

Eben MacBurney Byers was not, perhaps, a typical Radithor client. He could easily afford Radithor, for he was a wealthy man, an "ironmaster," chairman of the Byers Pipe Company in Pittsburgh. He owned homes in Pittsburgh; Aiken, South Carolina; and Southampton, Long Island. Byers competed throughout the world in various amateur sports—golf, trap shooting, horse racing—and won often enough to become a minor American celebrity. A bachelor, he was notorious for his many women companions. One lifelong friend was Mary Hill, also of Pittsburgh, who was considered a great athlete in her own right. When Byers, returning on a train from a Harvard-Yale football game in 1928, fell out of the upper berth and hurt his arm, his physician suggested that he try a patent medicine, Radithor. Byers, convinced of the medicine's efficacy, generously bought cases for many of his friends, including Mary Hill.[11]

In 1930, at age fifty, Eben Byers contracted a stubborn case of sinusitis, accompanied by tooth abscesses. By 1931 he had developed a full-blown necrosis resembling that of the dialpainters. His good friend Mary Hill was ill, too.[12]

During the twenties several federal agencies had become interested in radium medicines. As early as 1926 the Department of Agriculture's Bureau of Chemistry, then the enforcer of the Food and Drug Act, had warned of radium quackery; an estimated 95 percent of tested radium solutions contained no radium at all, and their danger was that users might not seek appropriate medical aid. The remaining 5 percent of radium solutions for sale were indeed radioactive. The bureau warned the public that the healing claims of these tonics were highly exaggerated and, further, that the radioactivity could be dangerous. In 1928 Radithor was identified by the new Food, Drug, and Insecticide Administration (FDIA) as one of the solutions actually containing radium. The FDIA had limited power to do anything about this. It could not prevent the sale of radioactive solutions, which was legal, but it could attack the fraudulent claims made on the solutions' labels. The Bureau of Chemistry began to gather evidence, collecting advertisements and finding physicians willing to testify to the ads' falsehoods.[13]

In the end, however, it was the Federal Trade Commission (FTC), not the enforcers of the food and drug laws, that moved against Radithor. The FTC had jurisdiction over false statements in advertising, and, spurred by the illness of the prominent Byers, it charged William J. A. Bailey with "unfair methods of competition in interstate commerce" through advertisements with "false and misleading statements." Bailey answered on March 8, 1930, by denying all the charges, and hearings began in August.[14]

Bailey's defense was based on claims that various government agencies were also promoting radium as a medicine. The radioactive waters at Hot Springs, Arkansas, were managed by the Department of the Interior. The physicians at the spa were under the authority of the Public Health Service, and the War Department ran a general hospital for veterans there. The U.S. Railroad Administration sponsored advertisements extolling the virtues of the radioactivity in the water. Saratoga Springs, in New York, was managed by the New York State government, which promoted its radium-laced waters. Ignoring these taunts, the Federal Trade Commission called witnesses to speak to the issue of Bailey's business.[15]

At this point Harrison Martland and Frederick Flinn reappear in our narrative. Flinn, the Columbia physiologist who worked for Connecticut's watch companies and New Jersey's U.S. Radium Corporation, testified against Radithor. A defender of radium for dialpainting firms, Flinn emerged as a leader in the fight against the medicinal use of radium; his criticism appeared in four articles published in major journals in the early 1930s. Meanwhile, Flinn continued to arrange settlements and supervise the medical care of dialpainters for the radium companies.[16]

Harrison Martland also noted radium's dangers as a medical treatment. After reviewing his own work in the field, he asserted that "no one ought to be allowed to give any radioactive substances inwardly by mouth or by intravenous injection." [17]

At the 1930 FTC hearings, various other physicians and scientists spoke for and against the medicinal use of radium, and several patients treated with radium attested to its healing abilities. For some reason, the matter was then put on hold for ten months. Hearings resumed in September 1931, when the Federal Trade Commission convened at Byers's Southampton mansion to take the testimony of the now dying sportsman-industrialist. Several operations to remove decaying bone had left him with practically no upper jaw; he retained only parts of the roof of his mouth and about half of his lower jaw. Frederick Flinn testified that Byers had a radium burden of 33.5 micrograms.[18]

At this point, William Bailey began negotiations to withdraw his objections to the FTC's charges. In October Byers's longtime friend Mary Hill died of radium poisoning. In December Bailey officially withdrew his refutation of the charges. He had already stopped producing Radithor—he moved appropriately enough into the advertising business—so in his case the cease-and-desist order issued by the FTC on December 19, 1931, was a formality. It did perhaps serve as a warning to other producers. Eben Byers died in March 1932.[19]

As a result of their mutual fight against radium medicines, Harrison Martland and Frederick Flinn began a professional rapprochement. Between 1930 and 1932 Martland sent Flinn patients for examination and samples for study. Flinn informed Martland about cancers among the dialpainters in his care. He suggested using a treatment devised for lead poisoning on radium victims that Martland saw put into use, making sure that Flinn got credit for the idea. The radium manufacturers were relying on Flinn to keep "the girls" safe, he wrote to Martland, and Flinn requested Martland's opinion about the dangers of inhaled radon. Martland let Flinn and U.S. Radium associates use his laboratory to measure radioactivity in dialpainters. Flinn continued to represent U.S. Radium and the Connecticut watch and clock manufacturers in matters pertaining to dialpainting. He and Martland disagreed over the appropriate tolerance standards for radium work, but their professional relations were cordial.[20]

There is a quandary raised by these scientists' condemnation of radium medicine: why did Flinn, who adamantly defended the luminous paint industry's use of radium, ardently attack the medicinal use of radium? And why did Martland, who withdrew from efforts to win compensation for dialpainters and who later acceded to dialpainting's resumption in New Jersey, also attack radioactive tonics? Unfortunately, Flinn's publications fail to indicate much about his attitudes toward industrial versus iatrogenic diseases. Little other material augments his publications, as Flinn ordered that all his personal papers be destroyed on his death.[21] Martland was more voluble in his publications and left personal papers, so we can attempt a reconstruction of his attitudes. The contradiction to be explained is Martland's withdrawal from activism on behalf of industrial victims of radium poisoning and his simultaneous advocacy on behalf of iatrogenic victims of radium poisoning.

An obvious answer might be that, as a physician bound by the Hippocratic Oath, Martland was sworn to prevent harm to patients and so felt duty bound to intervene in medicinal quackery. This may have accounted at least in part for his actions, although he never expressed himself as driven by such ethical motivations.

Another possibility is that Martland was influenced by the changing status of physicians. In the early twentieth century medical doctors were just consolidating their hold on their profession and were beginning to move into positions of social prominence;[22] to champion workers against their bosses might have adversely affected both of those trends. Despite the ideals of his profession, Martland might have been less interested in the welfare of working-class girls than in that of socially prominent patients. But, again, nothing in his papers indicates that this was the case.

Indeed, his work for the dialpainters, together with his later involvement in industrial health controversies over the toxicology of beryllium and benzene, earned him the reputation as a friend to working people. A colleague wrote in 1969 that Martland was "very partial to working people," that he "was usually the plaintiff's best witness in court," and that he never accepted a fee for his expert testimony but would appear only as a subpoenaed public servant. Further, the testimonial continues, as one of the earliest medical examiners, Martland shaped the duties and obligations of the office, contributing to a tradition in which "one of the primary functions of the Medical Examiner was to prevent the wastage of human life in industry."[23]

Friend of labor or not, Martland abandoned the dialpainters. It is possible that he did not understand how crucial his testimony was to their compensation cases, both his documentation of radium poisoning's etiology and his measurements of dialpainter body-burdens. His service to the dialpainters, however, had been widely praised and acknowledged as decisive in the eventual recognition of radium poisoning and the compensation of some dialpainters. Journalists labeled him the dialpainters' "star witness," and several dialpainters wrote to thank him for his help. Katherine Schaub informed him that he was in great measure responsible for the settlement. Also, Martland was fully aware of the difficulty caused by his refusal to assay dialpainters for radium and to testify about their body burdens. In a 1930 letter to Frederick Flinn, he noted that lawyers were continuously calling him to ask for referrals to someone who could diagnose radium poisoning. Martland told them that there were only two reliable institutions engaged in such work, "your's" (Flinn was still on staff at Columbia University) and Memorial Hospital (a noted research hospital in the forefront of radium therapeutics), but that neither institution would care to interest themselves in the dialpainters' concerns. In 1947 Martland wrote Frances Perkins, federal secretary of labor, that workers could not acquire competent toxicological examinations: there were few well-equipped medical examiners, leaving recourse only to industrial laboratories and universities, unlikely allies of labor because they "work chiefly in the interests of industry (the universities receive grants)." He hoped that state Labor Departments would pick up the slack.[24]

Martland's position as county medical examiner may have prevented his advocacy for the dialpainters. The medical examiner, like all other county officials, swore to "faithfully and impartially" discharge his duties. This may explain why Martland was willing to testify in the dialpainters' hearings of 1928, but only on subpoena for the fifty cents "customarily tendered" county officials, not as a recompensed expert witness. The dialpainters perceived it a favor that he testified essentially for free. Favor it may have been, but it also reflected Martland's disinclination to take sides as a paid consultant and his bent toward serving the public interest as an unbiased expert. His insistence on independence from partiality cut both ways, however. Although he used it to help the five dialpainters for whom he testified as a county official, he also used it to deny help to other dialpainters. "I cannot take sides in a civil suit," he wrote to a dialpainter in 1927. Around 1930 he pointed to a temporary Public Health Service consultantship as a bar to consultation with a dialpainter initiating another lawsuit.[25]

Martland's stand that as a neutral, government expert he could not take part in dialpainters' suits seems an extreme interpretation of a county employee's oath to "impartially" fulfill the duties of office. His county position may indeed have prevented him from serving as a paid consultant and compensated expert witness for the dialpainters. Still, tests for radioactivity in dialpainters and testimony about them in court were conceivably within the realm of his duty as medical examiner to investigate medical evidence of criminal acts. Nothing legally prevented Martland from speaking out about the dangers of radium. As an agent of the legal system, he believed himself responsible for recommending appropriate measures for prevention, compensation, and punishment of wrongdoers. In 1925, when he first concluded that radium had harmed the dialpainters, he notified the New Jersey Labor Department of his findings, asked that the department insist on preventive measures (including restricting the inhalation as well as the swallowing of

radium), and recommended that the dialpainters receive workers' compensation. He emphasized his "duty" to report to the labor commissioner and his
"impartial" stance in requesting compensation for the dialpainters. On the
other hand, Martland never reported the radium cases to the district attorney
for prosecution because he believed that U.S. Radium had not intentionally
harmed its employees.[26]

Also related to his refusal of further aid to dialpainters were his court-
room experiences. Martland hated the journalistic furor over his testimony.
Finding the public confrontations by paid consultants distasteful, he likened
scientists' participation in the American adversarial legal system to "the dis-
graceful wrangling of the hired medicolegal prostitutes and so-called experts
that prey on the court."[27]

If Martland rejected judicial hearings as unsuitable vehicles for settling
industrial health questions, he found legislative solutions equally inappropri-
ate. With New Jersey's faulty radium necrosis bill in mind, he complained that
compensation legislation was most often written by lawyers and laypeople,
not by "those who are acquainted with the real facts," physicians. "For this
reason," he continued, "many of the laws are vague in meaning, often quite
incomprehensible, full of loopholes, and in some cases containing jokers." He
objected specifically to the oft-suggested remedy of compensation acts providing for blanket coverage of all industrial diseases. These, he thought, were
likely to be abused because "political pull and graft" might unfairly establish
common diseases as of industrial origin.[28]

Of the three routes available in the American system to define and compensate industrial diseases—judicial, legislative, and administrative—Martland clearly preferred the administrative. Specifically, he envisioned administrative boards within Departments of Labor, comprised of medical specialists
who "could be called on if a new disease appeared to give their unbiased
opinion as to whether or not an occupational poisoning existed." In regard to
the dialpainters' suits, he wrote, "A competent body of medical authorities . . .
could have dispensed of the case in a day's time without the utter ridiculous-
ness of dragging a simple case along for years in the courts."[29]

Martland's preference for administrative law was inextricable from his belief that scientists could efficiently and fairly determine the fate of workers.
It seems from his writings that he thought the disputatious scientific debates
over radium poisoning were traceable to the public nature of the debates and
the cash nexus in which scientists were paid by various interests to promote
scientific interpretations based on their political utility. Earlier we considered
Alice Hamilton's support of the conference method and how it related to

her belief that professional scientists could impartially and prudently handle industrial health disputes. Martland shared this view. In asserting a preference for procedures empowering ostensibly objective medical scientists, he echoed both the Progressive Era's confidence in experts and scientists' confidence in their ability to define truth. He ignored struggles within the medical community over the toxicity of radium, or he blamed them on the requirements of an adversarial legal system. The concept of "truth" as influenced by perspective was not a part of Martland's world, and could not be, given the times in which he lived. Science, for him, was the neutral arbiter of controversy grounded in social difference; the scientific perspective offered hope that political struggles between social groups might be amicably and fairly settled.

An important motive, then, for Martland's abandonment of the dialpainters was his distaste for the juridical and legislative procedures with which the dialpainters' cases necessarily involved him. The iatrogenic radium poisoning cases were handled by federal bureaucracies under administrative law. Neutral federal experts heard testimony, weighed opinions, and balanced evidence, without the involvement of lay juries or legislators and largely beyond the journalistic eye. Federal bureaucrats were empowered to judge the safety and efficacy of Radithor, even if they could only act based on fraudulent advertising. Martland must have found participation in this kind of a system preferable to medicolegal wrangling in courts of law or to the political wheeling and dealing of legislatures. He preferred administrative venues for their dignity, their efficiency, and the authority they granted neutral experts, especially scientists.

Harrison Martland, like Alice Hamilton, was confident that scientific debate smoothly produced unanimity about appropriate policy. Unlike Hamilton, though, Martland spurned the involvement of interested parties (workers or their employers) or an uninformed lay public (journalists, legislators, attorneys). Still, had they ever spoken about it, Hamilton and Martland would have agreed that science was the most effective vehicle for the resolution of debates on industrial disease. Both relished the increasing influence of scientists over government policy.

Various government agencies better served iatrogenic victims of radium poisoning than industrial victims. Local government officials were generally hostile and federal agency representatives ambivalent about requests for thorough investigations of the dialpainters' charges. An eventual federal inquiry did confirm the workers' claims that their jobs were injurious to their health, but federal officials through the interwar years sought from the radium companies only voluntary reform of working conditions. In contrast, federal agents investigated and sought regulation of radium medicine before it had knowingly harmed anyone. Then, prodded by illness and death among patients using the radium tonic Radithor, a federal agency won discontinuance of its manufacture and sale. The federal government acted with far greater alacrity to help consumers than to assist workers.

WE SLAPPED RADIUM AROUND

LIKE CAKE FROSTING

Dialpainting in Illinois

A third set of cases involving industrial radium poisoning confirms what the evidence from New Jersey and Connecticut reveals: the radium companies' disregard for the health and safety of their employees. The companies were motivated by pressure to maintain a workforce, to protect their product, and to avoid expensive legal obligations. In Illinois, the Radium Dial Company's response to the New Jersey and Connecticut poisonings effectively delayed dialpainter legal action until 1934.

Dialpainters in Ottawa, Illinois, a small town southwest of Chicago, used friendship and kin networks to gather information and support each other in collective legal activity; in this, they echoed the efforts of their counter-parts in New Jersey. As in New Jersey and Connecticut, the dialpainters cases in Illinois brought in their wake changes in the laws concerning occupational diseases and workers' compensation. Unique among the three groups, the Ottawa dialpainters took advantage of legal recourse available through workers' compensation laws; they were the only workers to win state-sanctioned compensation for radium poisoning.[1]

The early history of Illinois dialpainting has already been told in the discussion of Swen Kjaer's 1925 investigation (Chapter 4) and his 1928–29 probe (Chapter 6). The Radium Dial Company of Ottawa was a subsidiary of the Radium Chemical Company, itself a subsidiary of one of the largest radium firms in the country, the Standard Chemical Company. Standard Chemical was based in Pittsburgh, where it purified radium for medicinal use as well as for luminous paint. Radium Dial moved around the country before settling in 1923 in Ottawa, where it set up operations in an old school. As elsewhere, dial-painters at the Ottawa plant were women, mostly young. Although management assured the dialpainters that radium was safe and would even improve their health, it learned through tests and physical examinations since 1925 that some dialpainters were radioactive, and some already had been harmed by radium body-burdens. This information was kept from the workforce.

Between Kjaer's 1929 visit and 1934, amazingly little notice was taken in Ottawa of the perils of radium poisoning. Ella Cruse and Margaret Looney, two of the affected workers known to Kjaer in 1929, died that year within a month of each other. Cruse succumbed to the jaw necrosis typical of early radium poisoning cases. Looney died of diphtheria and pneumonia, undoubtedly opportunistic infections, as she suffered from a jaw infection and was said to have "wasted away" for months before the onset of the diphtheria. In 1931 Cruse's family had accepted a $250 settlement from Radium Dial, approved by the Illinois Industrial Commission, although Cruse originally had sued for $3,750. Looney's sister reported that her family sought compensation but never received a settlement; the family also contested the death certificate's conclusion that Margaret had died of diphtheria and pneumonia. Despite this, there was no public outcry about radium poisoning in Ottawa.[2]

Catherine Wolfe started working for Radium Dial in 1922, when she was nineteen. Charlotte Purcell, age sixteen, began at the same time. By 1934 they were both married with young children; they became good friends and their families seem to have socialized. The good times ended when Wolfe, now Catherine Donahue, began to lose pieces of her jaw. Since at least 1925 she had experienced pains in the leg and hip, and in 1931 Radium Dial fired her—in her words, "I was told that my limping condition was causing talk and it wasn't giving a very good impression to the company and that they felt that it was their duty to let me go." Then cancer was discovered in Charlotte Purcell's arm, and the limb had be amputated.[3]

In April 1934 Purcell convinced a number of women who had worked for Radium Dial to consult with a Dr. Loffler, who was holding an informal clinic at a local hotel. Loffler told the former dialpainters that many of them were suffering from radium poisoning. An attorney was contacted, and about nine women signed a contract with him to sue the company under Illinois workers' compensation and occupational disease laws.[4]

On the lawyer's advice, Purcell and Donahue went to Radium Dial to see Rufus Reed, the dialpainters' supervisor, to legally notify him and the company about the women's illnesses and their determination to seek compensation and medical restitution. Donahue later reported, "He said he didn't think there was anything wrong with us at all." Reed also denied that radium poisoning existed: "He said there was nothing to it at all." Donahue's husband, Tom, spoke to Reed twice. The first time, he said, followed Purcell's return to Ottawa, "the first time with her arm off." "I saw him on the street," Tom Donahue related, "I told him the women were in a bad way, and that the doctors were finding that it was from the material in the paints they were using." A second encounter ended badly. Donahue wanted data from the radiological examinations of the dialpainters dating back to 1925: "This day I wanted to find out the name of them doctors, who was supposed to examine them women that was working there, that didn't give them a report. . . . I wanted to ask him why the report wasn't given to me, and he wanted to brush by me, and I said I only had another question to ask him, and I said we only wanted to help the women, and he started to swing at me and I swang at him. . . . He got excited." Reed had Tom Donahue arrested.[5]

In 1934, in addition to the nine women collectively considering legal action, another dialpainter, Inez Vallat, filed suit against Radium Dial, asking for $50,000 in damages. She alleged that the company was in violation of the state's Occupational Diseases Act, especially its requirement for the use of "reasonable and approved devices, means or methods for the prevention of . . . industrial or occupational diseases," and that, as a result, Vallat had contracted at least some of the symptoms of radium poisoning, listed as "anemia, rarefaction of the bones, alveoli of the jaws, and other bone complications and disorders."[6]

Illinois's Occupational Diseases Act made the state unique among those studied here. Its history demonstrates once again women reformers' interest and influence in regulating working conditions, but it also reveals the control exerted over occupational disease laws by manufacturers. In terms of measures to protect the workplace, Illinois entered the Progressive Era ahead of most other states, fell far behind in the early 1900s, and then advanced again to lead the nation in progressive legislation. Manufacturers learned to cooperate with reformers and to control, rather than fight, regulation.

Illinois in the Gilded Age led other states in occupational safety and health legislation, a dubious honor as often regulations required little or defied enforcement. Beginning in the 1890s labor leaders, progressive politicians, and reformers, especially women reformers and those associated with Hull-House, formed a coalition to lobby for safety and health reform.[7]

The safety and health movement coalesced first around the Sweatshop Act of 1893. In 1891 the Chicago Trades and Labor Assembly had resolved to undertake a survey of the prevalence and problems of sweated labor in the city. Florence Kelley of Hull-House joined the antisweating campaign and, with public interest aroused, was appointed by the Illinois Bureau of Labor Statistics to further study the sweatshops. In time she and the bureau recommended reform measures to the General Assembly, which authorized yet a third investigation to be guided by a legislative committee. Tutored by labor leaders and Hull-House residents, this committee proposed a bill that became the Sweatshop Act of 1893.[8]

The law's biggest impact was to establish a system of factory inspectors. By gubernatorial appointment Florence Kelley became the first chief factory inspector, a post that launched her on a career path that would eventually find her executive secretary of the National Consumers' League and a dialpainter advocate. The Sweatshop Act promised to regulate the conditions of labor in sweatshops, especially inadequate sanitation that threatened the health not only of workers but also of consumers of the products of sweated labor. The sanitation codes, however, were the act's least controversial measures because loopholes permitted much sweated labor to continue, authorized few inspectors to enforce the law, and assigned no penalties for most abuses. A more disputed part of the act was a limitation of women's and children's labor in sweatshops to eight hours. This evoked a forceful reaction on the part of manufacturers, who brought suit and saw the hours law voided by the Illinois Supreme Court in 1895 (in the famous Ritchie I decision).[9]

In 1908, in *Muller v. Oregon*, the U.S. Supreme Court upheld a law limiting the hours women could work. In Illinois, trade union women and members of the Women's Trade Union League decided to push for a new eight-hour bill. This measure, introduced in 1909, was opposed by the Illinois Manufacturers' Association and supported by the usual assortment of reform and women's organizations. Jane Addams lobbied hard for the bill. Little headway was made until the legislature passed a substitute act, the Ten-Hour Law of 1909, although it was immediately challenged in court by manufacturers. As with the *Muller* hearings, Louis Brandeis argued for the law, backed with a brief prepared by Josephine Goldmark and other women reformers in the National Consumers' League, now headed by Florence Kelley. The law was upheld in a decision known as "Ritchie II." [10]

Progress was made in passing other workplace safety and health regulations in the years around the turn of the century. In 1897 a Blower Law, won through the efforts of the Metal Polishers' Union, required the venting of dusts produced by grinding. The union agreed with the governor that with the new law Illinois's metal polishers were the best protected in the nation. In 1899 mining safety laws were revised. The new regulations, jointly proposed by miners and mine managers through the state Bureau of Labor Statistics, gave Illinois the country's most effective mining regulations. In 1905 railroad workers gained legislation requiring safety equipment on trains. [11]

Still, Illinois—by the early 1900s the third most industrialized state—had fallen behind other states in occupational safety and health. Factory inspectors criticized the dearth of protective legislation and proposed new bills to the General Assembly, without success. An executive branch proposal in 1907 for a comprehensive safety and health bill modeled on European law also failed. The Illinois Manufacturers' Association, created in response to the eight-hour provisions of the Sweatshop Act of 1893, continued to lobby effectively against most measures to improve health and safety in the workplace. [12]

Manufacturers and Progressives then switched from fighting to cooperating. The early-twentieth-century political philosophy, variously labeled "corporatism," "corporate liberalism," or "the associative state," which would operate to create the voluntary conferences of the federal Bureau of Labor Statistics in the 1920s, suggested that society's disparate parts were mutually dependent and thus had a mutual interest to weld their parts into a harmonious whole. In experiments with cooperation in Illinois, business learned that it could shape regulation to its needs. As Lizbeth Cohen observes, throughout the 1910s and 1920s only one reform statute seriously opposed by the Illinois Manufacturers' Association became law. But this did not mean that no protective legislation was passed—rather, manufacturers cooperated with reformers to write labor statutes. [13]

This short-lived coalition included physicians and workers as well as industrialists and reformers. Between 1905 and 1915 Illinois became a leader in progressive protective legislation for workers. Reformers began by lobbying for the creation of bipartisan commissions charged with recommending health and safety codes to the legislature. The first such commission established by statute was the Industrial Commission, composed of three employer representatives, three worker representatives, and three concerned citizens. The Illinois reform community was represented on this commission by Graham Taylor, a settlement leader. With the help of the Bureau of Labor Statistics, the Industrial Commission recommended legislation that was enacted as the Health, Safety, and Comfort Law of 1909. It provided for factory machine guards, sanitation, and ventilation and for an enlarged factory inspection department with increased powers of investigation. Also formed in 1909 was a Mining Investigation Commission, similarly structured; Graham Taylor served on this commission as well. A major mine disaster accelerated the commission's work, and in a special session the legislature approved more protective measures for miners. These actions catapulted Illinois to the front rank of states with industrial health and safety laws.[14]

Cooperative commissions also influenced the course of legislation focused more specifically on workers' health. In 1906 an Industrial Insurance Commission recommended that the state sponsor an investigation into the prevalence of industrial diseases in Illinois. With the approval of state lawmakers, a Commission on Occupational Diseases was organized in 1907 and instructed to propose legislation to control occupational diseases.[15]

Alice Hamilton was appointed by the governor to the Commission on Occupational Diseases. Hamilton was, of course, a Hull-House reformer, and her career in occupational health was advanced by her work for the state of Illinois. Her position on the Occupational Diseases Commission was short-lived, however, as she soon resigned in order to manage the actual investigation of workplaces. In 1909 the commission requested and received funds to to continue its studies for two more years. In a 1911 report to the legislature, Hamilton and other researchers revealed a wide variety of industrial diseases that threatened the health of many Illinois workers.[16]

The commission recommended principles it deemed important to the implementation of any occupational disease law. First, the law should expand existing requirements for sanitation and cleanliness. Second, women and children, who were believed to be especially susceptible to industrial diseases, should receive special protection. Third, the more dangerous industries should be more closely regulated. And fourth, effective administrative structures were necessary. A bill designed to these specifications was drafted but never formally proposed to the legislature; perhaps the commission recognized the impossibility of securing enactment of such a broad, utopian vision.[17]

Instead, the commission proposed what passed as the Occupational Diseases Act of 1911. This statute required that employers engaged in any work that might produce illness or disease "adopt and provide reasonable and approved devices, means or methods for the prevention of such industrial or occupational diseases as are incident to such work or process." Employees in dangerous jobs were to be examined monthly for signs of developing industrial diseases. Poisonous fumes and dusts were to be vented out of the workplace. The State Department of Factory Inspection was empowered to inspect workplaces and to stipulate health and safety devices and safe work practices; failure to comply could result in misdemeanor convictions and fines of up to $200. Employers in health-threatening industries were also required to post notices acquainting their employees with known dangers as well as copies of the Illinois law itself.[18]

Finally, employees were granted the right to sue any industry that willfully violated the act or failed to comply with its provisions for recovery of damages due to loss of health. By 1923, with support from the state Federation of Labor, industrial diseases were brought under the jurisdiction of the workers' compensation act, supervised by the Illinois Industrial Commission within the state Labor Department.[19] Although most occupational illnesses would be compensated following an administrative hearing, workers might still sue employers outside of workers' compensation, through the courts, for willful violation of mandated health practices.

An error in the wording of the original 1911 act—perhaps just a printing error—permitted the courts to conclude that the only industries it covered were those in which lead, brass, and zinc compounds were manufactured or employed. By 1934, however, when the Illinois dialpainters were filing suits, the language of the act had been clarified. In 1934 the Supreme Court of Illinois, in *Burns v. Industrial Commission*, inserted the missing word ("or") that rendered the disputed passage sensible. The act now covered all processes in which poisons were used in ways that might cause harm.[20]

Another liberal interpretation of the law by the Illinois Supreme Court, in the 1933 decision *Raymond v. Industrial Commission*, settled the problem that had haunted the New Jersey dialpainters: the status of statutes of limitations in cases where employees did not know they had been injured or made ill until after the stipulated time period had run out. The court ruled that the period of limitation could not be construed so as "to require the impossible" and "defeat the very purpose" of compensation; therefore, employees could only be required to notify their employers of their conditions when the facts of their injuries or illnesses became known to them.[21]

Thus, a coalition of women social reformers and labor leaders with the acquiescence of manufacturers birthed legislation that gave the victims of radium poisoning a chance at compensation. If well administered, the law might have prevented radium poisoning in the first place, but its implementation was ineffective. First, in Illinois enforcement of labor legislation was fragmented. A bill passed in 1917 to centralize labor administration left the Industrial Commission that oversaw workers' compensation separate from other labor agencies, so that factory inspectors did not have access to accident reports filed with the commission. Second, the commission had little power and seldom employed what little it did have. It left policy to an advisory board comprised equally of representatives of the Illinois Manufacturers' Association and the Illinois Federation of Labor, and, as they hardly ever agreed, little policy direction was given.[22]

These problems, which continued well into the 1930s, could be traced to a strong state Federation of Labor and a highly fragmented Democratic Party. During the Progressive Era, the Illinois federation was one of the strongest and most militant in the nation. Its workplace strength freed labor from the temptation to rely on politics to exert its will, although the federation was strong enough to influence both Republicans and Democrats when it chose. Thus, unlike other state federations, Illinois's entered the 1930s without an alliance with Democrats, and, indeed, the emergent Democratic machine in Chicago had closer ties to business than to labor. In turn, the state Democratic and Republican Parties remained fragmented later than elsewhere, and in the power struggles between factions, both labor and business wielded enough power to exercise vetoes over policy. This also prevented ties between labor and the Democrats. Finally, fragmentation and patronage politics had weakened civil service reform, too, and no autonomous administrative machinery had evolved to oversee labor issues or any other concerns. As a result, social experts, though present at the University of Chicago and at Hull-House, were unable to wield authority within administrative state boards.[23]

As a result, from the 1910s into the 1930s the preventive clauses of the Occupational Diseases Act of 1911 were not enforced and workers' compensation was in disarray. In Illinois, despite legislation requiring occupational disease prevention and compensation and a strong labor movement, radium dialpainters were endangered and had to struggle to win compensation.

In 1935 Inez Vallat's case came before the Illinois Supreme Court because of a challenge made by Radium Dial to the constitutionality of the Occupational Diseases Act. Among other complaints, the company's lawyers claimed that the phrases charging businesses with employing "reasonable and approved devices, means or methods" to prevent industrial diseases were "vague, indefinite, and did not furnish an intelligible standard of conduct." Thus, the act was unconstitutional because "it fails to set up an intelligible standard of duty and violates the due process clauses of the state and Federal constitutions." Further, the law allowed factory inspectors to determine what "devices, means or methods" employers were required to employ, and this "unlawfully confers legislative powers" on an administrative agency.[24]

The court, though insisting that the act, "designed to ameliorate harmful and dangerous working conditions and . . . humanitarian in purpose . . . should be sustained as a valid enactment if this can be done," agreed in substance with Radium Dial, writing: "In order that a statute may be held valid the duty imposed by it must be prescribed in terms definite enough to serve as a guide to those who have the duty imposed upon them. . . . but if the duty is imposed by statute through the use of words which have not yet acquired definiteness or certainty and which are so general and indefinite that they furnish no such guide the statute must be declared to be invalid." Moreover, leaving the definition of these terms to a "ministerial officer," that is, to factory inspectors, as the act stipulated, was "an unwarranted and void delegation of legislative power."[25]

With this decision, all the Illinois dialpainters' cases were dismissed for want of prosecution. Fortunately, a new Occupational Diseases act was quickly introduced, passed by the legislature and signed into law by March 1936. This statute, like the old one, made all industrial diseases compensable. Catherine Donahue, Charlotte Purcell, and thirteen other women refiled for compensation. Inez Vallat, though, had already died in February. A woman who took part in the new series of compensation cases remembered the course of her disease: "At first she had a tooth pulled, then a bandage on her jaw, and she hobbled around before she died. . . . She died a very horrible death."[26]

The Illinois Industrial Commission held hearings on the dialpainter complaints in early 1938. By this time Catherine Donahue was seriously ill. She weighed only seventy-one pounds and, a brother-in-law remembered, "was full of radium and dying by inches . . . she suffered agonies and Tom was nearly bankrupt buying medicines to try to relieve Catherine." Tom Donahue estimated that his wife's medical bills had exceeded $2,500. The Ottawa glass factory worker had mortgaged their home for $1,500 to help meet expenses. At the hearings Catherine collapsed when questioned about her chances for recovery and survival.[27]

Once again newspapers nationwide covered the plight of the dialpainters, many in the most lurid terms imaginable. "Ottawa's Doomed Women," "Ottawa's Living Dead" were sobriquets referring to the dialpainters in the Illinois suit. Again, tabloid papers emphasized the horrors of knowing one had a fatal disease and ignored the moral and legal issues of culpability. Marie Rossiter was "frightened to death. . . . There were ten of us girls in that one room. Four already are gone. Perhaps I'll be next." But in best "industrial martyr" style (the newspaper caption over a picture of a dialpainter holding her children), Rossiter added, "I want to live as long as I can for the sake of my little boy, Bill." One paper contrasted the Illinois and New Jersey cases: the latter "got lots of attention and sympathy," whereas the former were "suffering and dying in comparative obscurity." Certainly, the Consumers' League had helped advertise the plight of the New Jersey dialpainters.[28]

Catherine Donahue was the first dialpainter to have her case settled by the Industrial Commission. In April 1938 she was awarded a fair, if not staggering, sum of money: $3,320 for medical expenses, $2,395 for compensation, and a $378 annual pension for the future. She died in July. One day after her death Radium Dial filed an appeal of her award.[29]

Before the appeal was allowed, however, the company was required to post a bond, for two reasons. First, Radium Dial was self-insured, because, since 1931, the potential for compensation suits had made it impossible for the firm to purchase compensation insurance. Second, Radium Dial was no longer in business in Illinois.[30]

In the fall of 1934 a new company, Luminous Processes, had established itself in Ottawa to woo dialpainters away from Radium Dial. Several stories exist to explain the origins of this firm. The dialpainters remembered that a supervisor at Radium Dial treated them all to dinner and asked them to help him start a new company, promising them higher wages. A second account is that the president of Radium Dial, Joseph Kelly, was ousted in a stockholders' struggle and formed Luminous Processes in revenge. A third version is that Kelly, seeing his company's survival threatened by personal injury suits, helped his son, Joseph Kelly Jr., set up Luminous Processes.[31]

There may be some truth in all of these explanations; the corporate metamorphoses of the Illinois radium businesses are difficult to untangle. The

Radium Dial Company of Ottawa had been a subsidiary of the Radium Chemical Company, which in turn had been a subsidiary of Standard Chemical Company, the first radium extracting firm in America. When Belgian radium began to reach the United States in 1922, Standard Chemical struck a deal: over the next five years it would transfer its radium stores and its personnel to the Belgian concern. Standard Chemical was liquidated by 1928, but its assets in Illinois were purchased by a longtime official of both Radium Chemical and Radium Dial who organized a new firm called the Radium Service Company of America. The Radium Chemical Company then took over the sales of Belgian radium. It retained ownership of Radium Dial and created the Radium Luminous Corporation to sell luminous paint. Joseph Kelly, son-in-law of Joseph Flannery, who had founded Standard Chemical, was president of Radium Chemical and its subsidiaries, Radium Dial and Radium Luminous. The new firm in Ottawa, Luminous Processes, was presided over by Joseph Kelly Jr. Radium Dial was allowed to go out of business, but a 1936 move consolidated Radium Chemical and its subsidiary, Radium Luminous (and two other radium businesses) into a *new* Radium Chemical Company. It is not clear which Kelly directed the new Radium Chemical, but by 1973 it was Kelly Jr. Through these transformations the Illinois interests of one company, Standard Chemical (and its subsidiaries Radium Chemical and Radium Dial), had by 1928 been separated into two companies, Radium Service and Radium Chemical (and the latter's subsidiaries Radium Dial and Radium Luminous), and during 1934 into three separate entities, Radium Service, Radium Chemical (reorganized in 1936), and Luminous Processes. A Dun and Bradstreet investigation in 1973 revealed that all three corporations were affiliated through similarity of officers. Thus, the original Illinois assets were preserved, and the Radium Dial Company, against which the dialpainters were filing charges, was left outside the fold, practically valueless. In this way, radium company officials in Illinois protected themselves from financial loss for the harm done the dialpainters.[32]

In 1934 Luminous Processes took over Radium Dial's employees and business, and Radium Dial ceased operating in Illinois. Under these circumstances, the law required and the Insurance Commission insisted that Radium Dial post a $10,000 bond before it be allowed to appeal the Donahue award. Radium Dial claimed that was impossible because its assets only amounted to $22,000. The company challenged the bond in court but lost and posted the bond.[33]

Radium Dial then challenged Donahue's compensation award all the way to the U.S. Supreme Court, which in 1939 refused to hear the case. The Donahue award, about $5,700, was paid, leaving the remainder of the $10,000 bond (plus perhaps some of the company's alleged $22,000 in assets) to be spread among the fourteen other women suing for compensation. The records are unclear about how many women actually won such a settlement: Charlotte Purcell did receive $300, and at least three women withdrew their suits.[34]

If all the rest of the women received a financial settlement, and we add Ella Cruse, whose case was settled much earlier, the number of women obtaining some compensation, thirteen, is not much different than in other states. It is estimated that sixteen women won compensation in Connecticut, ten or eleven in New Jersey, and thirteen in Illinois. The amounts awarded in Illinois are less than the five large New Jersey settlements, but the Donahue award of $5,700 was similar to the $5,600 calculated to have been spent per recompensed employee by the Connecticut watch companies. Other compensated women in Illinois likely received less than $1,000.

A good number of dialpainters who were affected the worst, and the earliest, were compensated. By 1940 fifteen Connecticut women were known to have died from radium poisoning, and all but three were compensated; sixteen New Jersey women were known dead, and eleven received settlements. Of course, these figures neglect the dialpainters who may have died unknowingly from radium poisoning and whose deaths were not tied to the disease, as well as those who became sick but did not die before 1940.

Throughout the 1920s and 1930s women painted luminous dials with radium paint at Luminous Processes in Ottawa. During World War II luminous dials were once again in high demand, and the number of firms and dialpainters in the luminous dial business multiplied. Concern about the health of home front workers spurred government investigations of the safety and health of various defense industries, including dialpainting.

No official standards guided workplace radiation exposure until 1941, when military-preparedness experts began arranging for large-scale production of luminous dials for military equipment. The U.S. Navy secured the services of the radium expert Robley Evans by threatening him with induction into the navy if he did not quickly produce radium "tolerance level" standards. Robley put together a committee under the jurisdiction of the National Bureau of Standards. The Advisory Committee on X-Ray and Radium Protection included the physician Harrison Martland, the industrial hygienist Frederick Flinn, a radium industry official, a radium medicine representative, a naval officer, and, as chair, L. F. Curtiss of the National Bureau of Standards. As Evans has described its deliberations, the group reviewed records on twenty-seven deaths from deposited radium. Twenty had body-burdens above one microgram of radium and were injured; seven had body-burdens below half a microgram of radium and showed no injuries. Evans suggested that the tolerance level be set at 10 percent of the one microgram level, or 0.1 microgram of radium, and asked members of the committee if they would "feel perfectly comfortable if our own wife or daughter" were to accumulate such a level. "Unanimously," all found 0.1 microgram acceptable. Because radium in the body correlated with radon in exhaled air, a tolerance level for radon in expired air was set at 10^{-12} curies of radon per liter. Standards were also established for maximum permissible levels of radon in ambient air (10^{-11} curies of radon per liter) and for maximum permissible gamma ray exposure (0.1 roentgens per worker per day). These tolerances were not regulations but recommendations, as were the safety codes adopted by New York State and Massachusetts based on the National Bureau of Standards's regulations.[35]

The maximum radium body-burden for workers was often exceeded. During World War II Robley Evans studied dialpainters' radium accumulations and in 1943 revealed his data on 288 dialpainters working in nine plants in the eastern United States. "Under the best working conditions now existing in the dialpainting industry," he wrote, "about 15 per cent of the workers accumulated *more* than the tolerance dose of radium." In Illinois, two employees of the Division of Industrial Hygiene, State Department of Health, published a review of data they had compiled in cooperation with a nationwide study of the dialpainting industry by the U.S. Public Health Service's Division of Industrial Hygiene. There were now six luminous dial plants in Illinois engaged in dialpainting (a seventh assembled instruments containing radium dials). About thirty-six workers (the uncertainty is based on disparate numbers in the article) were examined for radon in their expired air, indicative of radium deposits in the body. About half of the dialpainters showed more than the tolerance dose of breath radon. By far, the greatest number of tested employees came from a plant identified as "E," which also had four employees who had worked for sixteen, seventeen, nineteen, and twenty years: this was most likely the Luminous Processes plant, which had hired away employees from Radium Dial, where they had worked since the early 1920s. About half of Plant E's employees demonstrated breath radon above the tolerance level.[36]

Despite these figures, and others demonstrating that the tolerance limits were often exceeded, the Illinois researchers argued that dialpainting was safely conducted in the state. Dialpainting became "the most feared occupation among women" because of "the publicity given to deaths from radium poisoning . . . in the last war." The publicity, not the deaths themselves, was responsible for dialpainting's bad reputation. Dialpainting, the hygienists argued, "is not any more dangerous than the use of benzol and the many other very toxic materials employed in industry." Given the high levels of illnesses caused by such "very toxic materials" as benzol (benzene), this was hardly reassuring.[37]

The authors overlooked key factors in promoting the health of dialpainters. For instance, they reported that of seven Illinois plants involved in radium dial manufacture, only two did not exceed the tolerance level for radon in air. Plant E (Luminous Processes?) was singled out for discussion. This factory did not provide ventilated worktables, as did more modern plants. With the windows open in summer, the air tested at 92 and 106 percent of the established tolerance. This proved, the hygienists argued, that ventilated worktables were unnecessary; workers could paint "safely" without expensive ventilation equipment. This perspective overlooked winter conditions, when the windows would be closed; neglected radon exposure in factories with even modern ventilation; and did not consider the five plants with radon levels above the tolerance standards.[38]

The authors' argument that dialpainting in Illinois was safe rested on criticism of the tolerance limits for radon, both breath radon and radon in the air, as too severe—too low by a factor of ten. The standards were set impressionistically rather than scientifically, they implied, maintaining that "tolerances should have reasonable scientific proof for their adherence before they are advanced as final." An increase in the tolerances on the order of a magnitude of ten would have considerable impact on "the physical requirements for control measures," the researchers argued; that is, allowing more radon in air and in breath would save money.[39]

The figure for expired air radon levels supported by the Illinois hygienists (and higher by a factor of ten than the Bureau of Standards figure) was currently under debate in New York State. Looser regulations were favored in a public hearing by Victor Hess (who formerly had worked for U.S. Radium in New Jersey), Frederick Flinn (likewise connected to U.S. Radium and other dialpainting companies), and several manufacturers. In the opposite camp was Robley Evans, who was "violently opposed" to the higher figure, "because it kills," he wrote Harrison Martland. "I think you better join me in this little battle," Evans advised Martland, "or Flynn [*sic*] may win." In a letter to the New York State Department of Labor, Evans repeated that there was no doubt that the higher expired air standard "kills": there was, to his mind, "no scientific doubt whatsoever but that such a value can and does produce death from chronic radium poisoning." In response to Evans's concern, Martland wrote, "I am very suspicious about the New York State Labor Code . . . and wonder if the U.S. Radium has done some underhanded work."[40]

The only danger stressed in the Illinois Department of Health study was the poor work and hygienic habits of the dialpainters, who would get paint on their hands and then put their hands in their mouths. Getting paint on the hands was deemed unavoidable. To prevent radium from entering the mouth, employers were urged not to hire nail-biters or women with the "habit of swiping the corners of the mouth or the lips with the hand." Jewelry and "talon type" nails should be avoided because of the difficulty of removing embedded radium, although the study noted that "it may be difficult to enforce this regulation among our present-day girls." Head coverings hiding all of the hair were also desirable, although, once again, "it is . . . difficult to secure the co-operation of girls in this measure." Proper hand washing was stressed as perhaps the most important health precaution, but "it is difficult to convince girls as to the necessity for this thorough practice." Thus, according to the Department of Health study, the workers themselves were primarily responsible for endangering their lives.[41]

The study's insistence that employees were to blame for their own contamination makes sense only in that it might help explain why some women accumulated more radium than others in less time, if we assume that poor work habits led to the ingestion or inhalation of radium. The Illinois data show little correlation between radium body-burden and length of employment. Although the dialpainters employed longest at Plant E did have high body-burdens, theirs were exceeded by two women who had worked only two years; high burdens were also found in two women who had worked for a year or less. This meant that although a woman might work for a long time without accumulating a great deal of radium, she might also exceed the tolerance level in but a few months. Blaming dialpainters for their radium accumulations permitted employers to describe the job as safe despite some women's high accumulations.[42]

Other factors might explain the variation. Some women may have excreted radium more efficiently than others, thus permanently depositing less in their bodies. A woman who painted faster than another would use more paint each day and thus be exposed to more radium. These workers may have had different painting techniques, resulting in more or less radium paint on their hands or dust in the air. Despite the data it provided on how different kinds of dials demanded different techniques, the Illinois study did not at-tempt to make a correlation between type of job and radium accumulation. Instead, it grasped the ideas most congenial to luminous dial producers. Tol-erance levels were too low and too expensive to meet. Women accumulating radium were themselves at fault, not conditions in the plants.

The Illinois hygienists also shared the radium industry's position on man-dated safety and health codes. They thought that state codes defining con-ditions in the workplace were undesirable. "Recommended standards" were preferable to a code, as they "could contain much that could not be en-forced in a code." Recommended (but not enforceable, like a code) standards would also be educational, and "education in the methods of control is more necessary than a code which is only the result of compromise between all interests."[43]

If education rather than regulation was the best solution, and if workers were largely to blame for their radium body-burdens, then employee educa-tion would seem to have been a logical priority. Yet apparently dialpainters were not told about the dangers of radium. One employee of the war years, who was hospitalized and examined by radiation experts in 1952, reported: "That was the first time I knew that radium might be dangerous. The com-pany had never told us a thing." Luminous Processes employees interviewed by a journalist in 1978 had also been left ignorant of radium's dangers. They were told that eliminating lippointing had ended earlier problems. They worked in unvented rooms. They wore smocks that they laundered at home. Geiger counters could pick up readings from pants returned from a dry cleaner and from clothes stored away in a cedar chest. Workers reported that their calluses would glow in the dark from deposited radium and that, for fun, they would paint their nails with the luminous paint. "We slapped radium around like cake frosting," remembered one employee. Pearl Schott recalled: "You couldn't work in that plant without getting covered with the stuff. Some-times I'd get up in the night and look in the mirror and my hair would be glowing." She suffered from loose teeth, an ill-defined "blood disease" (per-haps anemia), gastrointestinal bleeding, and cataracts—all possible effects of overexposure to radiation. Schott mentioned the gastric symptoms to a Lumi-nous Processes executive: "A man from the Luminous offices in New York was out here. He used to check on safety in the plant and one time he asked me, 'Pearl, how are you?' I said I had massive gastro-intestinal bleeding and do you know what he told me? He said, 'Don't worry about it, lots of people have that.' . . . Once I asked him if gamma rays—that's what radium gives off—

if they were dangerous. And he said, 'No, no, gamma rays won't do you any harm.'" Coworker Eleanor Eichelkraut reported: "It was fun to work there. We girls would all sit around big tables, laughing and talking and painting. It never entered anyone's mind that we were doing something dangerous. Why didn't they say something?" Luminous Processes, then, was as contemptuous of its workers' health in the 1940s and 1950s as Radium Dial had been in the 1920s and 1930s.[44]

Inadequate protection of employees from radiation continued at Luminous Processes into the 1960s and 1970s. In 1957 the state of Illinois had been charged by the federal government with the regulation of radium, but not until 1975 did the state grant the Public Health Department the necessary authority. Regulations were then established to control the handling of radium. At about this time, Luminous Processes switched from using radium in its luminous paints to using tritium, a radioactive form of hydrogen. Tritium was regulated not by the state but by the federal government through the Nuclear Regulatory Commission (NRC). On inspecting the Luminous plant in Ottawa, the NRC found excessive levels of radiation in the workplace; the plant's radiation level was 1,666 times the NRC allowable level. After three visits in 1976, the commission fined Luminous $3,250. Two more visits in 1977 and yet another in 1978 failed to find substantial improvement, and the NRC suspended the license of Luminous Processes, threatening to revoke it unless the company showed cause why it should not.[45]

By 1980 ten women who had worked at Luminous Processes in Ottawa, alarmed at seemingly high rates of breast cancer and foot tumors among them, filed suit against their former employer. Of 463 dialpainters hired in Ottawa after lippointing of brushes ceased, 9 were diagnosed with breast cancer, which, according to the *Wall Street Journal*, was double the expected rate. Sixty-four deaths among dialpainters who had worked since the late 1960s included twenty-eight from cancer, over two times the expected level. But badges worn between 1964 and 1975 showed that the women's average annual radiation exposure was less than half that of the maximum allowed by safety standards, calling into question either the veracity of the badges or the standards if the cancer rate was indeed elevated.[46]

Luminous Processes responded by closing its Ottawa plant, laying off the few dozen women employees, and leaving town. The federal government ordered Luminous to clean up the plant. The company gave away thirty-five dialpainting desks to a local Catholic school, but when it was discovered that the desks were contaminated with radium the company was forced to take them back. Over two and a half years, Luminous promised and reneged on several decontamination plans, then was ordered by the circuit court to clean up the Ottawa plant. The firm claimed that it was willing to do so but lacked the necessary funds.[47]

To escape financial liability for environmental pollution and industrial diseases, Luminous Processes shuffled corporate assets into other holdings, in much the same way that Radium Dial had in the 1930s. Luminous made a gift of two of its foreign subsidiaries, in Scotland and in Canada, to the Radium Chemical Company president, Joseph Kelly. A third subsidiary, in Mexico, had been sold the year before to local businesspeople, and a second U.S. plant, in Alabama, was closed.[48]

Meanwhile, Luminous was charged with inadequate radiation safeguards at a third U.S. plant, in Athens, Georgia, which had been operating since 1954. There, the federal government cited the company for excessive radiation twice in 1976, once in 1977, and twice more in 1978. Finally, regulators told Luminous Processes that its license would not be renewed. The company started several cleanup campaigns but completed none, and in 1981 the state of Georgia filed suit against Luminous Processes for $5.75 million in damages.[49]

Illinois also went to court against Luminous Processes and against Radium Chemical as well. The latter firm had but few assets that could be confiscated by the state to use in cleaning up the Ottawa facility; Radium Chemical's main business was supplying radium products to hospitals, but other types of radioactive compounds often replaced radium now in medical practice.[50] Illinois prosecutors settled out of court for the highest sum they figured they could get, about $62,000. The state legislature approved in 1983 the allocation of $2 million to cover decontamination costs. The Georgia plant was finally cleaned up using federal Superfund money.[51]

By 1925 radium was known to endanger dialpainters' lives. In response, sometime around 1926 and 1927, the technique of pointing the paintbrush between the lips ended. This lowered radium accumulation enough so that no case of cancer definitely attributable to radium has been found in dialpainters hired after 1927.[52] Still, dialpainters developed cancers and they continued to accumulate radium. Dialpainting remained a dangerous profession, and over the years luminous dial manufacturers consistently resisted attempts to make it safer.

As the dangers of low levels of radiation became understood, communities surrounding radium facilities became alarmed at the possible *public* risk from environmental radiation pollution. When workers had been endangered, government agencies did little to ensure safe working conditions; some investigators demonstrated greater concern for manufacturers' costs than for dialpainters' health. When neighborhoods were deemed endangered, however, the government acted with more alacrity and forcefulness, putting dialpainting firms out of business.

During the 1980s residents of towns surrounding Orange, New Jersey, discovered high levels of radon in their homes. Such contamination is usually caused by radon from bedrock seeping into tightly insulated houses and concentrating there. But in some radon-contaminated homes in Montclair, Glen Ridge, and West Orange, New Jersey, the gas had a different source. It came from the sand used as fill for the house sites. The sand was probably residue from the radium extraction process at U.S. Radium, and it still contained enough radium to produce the radon that filled the houses. Sixty-three residents joined together to file suit for damages against the Safety Light Corporation, a successor corporation to U.S. Radium.[1]

Also bringing suit was T&E Industries, which had purchased the U.S. Radium site in East Orange. The facility had changed hands several times since U.S. Radium first sold it. T&E was forced to relocate when state and federal environmental officials discovered radium still present at the former U.S. Radium plant. Besides contaminating the buildings, U.S. Radium had used radium tailings as fill around the factory. T&E sought $2 million in damages from Safety Light plus millions more to clean up the site. Safety Light argued that it was not liable because its predecessor, U.S. Radium, had taken the appropriate safety precautions dictated by the period's state of knowledge about radium and its dangers. "Nobody knew the dangers" of dumping radium-laced fill during the years U.S. Radium and its successor were in business, claimed Safety Light's attorney (U.S. Radium ceased operations at the Orange site around 1925, and Safety Light sold the property in 1943). This was the same argument used by U.S. Radium to evade liability for dialpainter illnesses.[2]

In 1991 the New Jersey Supreme Court found that U.S. Radium was "forever" liable for radium at the East Orange site. One judge, speaking for the court, cited the "abnormally dangerous activity doctrine," which stipulated that when persons engaged in dangerous activities they must bear the cost of accidents. Further, U.S. Radium's successor corporation, Safety Light, failed to disclose a dangerous condition to the buyer of the property. The court assumed that U.S. Radium either knew or should have known the dangers of radium contamination. Although company officials could not have understood the full effects of radium and radon as we do today, they recognized that "radium has always been and continues to be an extraordinarily dangerous substance." According to the New Jersey Supreme Court, U.S. Radium had "constructive knowledge" about radium's dangers. The corporation's "liability for the harm caused by abnormally dangerous activities does not necessarily cease with the transfer of property," the court concluded.[3]

Once again, a double standard seems to have prevailed in radium damages cases. Employers were seldom found liable for harm done by radium to workers, but government agencies have been prepared to hold radium businesses liable for harm done to patients, and in the case of T&E Industries a New Jersey court held a radium business liable for harm done to a property-owner. Why was the industrial hygiene movement of the 1910s and 1920s unable to win greater business accountability for the health of workers?

Businesses successfully resisted reformers' push for government regulation of the radium industry. The industry stood to lose money to liability suits if workers' claims proved true. It stood to lose business in the radium internal medicine market if radium treatments proved dangerous. Dialpainting would become more expensive if more precautions were necessary, leading to increased costs and prices and perhaps limiting the sale of luminous watches.

Personal, as well as financial, motives prompted industry officials to deny the possibility of radium poisoning among their operatives. Many officers of the various radium companies were themselves former or still active physicists, chemists, or physicians who had worked with large quantities of radioactive substances. They must have worried about their own health as suspicion grew about the dangers of radium work. Denial might have been a mechanism to deal with their own fears. Denial of harm to patients or dialpainters would have followed their refusal to consider the damage they themselves might have sustained.[4]

Charles Viol, a physicist with Standard Chemical, was exposed to large quantities of radium. He suffered from radium burns on his hands beginning around 1912, when he was twenty-six. In a few years these burns developed into a malignancy on one hand and led to the amputation of one finger. Later, a lump was removed from his forearm and then one from his armpit. The malignancy continued to spread up to and through his chest, and he died in 1928 just before his forty-second birthday. At U.S Radium, physicist Sabin von Sochocky knew that his body contained enough radium to distort laboratory readings; his fingers were also severely affected. Von Sochocky suffered from anemia for years before succumbing to his radium-caused death in 1928. His clerk and laboratory assistant, Victor Roth, died in 1927 from anemia; his ill health had begun around 1924. The chemist Edwin Lemen, employed at U.S. Radium since 1920, died from anemia in 1925. The physicist Howard Barker, who had worked with radium since 1913, joined the U.S. Radium Corporation in 1923 and quickly rose to vice president. He died of lung cancer with cerebral metastases in 1938, most likely a victim of his work. Dr. Joseph Bissel, founder of the New York Radium Institute, died of radium-related causes. Dr. Douglas Moriarata, of Saratoga Springs, lost fingers to radium use, although his death from diabetes was likely unrelated to his profession. Such illnesses and deaths among colleagues and coworkers may have contributed to the radium industry's denial of possible harm from internally deposited radium.[5]

These practical, financial and personal, psychological reasons for resisting the conclusion that radium was harmful did not preclude all reform. Radium businesses in this period did improve their safety records,[6] and they eventually did create healthier, if not healthy, workplaces. Although they argued with evidence that radium was dangerous and opposed regulation of their workplaces, industry officials did meanwhile forbid lippointing, and this improved the safety of dialpainting: no cancers definitely attributable to radium can be identified in dialpainters working after lippointing ended. But radium companies failed to protect dialpainters from hazards other than from lippointing; dialpainters continued to accumulate radium. While epidemiologists debated whether cancers other than the bone cancers and the "head cancers" might be caused by radiation, dialpainters continued to worry about their excessive radium body-burdens.

The existing medical literature justified the industry's stand. In a sense, though, this was a self-fulfilling prophecy, as that literature had been largely produced by those with a financial interest in radium. Further, dialpainters imbibed far greater amounts of radium than most patients. The radium companies reported that patients had been safely treated (ostensibly) with up to one milligram of radium element, although cumulative doses of around 300 micrograms were more common. Researchers would eventually attempt to calculate the quantity of radium element consumed by the dialpainters. Such calculations were complicated by the variable formulas of different grades of luminous paint and by the varying work habits of dialpainters, both in the number of watches painted a day and in the number of times the brush was pointed per watch. Harrison Martland, the scientist most often credited with discovering radium poisoning, gave a broad estimate of 15 to 215 micrograms of radium element ingested a week, with much less than that permanently deposited. The cumulative effect of these doses, which would lead to the frightful illnesses of the dialpainters, had never before been considered. As little as half a microgram of deposited radium ultimately proved to be dangerous.[7]

Controlling research on industrial health issues became a primary corporate concern, and in the period beyond this study, business would continue, with great success, to control the conclusions of industrial health scientists. Among the institutions created by the industrial hygiene movement of the 1910s and 1920s were government agencies responsible for the investigation, regulation, and compensation of industrial health, including workers' compensation agencies, as well as university research centers for the study of industrial hygiene. The government agencies with an interest in industrial disease were state Labor Departments and health boards and the federal Labor Department and Public Health Service.

With the exception of the Illinois Industrial Commission in the 1930s, government agencies responded slowly to the problems posed by the discovery of radium poisoning among the dialpainters. We have seen evidence that public agencies in New Jersey, Connecticut, Illinois, and at the federal level were more concerned with maintaining good relations with business than in helping dialpainters discover the cause of their illnesses. Medical industrial hygiene experts also responded more to business concerns than to workers' needs. Both Cecil Drinker of Harvard University and Frederick Flinn of Columbia University aligned themselves with radium company interests and resisted alliances with the dialpainters. Harrison Martland refused to help most of the dialpainters with diagnoses and testimony to back their suits against their former employer.

Few individuals in government and medicine operated independently of business. A review of responses by industrial hygiene institutions to radium poisoning in the workplace demonstrates that without continuing pressure from outside of business and government, manifestations of Progressive concern about industrial diseases were largely ineffective. The outside pressure was applied by the Consumers' League.

A new industrial disease, like radium poisoning in the 1920s, presents three political problems to those affected: proving the disease exists, compensating its victims, and preventing future cases. The Consumers' League contributed to progress made on all three fronts. Without the league's intervention, the dialpainters would never have established the cause of their illnesses and deaths. The discovery of radium poisoning was as much a political as a medical process. League reformers pressured Drinker to release his dialpainting data. They forced Flinn to back away from his denial of radium poisoning.

Martland's reputation as the discoverer of radium poisoning rests on data funneled to him by the Consumers' League. The league thus prodded state and federal officials and medical authorities to act.

The Consumers' League also helped win compensation for victims of industrial radium poisoning. It played a direct role only in New Jersey, where individual plaintiffs received the highest monetary awards among the states examined. Indirectly, the league influenced the Connecticut settlements. The Connecticut watch companies wished to avoid the publicity utilized by league reformers in New Jersey, so that even though radium poisoning was likely compensable under Connecticut's workers' compensation system, the companies preferred private settlements. Moreover, by settling out of court the watch companies could avoid admitting legal liability. The Illinois settlements took place a decade after those in New Jersey and Connecticut, in a far different political climate. No Consumers' League or other women's reform groups became involved, but again, indirectly, women reformers can be credited with some role in the settlements. The Consumers' League leaders Alice Hamilton and Florence Kelley, linked to other female reformers through the Hull-House settlement, were instrumental in the passage of legislation during the first decades of the twentieth century that made industrial diseases compensable in Illinois.

The league was responsible, too, for the small progress made in preventing future cases of industrial radium poisoning. The Consumers' League of New Jersey pressured the state labor commissioner to forbid U.S. Radium to continue painting dials in the state. No other state prohibited dialpainting; no other state had a Consumers' League committed to industrial health reform. Further, publicity manipulated by the league acquainted dialpainters and their employers throughout the country with the dangers of lippointing paintbrushes, initiating in-house reform by the dialpainting firms. Finally, the Public Health Service, under pressure from the Consumers' League, studied dialpainting and in 1933 suggested safe practices for producing luminous watches. Certainly these recommendations did not bear the authority of government-enforced regulations and they may have been ignored; yet even this minimal government action to prevent radium poisoning would not have been taken without league intervention.

The accomplishments of the Consumers' League reformers should not be exaggerated. They made mistakes—as in the radium necrosis bill pushed through the New Jersey legislature—and their successes were largely outweighed by their failures and oversights. The league never addressed the workers' need for sympathetic scientists to corroborate their concerns about workplace health and safety; research into industrial diseases is still largely controlled by businesses, state agencies, and universities subsisting on business philanthropy and government grants, institutions that historically have not allied themselves with workers' interests. Efforts to make industrial diseases compensable in New Jersey were somewhat successful, but workers still have to fight for compensation—witness the recent battles over compensation for silicosis—and compensation payments remain low and are paid over a relatively short term.

The illnesses of iatrogenic victims of radium poisoning offer a good contrast to the industrial cases. Government officials and medical personnel responded with greater alacrity to the discovery of radium poisoning among wealthy medical patients. Although they offered little or no help to the dialpainters, Martland and Flinn crusaded against the medical use of radium. Whereas federal agencies responded to industrial radium poisoning with a voluntary and toothless conference, the Federal Trade Commission sued and shut down a producer of radium for internal medical use. Medical and governmental experts, then, were far more willing and able to act for wealthy consumers than for working-class dialpainters.

Of the three groups closely studied here—medical scientists, government experts, and social reformers—only the Consumers' League continued to fight for workers' health beyond the 1920s. Physicians and other scientists founded industrial hygiene as a broad academic and professional field incorporating all threats to workers' health. Industrial hygiene did not survive as a specialty past the twenties but broke up into several component specialties. Toxicology was one offshoot. Toxicologists studied the actions of various substances on human physiology; the field retained little mark of its original interest in the effects of industrial materials on workers. Two other new fields were physiology and pathology, but practitioners like the physiologist Frederick Flinn and the pathologist Harrison Martland demonstrated scant concern for or little long-term commitment to workers' well-being. Industrial medicine, another offshoot, became a specialty in which medical experts worked under management in industry to prevent the hiring of unhealthy workers and to certify the continued health of those employed. It was no longer a research branch of medicine but a service to industry.

Industrial hygiene departments in universities disappeared; at Harvard, an interesting metamorphosis took place. Perhaps because of his oft-overlooked insistence on inhaled radium as a source or radium poisoning in dialpainters, Cecil Drinker continued research into the effects of inhaled dusts. His brother, Philip Drinker, an engineer, joined him in this effort, and while designing an experiment to study inhaled dusts in cats, Philip invented a device to control cats' inhalations. When the polio epidemic began, this device became the iron lung that kept many polio patients alive. Meanwhile, funds for research into workers' diseases, never free-flowing, waned. Industry, however, was willing to spend money for preventive measures. Philip Drinker became adept at designing ventilation systems to remove unwanted dusts and fumes from factory air, and by 1932 the Industrial Hygiene Department at Harvard was enveloped as a subdivision in his new department of Public Health Engineering. A course on industrial diseases was taught in the Industrial Medicine Department until 1936, when, following Alice Hamilton's retirement, industrial medicine was also subsumed under Public Health Engineering.[8]

In government, industrial hygiene had been institutionalized within Departments of Labor at the state level. Following an interesting turf fight at the national level (beginning in the 1910s and culminating in the 1930s), industrial health was increasingly removed from the Labor Department's jurisdiction and placed within the Public Health Service. Less dedicated than the Labor Department to improving workers' lives, the Public Health Service tended to focus on dangers to consumers from contaminated goods rather than dangers to workers from dangerous processes. Both federal and state governments left workplace safety and health largely to voluntary efforts by industry.[9]

From the 1920s on the Consumers' League continued to fight for industrial health and safety. Reformers maintained their Progressive tactics: investigation, education, and legislation. In New Jersey, concurrent with its political work in the 1920s to amend the compensation law to include radium poisoning, the league fought to include silicosis as well. In 1930 it studied mercury poisoning in a Belleville hat factory and brought a new, nonmercury felting formula to the attention of the New Jersey Labor Department. Soon afterward, Consumers' League reformers took up the plight of child agricultural workers who were exposed to pesticides containing arsenic. In 1932 Katherine Wiley discussed "the calamity in South Jersey" of workers dying of silicosis (probably in the glass industry) and spoke hopefully of further efforts to pass a silicosis bill.[10] In 1933 she described her "ten year fight" to conquer industrial diseases at the New Jersey league's annual meeting.

The Consumers' League of New Jersey continued its work in industrial health reform following Wiley's departure in 1935. The Great Depression had hurt its finances, so the league could no longer pay Wiley full-time after 1933; she became interested in police work, was the first policewoman in her county in New Jersey, and eventually advanced to national police work. In 1937 the New Jersey league helped a radium dialpainter obtain medical care. It continued to fight for fairer laws dealing with compensation and prevention of industrial diseases. In 1944, with Alice Hamilton newly installed as president of the National Consumers' League, the New Jersey league launched an effort to win a compensation bill covering all industrial diseases, not just those listed on the state's schedule. Hamilton spoke at the New Jersey league's annual meeting in 1946, urging the struggle on, and the league arranged for her to address a New Jersey Welfare Council that fall. Networking brought endorsements for a full-coverage law from the Welfare Council and from state branches of the League of Women Voters, the National Conference of Jewish Women, and the AFL-CIO, among others. The Consumers' League then organized an Inter-Organization Committee to lobby in the state legislature for a full-coverage law.[11]

As in so many reform movements, a tragedy helped the cause along. A Westinghouse employee, Dorothy Burns, brought suit in the mid-1940s against her employer for a new industrial disease, beryllium poisoning. Beryllium was used to coat fluorescent tubes, and beryllium dust led to lung disease. Harrison Martland once again cooperated with the Consumers' League's industrial disease efforts and described beryllium lung disease after the autopsy of a deceased electrical worker. A former Hamilton student, Harriet Hardy, became involved and undertook the definitive industrial and epidemiological research. More suits were filed against Westinghouse and other producers of fluorescent tubes. Echoing the radium cases, the Consumers' League of New Jersey organized a voluntary Beryllium Conference in 1947 to which it invited state officials, business representatives, legislators, and physicians.[12]

The conference endorsed the idea of a full-coverage industrial diseases bill. Such a bill was introduced into the state legislature along with one to extend the time period within which an employee could discover an illness and file for compensation. With the help of the Inter-Organization Committee, the AFL-CIO, Harrison Martland, Alice Hamilton, Paul Kober (an engineer dying of beryllium poisoning), and the governor of New Jersey, the bill was protected from crippling amendments. It was, however, rewritten to satisfy the New Jersey Manufacturers' Association, which demanded that silicosis, asbestosis, and occupational hernias be excluded from full coverage. The revised measure passed and was signed into law in 1949. The league fought against the restrictions in the statute, and they were repealed by 1951.[13]

The Consumers' League continued to assist those who worked with radioactive substances. In 1958 the New Jersey league helped pass legislation to control radioactive substances in the state. The National Consumers' League remained involved in regulating radioactivity in the workplace into the 1960s. In the early sixties the Consumers' League worked with labor unions to require public reporting of radioactive "spills" in which both workers and the public were exposed to radiation. Opposing the league and the unions was none other than the U.S. Radium Corporation, then situated in Morristown, New Jersey. C. W. Wallhausen, vice president, wrote to the Atomic Energy Commission protesting regulations that required incident reports to be made public. "Such a change," he declared, "would serve no useful purpose, but could . . . be detrimental to the entire program of peacetime use of atomic energy. . . . It is certainly true that the public is entitled to know of incidents, but we wonder if the public has enough of an understanding of radioactivity to properly use such information and not become even more confused than it now is." Public "confusion" would present "an extremely expensive proposition for the licensee involved." U.S. Radium recently had opened its records on the early dialpainters to three groups studying radiation exposure. Apparently, charged corporate officials, one of these groups passed on information to a former employee, who then filed a suit for one million dollars "for injuries which could have been, but were not necessarily, the result of radiation exposure." The company settled out of court for $5,000. "As a result of our kind-heartedness," the vice president noted, "we may be faced with additional law suits." He argued against public reporting of "incidents": "Reporting of incidents . . . are many times misinterpreted by the general public and can result in claims for injuries, generally always imagined, and . . . expensive to disprove." Thus, corporate welfare was defended against the right of workers and the public to know of threats to their health.[14]

The Consumers' League, then, linked the industrial hygiene movement of the early twentieth century with the industrial health movement of the 1960s. To union and government officials concerned with radioactivity and other workplace dangers, the league consistently made reference to its work with and for the New Jersey dialpainters. Their experience in New Jersey during the 1920s taught league members how to word legislation, how to manipulate medical personnel, how to pressure public officials.[15]

The league's continuous fight for industrial health and specifically its work for the dialpainters highlight the necessity for constant oversight by worker or citizen groups of industry and government regulation for health and safety in the workplace. Industries have consistently opposed regulation. Government has frequently responded to industry pressure by neglecting regulation altogether, ceding health and safety to "voluntary" efforts on the part of industry or writing regulations that satisfy industry requirements. Physicians and other health scientists have seldom supported workers in political struggles over workplace regulation. Workers' organizations alone have generally not had the power to force effective health and safety measures. Reformers, then, as in the dialpainters' cases, may play a pivotal role in efforts to improve health and safety in the workplace by throwing the weight of public pressure on the side of workers. The Consumers' League, although it made mistakes and won only limited concessions from government and industry, helped dialpainters assure that the existence of radium poisoning was acknowledged, helped some dialpainters win compensation, and helped prevent many illnesses among those who worked with radium and other radioactive substances. In this vein, the New Jersey dialpainters' lawyer, Raymond Berry, wrote the president of the New Jersey Consumers' League, "I never realized, until these cases arose, the great importance of a fact-finding body distinct from any governmental function and unallied with any particular interest except that of social welfare." [16]

This study was undertaken in response to concerns about dangers faced in today's workplaces. Reform efforts in industrial health began over a century ago and reached an apogee during the industrial hygiene movement of the 1910s and 1920s. By considering not just the reforms of that period but how those reforms worked in practice, we have traced the failures of industrial health reform to a faith in the autonomy of "experts" in both government and medicine. Workers have always doubted that government agencies could act independently of business, but science has not been understood as another political realm. Workers must fight business influence on government agencies, but to cope effectively with industrial diseases, workers must find their own medical experts and finance their own industrial disease investigations. Alternately, workers must conceive of themselves as experts who better than many recognize occupational dangers and struggle to become part of institutions overseeing safety and health in the workplace.

We have considered only briefly working-class organization around the issue of industrial health and safety. Dialpainters never organized formally in response to radium poisoning. One reason for this is that by the time they began to get sick, most of them no longer worked as dialpainters. When New Jersey women discovered their illnesses, their jobs had already migrated to Connecticut and Illinois. Second, because most dialpainters became ill when they were young (in their twenties), they did not have the personal resources that longer life might have brought them. Third, the affected women were sick, they were scared, and eventually they knew they had only a short time to live. Organizing a workers' group must have been low on their list of priorities.

In some instances dialpainters did act in concert. At least one meeting was held at a dentist's office for a number of women to consider the work origins of their illnesses. Theodor Blum was consulted by two of his patients about the similarities of their illnesses, leading to his paper on "radium jaw." A pair of dialpainters in Illinois together confronted their supervisors, demanding compensation. Dialpainters acted in groups to sue their employers for compensation. In building cooperative networks, dialpainters were helped by the kinship and neighborhood networks that had brought many of them to their jobs. Progress on the industrial radium poisoning front could not have been made without the dialpainters' collaborative efforts.

Still, the dialpainters mostly relied on health scientists to establish the cause of their illnesses, on business and government for compensation and preventive measures, and on middle-class reformers to push industrial health institutions to work on the women's behalf. Workers today, in addition to supporting their own medical experts and organizing around their own expertise in industrial health, may consider alliance with middle-class reform organizations. In the case of the dialpainters, the Consumers' League pressured state and medical experts to serve the workers' ends.

Many dialpainters survived into old age, but those with high body burdens, including most of the women introduced here, succumbed to radium poisoning in the 1920s and 1930s. Again, we know most about the life of Katherine Schaub, whose cousin and fellow dialpainter was one of the earliest victims of radium poisoning and whose complaints to the New Jersey Health and Labor Departments led to early government consideration of dialpainting's dangers. Her story will have to stand in for the rest.

Following the 1928 settlement, Schaub made plans. First, she contributed money toward her parents' house mortgage and tax bill, because "I could find, I knew, no greater happiness than that which would be mine by making the folks happy." Then she went shopping. She bought "the coat I had always wanted and a tan felt hat to match . . . four silk dresses, a white flannel sport skirt, a rose-colored sweater, two silk blouses, lingerie, stockings, shoes and a purse"—the doctors had given her only a year to live, and she was determined to "live like Cinderella as the princess at the ball." She bought an automobile and moved to a rural resort for "real country life." She decided to continue her education and enrolled in a home study course from Columbia University, taking college preparatory and writing courses. And she worked on her autobiography, a portion of which was published.[17]

Schaub struggled with ill health and depression. She sought to soothe her fears in part through religious faith; she once spoke of wishing to visit religious shrines "with the faith and the hope that something can be done for me by prayer." Schaub also attempted to allay her anxiety by "seeking forgetfulness," and this worked to some extent; her health improved overall, but a relapse in early 1929 put her in the hospital for four weeks. A few months later churlish newspaper reporters observed with suspicion that the women "doomed to die" were still alive. At times Schaub's worries seemingly overcame her. She wondered about "this craving for alcohol of mine" and believed that sunshine, prescribed as part of an experimental treatment by the medical board established in the settlement, made her "very nervous" and "full of fears."[18]

Although labeled "hysterical" in several medical documents, Schaub perhaps faced life with more realism than others. She wrote scornfully of former dialpainters who "won't own up. . . . They know it isn't rheumatism they've got. . . . God—what fools—pathetic fools! . . . Afraid of being ostracized!" But Schaub, herself twenty-six years old, understood that they feared "losing their boyfriends and the good times." In a 1928 newspaper article, she lamented: "While other girls are going to dances and the theater and courting and marrying for love I have to remain here and watch painful death approach."[19]

By the summer of 1930 Schaub had outlived her prognosis but also had used up all of her $10,000 settlement. She returned to the Orange area to live with her family for a brief time. She had been suffering for months from severe pain in her left knee. While visiting friends in New Jersey, she stumbled going up a stairway and when she fell the bone above her knee fractured. X rays revealed cancer of the bone. In Harrison Martland's first paper on cancer among radium dialpainters, Katherine Schaub would become the sixth known case.[20]

Wearing a cast and then a brace, Schaub returned to the countryside. She still received her annuity from U.S. Radium. To stretch this, she reasoned that her mental and physical health was improved by country life and that her disability necessitated taxi rides, so she charged her living and transportation expenses to her former employer. (The company protested.) In 1931 Schaub moved to a private sanitorium in New Jersey. There she completed the published portion of her autobiography with hope that she would regain her "lost health"; but if this was denied her, "I have been granted another priceless gift—I have found happiness."[21]

We know less about Schaub after this. The medical committee charged with overseeing the dialpainters' health costs continued to devise treatments that might help them. By 1930 Schaub, not wishing to be used in experiments any more, resisted further treatment. By October 1932 doctors urged amputation because of a large tumor in her leg, but she refused. Schaub was almost thirty-one at her death on February 18, 1933. She was buried in Holy Sepulcher Cemetery, Newark.[22]

NOTES

ABBREVIATIONS

AALLP   American Association for Labor Legislation Papers, Labor Management Documentation Center, New York State School of Industrial and Labor Relations, Cornell University, Ithaca, N.Y.

AFPSR   American Fund for Public Service Records, Rare Books and Manuscripts Division, Astor, Lenox and Tilden Foundations, New York Public Library

AHP   Alice Hamilton Papers, Schlesinger Library, Harvard University

AMADIR   American Medical Association Department of Investigation Records, Chicago

ANL   Argonne National Laboratory Center for Human Radiobiology, Argonne, Ill.

CLMP   Consumers' League of Massachusetts Papers, Schlesinger Library, Harvard University

CLNJP   Consumers' League of New Jersey Papers, Special Collections and University Archives, Rutgers University Libraries, New Brunswick

EKP   Edward Bell Krumbhaar Papers, Library of the College of Physicians of Philadelphia

FDAR   U.S. Food and Drug Administration Records, Record Group 88, National Archives, Washington, D.C.

FTCR   U.S. Federal Trade Commission Records, Docket 1756, National Archives, Archives II, College Park, Md.

HMP   Harrison Martland Papers, Special Collections and Archives, University Libraries, University of Medicine and Dentistry of New Jersey, Newark

HUIHDR   Harvard University Industrial Hygiene Department Records, Archives of the Francis A. Countway Library of Medicine, Harvard Medical School, Boston

HUPDR   Harvard University Physiology Department Records, Archives of the Francis A. Countway Library of Medicine, Harvard Medical School, Boston

JAMA   *Journal of the American Medical Association*

LWVNJP   League of Women Voters of New Jersey Papers, Special Collections and University Archives, Rutgers University Libraries, New Brunswick

NA   National Archives, Washington, D.C.

NCLR   National Consumers' League Records, Library of Congress, Washington, D.C.

NJDPH   New Jersey Department of Public Health

NPL   Newark Public Library

NYPL   New York Public Library

RBP   Raymond Berry Papers, Library of Congress, Washington, D.C.

SKP   Swen Kjaer Papers, Center for Human Radiobiology, Argonne National Laboratory, Argonne, Ill.

USBOL   U.S. Bureau of Labor
USBOLS   U.S. Bureau of Labor Statistics
USDOE   U.S. Department of Energy
USDOL   U.S. Department of Labor
USPHSR   U.S. Public Health Service Records, Record Group 90, National Archives,
    Washington, D.C.
USRC   U.S. Radium Corporation
USRCR   U.S. Radium Corporation Records, Center for Human Radiobiology,
    Argonne National Laboratory, Argonne, Ill.

A NOTE ABOUT CITATIONS

The U.S. Department of Energy and its subsidiary, the Argonne National Laboratory, in 1995 became concerned about possible objections by persons represented in its medical files to public use of their names and medical data. By agreement with the USDOE, the author has cited public sources for all medical information relating to specific individuals and has cited medical files held by the department only when the medical files were used to corroborate public sources or when using nonmedical information. In the small number of cases where no other alternative existed, citations refer readers to earlier publications by the author. Original sources may, of course, be located in those publications.

INTRODUCTION

1. In the late nineteenth and early twentieth centuries over three-quarters of teenage girls worked for wages, and they made up the bulk of women in the workforce. For women employed in sex-typed jobs where they worked only with other women, the factories were, one historian suggests, "a preponderantly adolescent world." Statistics for 1908 on factories in five cities reveal that approximately half the women workers were between sixteen and twenty-four years of age; about three-quarters of them were under twenty-five.

The increased number of women working during World War I proved to be a temporary acceleration in the long-term trend of increased paid employment for women. As the percentage of women going into domestic service declined (from 65.6 percent of nonagricultural workers in 1870 to 18.7 percent in 1920), the greatest increases in paid employment for women were in professional jobs such as teaching and nursing, in white collar jobs such as bookkeeping and typing, and in service jobs such as clerking in department stores, but these were generally reserved for daughters of native-born, wealthier families, and manufacturing jobs presented the best opportunities for women who otherwise faced domestic duties. (The dialpainters seem to have come from fairly well-off working-class families, and some followed their dialpainting employment with clerical jobs in offices or positions in department store sales.) The percentage of women working in manufacturing declined between 1870 and 1930, from 24.9 to 19.2 percent of nonagricultural workers, although in absolute numbers their employment in manufacturing rose by a factor of five.

Increasingly, manufacturing work meant factory work: between 1910 and 1920, while the number of women in manufacturing remained stable, the number working in factories increased by one-third—from 52 percent to 66 percent. See Katzman, *Seven Days a Week*, 48, 51–54; Tentler, *Wage-Earning Women*, 59; and Kessler-Harris, *Out to Work*, 122–27.

On invention of the luminous wristwatch, see "A Short History of Elgin," vol. 8, Radium Archives, ANL.

2. See, e.g., Sharpe, "The New Jersey Radium Dial Painters." The most significant work to date on the dialpainters is Nugent, "The Power to Define a New Disease," and Nugent Young, "Interpreting the Dangerous Trades." See also Cloutier, "Florence Kelley and the Radium Dial Painters," and Baron, "Watches, Workers, and the Awakening World." Other secondary works by Sharpe include "Radium Osteitis with Osteogenic Sarcoma."

3. Richard Gillespie shares my understanding of the politics of the discovery of industrial diseases, employing the phrase "accounting for illness" to label the complex political negotiations over the intertwined issues of recognition, compensation, and prevention. See his "Accounting for Lead Poisoning," esp. 305-6. Similar reasoning may be found in Rosner and Markowitz, *Deadly Dust*, 62–74, 89–91, 98–101.

4. See Montgomery, *Worker's Control in America*; Edwards, *Contested Terrain*, esp. chap. 4, and Gordon, Edwards, and Reich, *Segmented Work, Divided Workers*, esp. chap. 4. See also Rosner and Markowitz, *Deadly Dust*, 52–58.

5. See Weinstein, *The Corporate Ideal*; Wilson, *Herbert Hoover*; and Hawley, "Herbert Hoover."

6. Several articles in Rosner and Markowitz, *Dying for Work*, consider the issue of science and neutrality. See Nugent, "The Power to Define a New Disease"; Zwerling, "Salem Sarcoid"; and Graebner, "Hegemony through Science."

7. See Rosner and Markowitz, "Research or Advocacy?."

8. Considering the trade-off between jobs and health, see Kazis and Grossman, *Fear at Work*, and Hays, *Beauty, Health, and Permanence*, esp. 297–311, 313–15. An important dissertation on the social debates over industrialization's effects on health, one that has been generally overlooked (perhaps because of organizational weaknesses), is Donegan, "For the Good of Us All." On reactions to increased accidents among workers that focus on reform, including workers' compensation, see Lubove, *The Struggle for Social Security* and "Workmen's Compensation"; Weinstein, *The Corporate Ideal*, chap. 2, and "Big Business"; Brandes, *American Welfare Capitalism*, chap. 19; Asher, "Business and Workers' Welfare"; and Leiby, *Carroll Wright and Labor Reform*. For a discussion of law, see Friedman and Ladinsky, "Social Change and the Law of Industrial Accidents." For an emphasis on workers, see Dubofsky, *When Workers Organize*; Montgomery, *Workers' Control in America* and *The Fall of the House of Labor*; and Brody, *Steelworkers in America*, 156–66.

9. Industrial health history is not an altogether neglected field. The best of older work was done by George Rosen and Henry Sigerist—see, e.g., Rosen, *The History of Miners' Diseases* and "On the Historical Investigation of Occupational Disease," and Sigerist, "Historical Background of Industrial and Occupational Disease." A few general texts have appeared, valuable for details more than for interpretation—see, e.g., Selleck and Whittaker, *Occupational Health in America*; Teleky, *History of Factory and Mine Hygiene*; and Hazlett and Hummel, *Industrial Medicine in Western Pennsylvania*. Two useful autobio-

graphical texts are Hamilton, *Exploring the Dangerous Trades*, and Hardy, *Challenging Man-Made Disease*. Two important compilations of recent work are, for American history, Rosner and Markowitz, *Dying for Work*, and, for European history, Weindling, *The Social History of Occupational Health*. Other fruitful new work includes Rosner and Markowitz, *Deadly Dust*; Cherniack, *The Hawk's Nest Incident*; and Derickson, *Workers' Health, Workers' Democracy*.

10. The classic question on the fate of Progressivism was framed in Link, "What Happened to the Progressive Movement?."

11. No general work on the Consumers' League has appeared since Goldmark, *Impatient Crusader*. See also Athey, "The Consumers' League and Social Reform"; Wolfe, "Women, Consumerism, and the National Consumers' League"; and Sklar, *Florence Kelley and the Nation's Work*.

12. My understanding of women, politics, and the Progressive Era has been shaped by Lebsock; see her "Women and American Politics." See also Muncy, *Creating a Female Dominion in American Reform*, and Koven and Michel, "Gender and the Origins of the Welfare State."

13. See Nugent Young, "Interpreting the Dangerous Trades," and Sicherman, *Alice Hamilton*.

14. For similar conclusions about business, see, e.g., Zwerling, "Salem Sarcoid"; Graebner, "Hegemony through Science"; and Rosner and Markowitz, *Deadly Dust*.

15. These figures have been drawn from Stebbings et al., "Mortality from Cancers," 436–37, and Rowland and Lucas, "Radium-Dial Workers," 232, 234–36.

16. Stebbings et al., "Mortality from Cancers," 435, 446–58; Rowland and Lucas, "Radium-Dial Workers," 235–36.

17. Rowland et al., "Current Status of the Study of 226-Ra and 228-Ra," 160, citing "Radium Poisoning," USBOLS; Martland and Humphries, "Osteogenic Sarcoma"; Martland, "The Occurrence of Malignancy," 2440, 2443. The Connecticut figures are from my compilations—consult Chapter 5 for data.

18. Stehney et al., "Survival Times," 333ff.

19. Polednak, "Fertility of Women after Exposure"; Hunt, "A Brief History of Women Workers," 282.

20. Rowland and Lucas, "Radium-Dial Workers," 18 (percentage calculated from table 3, p. 238); Stebbings et al., "Mortality from Cancers," table 2, p. 442.

21. U.S. National Institute of Occupational Safety and Health, President's Report on Occupational Safety and Health, 1972. For criticism about the compilation of occupational health statistics, see the summary in Berman, *Death on the Job*, 38–53, and Gordon, Akman, and Brooks, *Industrial Safety Statistics*.

CHAPTER ONE

1. The damages suit is recorded in numerous newspaper accounts—see, e.g., clippings in Schaub medical file, ANL, and references in EKP. For Schaub's letters to Martland, see HMP. A good secondary account of Schaub's illnesses is in Sharpe, "Radium Osteitis

with Osteogenic Sarcoma." The records of the dialpainters' attorney, Raymond Berry, are in RBP. The autobiographical fragment is Schaub, "Radium."

2. Schaub's birth certificate, New Jersey Bureau of Vital Statistics, Trenton. The description of the dialpainters is largely impressionistic, based on my study of the medical files of forty-three Orange dialpainters located in ANL. The files were not picked at random but on the basis of names of women discovered elsewhere in my research. Most of the files contain birth and death certificates, which give some information on parents (places of origin and occupations). Knowledge of home ownership is generally from accounts of families that lost their homes because of the large medical bills of their daughters, wives, or mothers.

3. Again, these conclusions are based on an impressionistic reading of the dialpainters' medical files in ANL.

4. Schaub medical file, ANL; Schaub, "Radium," 138.

5. Schaub, "Radium," 138; Schaub medical file, ANL; *New York World*, May 13, 1928, 1N–2N, and clipping, *New York Sun*, April 17, 1928, reel 2, RBP (Schaub's physical description); memo of conversation with Mae Kluver, March 29, 1972, Schaub medical file, ANL ("nightlife" quotation). Many photographs of Schaub survive in newspaper clippings.

6. On von Sochocky, see Wall, "Radioactivity in Industry," and von Sochocky, "Can't You Find the Keyhole?," 24–27, 106–7; see also von Sochocky medical file, ANL. On the founding of the Radium Luminous Materials Corp., see memo by Dr. James Stebbings, May 13, 1985, filed with vol. 59, Radium Archives, ANL, and Lang, "A Most Valuable Accident," 53–54.

7. "A Short History of Elgin," vol. 8, Radium Archives, ANL.

8. "Historical Outline" of USRC, enclosed in C. B. Lee, USRC, to Harris, Commissioner Harris Correspondence Files, ANL.

9. "Necrosis Cases Declared Growing," *Newark Evening News*, June 29, 1925, pp. 1, 4; Swen Kjaer, Report, April 8, 1925, SKP.

10. On the stereotyping of working women, see Kessler-Harris, *Out to Work*, esp. 139. Wages depended on the number of watches painted. That number could vary from the 72 watches a day reported by Albina Larice (testimony in Chancery Court, April 25 or 26, 1928) to the 400 to 500 watches a day claimed by Edna Hussman (testimony in Chancery Court, April 12, 1928), reel 1, RBP. Quinta McDonald said that she painted 240 to 288 watches a day (testimony in Chancery Court, April 12, 1928); Grace Fryer, about 250 (testimony in Chancery Court, April 12, 1928); and Katherine Schaub, about 300 a day (testimony in Chancery Court, April 25, 1928)—all in reel 1, RBP. A reasonable average would seem to be 250 to 300 watches per dialpainter.

The price paid per watch varies in different accounts. The highest figure is presented by Sharpe ("The New Jersey Radium Dial Painters," 562), who claims that the women were paid $1.44 for twenty-four watches, or 6 cents a watch. A clipping from the *Newark News* of August 5, 1932 (quoted in a memo in Mae Canfield medical file, ANL) asserts that in May 1917 women were paid 87 cents for twenty-four watches, or 3.6 cents a watch. Mae Canfield reported that she was paid 1.5 to 2 cents a watch (Pre-trial examination of plaintiff, January 17, 1929, reel 2, RBP). For reasons made clear below, I believe the last figure.

The total earned also varies. According to the *New York Times* (June 20, 1925, pp. 1, 6), Grace Fryer earned between $35 and $50 a week. The *Newark News* (clipping cited above) is quoted as saying that one dialpainter earned $22.50 a week. The Consumers' League claimed that dialpainters averaged $25–$35 a week from October through December and $18–$20 a week during the rest of the year ("Data collected by the Consumers' League of New Jersey," in vol. 28, Radium Archives, ANL). In "Radium Poisoning," 1220, it was the opinion of the USBOLS that "the earnings of the workers were good, ranging ordinarily from $17 to $35 per week in one studio and said to have reached as high as $42." Actual payroll figures are available for the USRC (then the Radium Luminous Material Corp.) for the year 1920. Annual earnings varied from a high of $1,482 [$28.50 a week] for the forewoman Anna Rooney to a low of $243.48 [$4.68 a week] for Katherine Schaub. Assumably the forewoman was paid a bonus for her supervision, and women like Schaub were not working full-time (Schaub may have been helping to set up dialpainting studios elsewhere and was being paid by other companies). On the average, women earned almost $14 a week; those in the middle range, making $700 to $1,000 a year, earned between about $13.50 and $19.25 a week. "Payroll Report, January 9, 1921" (vol. 5) and "U.S. Radium Corporation Payroll List, January 1921" (vol. 7), Radium Archives, ANL. I assume that a *full-time* worker averaged about $20 a week.

Pay may have varied over time, too, rising during the war and falling with the recession that followed it. However, from the figures above, it would seem that an average worker painted 250 watches a day, at 1.5 cents a watch, yielding almost $21 a week, but also that because full-time employment was rare, dialpainters generally earned about $14 per week. I have compared these earnings to those cited in "Women in New Jersey Industry," which reports that in the last four months of 1922 median income for New Jersey women was around $15, with the highest paid earning about $23 a week. The median yearly pay was $811.

Statistics on luminous watch production are from "Radium Poisoning," USBOLS, 28. Statistics on Orange employees are from Industrial Medicine Inspector Leo Tobias to Commissioner of the Department of Public Health, New York City, June 4, 1928, Commissioner Harris Correspondence Files, ANL.

11. Photographs, ANL, and Schaub, "Radium," 138 (description of studio); Medical Files, ANL (information on friends and relatives); Florence Wall to Michael Bishop, November 6, 1977, vol. 59, Radium Archives, ANL, and quotation in Lang, "A Most Valuable Accident," 78 (descriptions of dialpainters); Schaub, "Radium," 138 (last quotation).

12. Schaub, "Radium," 138; Lang, "A Most Valuable Accident," 78.

13. Schaub, "Radium," 138; Lang, "A Most Valuable Accident," 83–85, 88. For accounts of glowing in the dark, see Edna Hussman, testimony in Chancery Court, January 12, 1928, reel 1, RBP; Schaub, "Radium," 138; Castle, Drinker, and Drinker, "Necrosis of the Jaw," 375.

14. Schaub, testimony in Chancery Court, April 25–26, 1928, reel 2, RBP.

15. Memo regarding interview with Warren Holm on December 4, 1923, January 23, 1924, vol. 1, Radium Archives, ANL (source of lippointing technique); unidentified newspaper clipping, Radium Clipping File, New Jersey Room, NPL (story about playbox sand).

16. Grace Fryer, testimony in Chancery Court, April 12, 1928, reel 2, RBP.

17. Interview with Florence Wall, April 24, 1986 (tape in the author's possession); Elizabeth Hughes, testimony in Chancery Court, April 25–26, 1928, reel 2, RBP.

18. Mae Canfield and Anita Burricelli medical files, ANL.

19. Satre, "After the Match Girls' Strike," 8–9; Lee, "The Eradication of Phossy Jaw," 2–4.

20. Satre, "After the Match Girls' Strike," 28–30; Lee, "The Eradication of Phossy Jaw"; Ward, "Phosphorous Necrosis," 314; Andrews, "Phosphorous Poisoning." This story is also briefly recounted in Nugent Young, "Interpreting the Dangerous Trades," 12–13; Hamilton, *Exploring the Dangerous Trades*, 116–18; and Sicherman, *Alice Hamilton*, 154–56. Passage of the prohibitive tax on white phosphorous matches was a rare early use of taxation to create a national police power and the only time this power was used to control an industrial disease.

21. On phosphorous poisoning among match workers in New Jersey, see "Report to the International Labor Office on Phosphorous Poisoning in the United States," 1909, reel 64, AALLP. See also Andrews, "Phosphorous Poisoning." For fireworks manufacturing in New Jersey, see Ward, "Phosphorus Necrosis," 316.

22. Bates, *Mercury Poisoning*, 42, 44–68, 69.

23. Ibid., 16–24.

24. These articles are cited in ibid., 69–78.

25. Ibid., 8, 32, 83.

26. "Fifty Years of Occupational Health"; Derickson, *Workers' Health, Workers' Democracy*, 52–55. The list of colloquial names for industrial diseases is from "Memorial on Occupational Diseases," 128, AALLP.

27. Milles, "From Workers' Diseases to Occupational Diseases," 55, 61, 63 (on the narrowing of industrial health as a field); Rosner and Markowitz, *Dying for Work*, xii (on the earlier broad conception of workers' diseases) and xii–xiii (on the AFL's recognition of tuberculosis as a workers' disease); "Memorial on Occupational Diseases," 137–38 (1908 census data). The 1934 death rate statistics are from J. Whitney, *Death Rates by Occupation* (1934), cited in Teleky, *History of Factory and Mine Hygiene*, chap. 11. For diseases due to dust inhalation, see generally Hoffman, *Mortality from Consumption*.

28. Bremner, "Brief," 11. The denial of life insurance to asbestos workers is documented in Hoffman, "Mortality from Respiratory Diseases," USBOLS, Bulletin 231 (1918), cited in Kotelchuck, "Asbestos," 195.

29. On silicosis, see Rosner and Markowitz, *Deadly Dust*, 9, and chap. 2. On new chemical hazards beginning with World War I, see Hamilton, *Exploring the Dangerous Trades*, 183–84, 294–95, and "The Growing Menace of Benzene," 627; Haynes, *The American Chemical Industry*, 3:353–68; Chandler, *Strategy and Structure*, 78–90; Nugent Young, "Interpreting the Dangerous Trades," 69–71.

30. Dr. Henry Kessler, "Can Occupational Diseases Be Controlled?," Address at the Ninth Annual Eastern Safety Conference, ca. 1935, pp. 1, 3, reel 64, AALLP.

31. Hamilton, *Exploring the Dangerous Trades*, 294–96; Hunt, "A Brief History of Women Workers," 277–80; Dr. Henry Kessler, "Can Occupational Diseases Be Controlled?," 1–3; Hamilton, "The Lessening Menace of Benzol" and "The Growing Menace of Benzene"; "State Reporting of Occupational Diseases," 8; Hamilton, "Women Workers and Industrial Poisons."

32. Rosner and Markowitz, " 'A Gift of God,' " 344–46; Nugent Young, "Interpreting the Dangerous Trades," 158–60.

33. Hamilton, *Exploring the Dangerous Trades*, 294–99, and "The Growing Menace of

Benzene," 627; Consumers' League of Massachusetts, *Some Types of Industrial Poisoning* (1929), cited in "State Reporting of Occupational Diseases," 17–18.

34. Hamilton, *Industrial Poisons*, 8–9, "Lead Poisoning in American Industry," 8–21, "Women in the Lead Industries," and "TNT as an Industrial Poison," 108. W. Gilman Thompson (*The Occupational Diseases*, 60) also believed that women were more susceptible to some poisons. Modern treatments of the subject of women and industrial diseases often consider the possibility of women's greater susceptibility. See Hunt, "A Brief History of Women Workers"; Hunt, Lucas-Wallace, and Marsch, *Work and the Health of Women*; Chavkin, *Double Exposure*; Stellman, *Women's Work, Women's Health*; and Hricko and Brunt, *Working for Your Life*.

35. That women may have been more exposed than men to chronic levels of benzene is suggested in Hunt, "A Brief History of Women Workers," 278.

36. "State Reporting of Occupational Diseases," 3, 12.

37. Graebner, *Coal Mining Safety*, 3–4; Levenstein, Plantamura, and Mass, "Labor and Byssinosis," 213; Nugent, "Organizing Trade Unions to Combat Disease," 431; Derickson, *Workers' Health, Workers' Democracy*, 47–48 (quotation on silicosis) (Derickson is striving to make exactly the opposite point, that workers did organize around health issues); James Maurer, "The Increase of Accidents and Occupational Diseases," Address at the First National Labor Health Conference, June 18–19, 1927, box 20, AFPSR (last quotation).

38. Weindling, "Linking Self-Help and Medical Science," 9–10; Milles, "From Workers' Diseases to Occupational Diseases," 57–58; Levenstein, Plantamura, and Mass, "Labor and Byssinosis," 209; Donegan, "For the Good of Us All," 327–28, 355–57, 368, 384, 389.

39. Montgomery, *Workers' Control in America* and *The Fall of the House of Labor*, esp. chap. 3; Donegan, "For the Good of Us All," 197–200, 334; Markowitz and Rosner, "Death and Disease." For an analysis of the antebellum working-class belief that labor was property, see Wilentz, *Chants Democratic*, esp. 17, 242–43.

40. On mutual assistance groups, see Derickson, " 'To Be His Own Benefactor,' " 14, and *Workers' Health, Workers' Democracy*, xi–xiii, 17–22, 27; Asher, "The Limits of Big Business Paternalism," 21; and Markowitz and Rosner, "Death and Disease," 115–16. On limiting the length of the workday, see Bale, "America's First Compensation Crisis," 47; Corn, "Protective Legislation for Coal Miners," 76–78; Levenstein, Plantamura, and Mass, "Labor and Byssinosis," 209; and Donegan, "For the Good of Us All," 178–79, 183, 234–35, 244–46, 314–17. On protective legislation as a health and safety strategy, see Levenstein, Plantamura, and Mass, "Labor and Byssinosis," 209, and Donegan, "For the Good of Us All," 191–93, chap. 9. On union actions to win health and safety improvements, see Asher, "Industrial Safety and Labor Relations," 127–28. A useful, new look at corporate welfare programs that hints at their health and safety components is Cohen, *Making a New Deal*, chap. 3, 158–212.

41. For examples of employees before the 1930s seeking legislation to control working conditions, see Donegan, "For the Good of Us All," 183, 201, 315–16, 324; Derickson, *Workers' Health, Workers' Democracy*, xi, 59, 156–58; Asher, "Failure and Fulfillment," 199–201; and Graebner, *Coal Mining Safety in the Progressive Period*, 1–4, 139. See also Corn, "Protective Legislation for Coal Miners," 67; Buder, *Pullman*, 141–42, and Montgomery, *The Fall of the House of Labor*, 6, 165–69, 178, 194–99, 291, 357. On labor debates about this government strategy, see Grob, *Workers and Utopia*, 183–85; Lubove, *The*

Struggle for Social Security, 14–17, 85–86, 221; Yellowitz, *Labor and the Progressive Movement in New York,* 93, 102, 137–38, 183; Green, *The National Civic Foundation,* 284–86; Greenbaum, "Social Ideas of Samuel Gompers"; Weinstein, *The Corporate Ideal,* 120–21; Montgomery, *The Fall of the House of Labor,* 6, 165–69, 178, 194–99, 263–81, 291, 357; Bernstein, *The Lean Years,* 225–26; Orloff, "The Political Origins of America's Belated Welfare State," 55–57; Weinstein, "Big Business and the Origins of Workmen's Compensation," 157–59; Lubove, "Workmen's Compensation and the Prerogatives of Voluntarism," 259–61; and Foster, "The *Western Dilemma,*" 278.

42. Galishoff, *Safeguarding the Public Health,* xiv, 35, 39, 135, 141, 162–63.

43. On Bureaus of Labor Statistics and Departments of Labor in general, see Fones-Wolf, "Class, Professionalism"; Clague, *The Bureau of Labor Statistics;* and Grossman, *The Department of Labor.* For New Jersey developments, see Ethel Hanks Van Buskirk and Alfred W. Biggs, "Factory Inspection in New Jersey," typescript report, December 1930, reel 65, AALLP; "Development of Labor Laws in New Jersey and Events Leading Up to the Reorganization of the Department in 1904," typescript report, ca. 1914, and Lillian Erskine, "Code Formulation and Investigation in Department of Labor — New Jersey," typescript report, April 19, 1928, both on reel 65, AALLP; New Jersey State Library, "Legislative History of Workers' Compensation Act," April 12, 1967; and see New Jersey Bureau of Statistics of Labor and Industry, *Annual Reports,* esp. for 1881–82, 1887, 1889–95, 1901–2, 1905–6. Secondary works include Berkowitz, *Workmen's Compensation;* Ritchie, "An Appraisal of Workmen's Compensation Legislation"; Mahoney, "New Jersey Politics after Wilson"; Burch, *Industrial Safety Legislation in New Jersey;* and Newman, *The Labor Legislation of New Jersey.*

44. Fones-Wolf, "Class, Professionalism," 38–43 (quotation, 40).

45. Galishoff, *Safeguarding the Public Health,* 137–38; Van Buskirk and Biggs, "Factory Inspection in New Jersey"; "Development of Labor Laws in New Jersey and Events Leading Up to the Reorganization of the Department in 1904."

46. Mahoney, "New Jersey Politics after Wilson," chap. 7; New Jersey Bureau of Statistics of Labor and Industry, *Annual Reports,* in which health and safety reports cease after 1906; Van Buskirk and Biggs, "Factory Inspection in New Jersey." On the end of Progressivism in New Jersey, see Mahoney, "New Jersey Politics after Wilson," Hirst, "James Fairman Fielder," 185; Ritchie, "An Appraisal of Workmen's Compensation Legislation."

47. Mahoney, "New Jersey Politics after Wilson," chap. 7; Galishoff, *Safeguarding the Public Health,* 138–40.

48. On New Jersey's liability law, see Ritchie, "An Appraisal of Workmen's Compensation Legislation," 1. On employers' liability laws and workers' compensation as national movements (and workers' opinions of them), see Weinstein, "Big Business and the Origins of Workmen's Compensation," 157–59; Lubove, "Workmen's Compensation and the Prerogatives of Voluntarism," 259–61; Foster, "The *Western Dilemma,*" 278; Lubove, *The Struggle for Social Security,* 85–86; 221; Yellowitz, *Labor and the Progressive Movement in New York,* 137–38; Weinstein, *The Corporate Ideal,* 120–21; Montgomery, *The Fall of the House of Labor,* 263–81; Bernstein, *The Lean Years,* 225–26; and Orloff, "The Political Origins of America's Belated Welfare State," 55–57.

49. Ritchie, "An Appraisal of Workmen's Compensation Legislation," 11, 13–14; Van Buskirk and Biggs, "Factory Inspection in New Jersey."

50. Galishoff, *Safeguarding the Public Health*, 139–40; Ritchie, "An Appraisal of Work-men's Compensation Legislation," 27; Hamilton, "Lead Poisoning in Potteries, Tile Works, and Porcelain Enameled Sanitary Ware Factories," cited in "Lead Poisoning in New Jersey," March 31, 1913, reel 70, AALLP.

51. Erskine, "Code Formulation and Investigation in Department of Labor—New Jersey."

52. Mahoney, "George Sebastion Silzer," 194–96.

53. New Jersey State Library, "Legislative History of the Workers' Compensation Act," April 12, 1962. Whorton, *Before Silent Spring*, provides a good discussion of the development of our understanding of chronic poisoning.

54. "Andrew Francis McBride, M.D.," in Nelson and Shriner, *History of Paterson and Its Environs*, 464; W. S. Myers, *The Story of New Jersey*. See also Golin, "Bimson's Mistake"; Dodyk, "Winders, Warpers, and the Girls on the Loom," 79–80. On the New Jersey appointment, see Van Buskirk and Biggs, "Factory Inspection in New Jersey."

55. "Andrew Francis McBride, M.D.," in Myers, *The Story of New Jersey*, 436; Mahoney, "George Sebastian Silzer," 194.

56. "Radium Poisoning," 39; C. B. Lee, "Historical Outline," and President of USRC to Commissioner of Health, New York City, June 18, 1928, both in Commissioner Harris Correspondence Files, ANL. Documentation that Orange dialpainters taught dialpainters elsewhere to lippoint is from transcript of interview with Agnes Kelley, May 19, 1934, and letter (probably from Frederick Flinn) to USRC, both in vol. 4, Radium Archives, ANL.

57. Leo Tobias to Commissioner, June 4, 1928, Commissioner Harris Correspondence Files, ANL (employment statistics); Schaub and Rudolph medical files, ANL; Shaub, "Radium," 138, and "Transcript of Pleadings for Trial," reel 1, RBP (location of Luminite Corp.).

58. Rudolph medical file, ANL; "Radium Poisoning," Case 1, 40–41; Walter Barry, testimony in Chancery Court, January 4, 1928, reel 1, RBP; Swen Kjaer, "Notes on Cases," SKP; Katherine Schaub, testimony in Chancery Court, April 25–26, 1928, reel 2, RBP.

59. "Radium Poisoning," 1222; Martland, "The Occurrence of Malignancy," 2443; Clark, "Glowing in the Dark," 49.

60. Amelia Maggia medical file, ANL; "Radium Poisoning," Case 2, 41–42; St. George, Gettler, and Mettler, "Radioactive Substances in a Body Five Years after Death"; Martland, "The Occurrence of Malignancy," 2443.

61. Erskine's testimony, reel 2, RBP; "Notes on Cases," vol. 13, Radium Archives, ANL; "Radium Poisoning," 1220–21; John Roach to Leonore Young, August 30, 1923, reel 3, RBP. Other details about Lillian Erskine are from letterhead, New Jersey State Industrial Safety Museum, Jersey City, "under jurisdiction" of the New Jersey Department of Labor, reel 29, AALLP; see also Sicherman, *Alice Hamilton*, 177, n. a, which cites Erskine as a special investigator of industrial diseases for the New Jersey Department of Labor and the AALL, with a "long career as an industrial hygienist."

62. "Radium Poisoning," 40; John Roach, Deputy Commissioner of Labor, to Leonore Young, Orange Department of Public Health, August 30, 1923, reel 3, RBP; M. Szamatolski to New Jersey Department of Labor, January 30, 1923, reproduced in "Radium Poisoning," 40.

63. Schaub, "Radium," 138; Memo, July 19, 1923, Leonore Young File, reel 3, RBP; Leonore Young, testimony in Chancery Court, January 12, 1928, reel 1, RBP.

64. Kuser medical file and clipping therein from *Newark Ledger*, May 31, 1928, ANL; *New York World*, May 31, 1928, 1; "Radium Poisoning," 1222.

65. Quinlan medical file, ANL; "Radium Poisoning," Case 3, 42; "Notes on Cases," vol. 13, Radium Archives, ANL.

66. Leonore Young, testimony in Chancery Court, January 12, 1928, reel 1, RBP; Kuser medical file, ANL; Leonore Young to John Roach, February 2, 8, 1924, Young File, reel 3, RBP; Walter Barry, testimony in Chancery Court, January 4, 1928, reel 1, RBP; Carlough medical file, ANL; "Radium Poisoning," Case 9, 45.

67. Schaub, "Radium," 138; Barry, testimony in Chancery court, January 4, 1928, reel 1, RBP; Genevieve Smith Kernan and Josephine Smith Farrel medical files, ANL; "Memorandum re Fryer et al. against U.S.R.C.," reel 1, USRCR.

68. Barry, testimony in Chancery Court, January 4, 1928, reel 1, RBP.

69. Leonore Young, testimony in Chancery Court, January 12, 1928, reel 1, RBP; Williams, "Preliminary Note on Observations."

70. Interview with Young, Young file, reel 3, RBP.

71. Vice President of USRC (Barker) to American Mutual Liability Insurance Co., January 19, 1924, reel 7, USRCR; "Radium Poisoning," 43; "Memorandum on Fryer et al. against U.S.R.C.," reel 1, USRCR.

72. Young, testimony in Chancery Court, January 12, 1928, reel 1, RBP.

CHAPTER TWO

1. On the relation between the "theoretical nihilism" of contemporary medicine and the propensity of the public to use patent medicines, see James Harvey Young, *The Medical Messiahs*, esp. 25–26. Young touches on some of the topics of this chapter: radium quackery and the promotion of patent medicines in medical journals and through personal educational visits with physicians—see 26–27, 101, 121, 165, 169. For a comprehensive history of medical quackery in America, see *The Medical Messiahs* in its entirety and its companion volume on the nineteenth century, *Toadstool Millionaires*.

2. Grigg, *The Trail of Invisible Light*; Brecher and Brecher, *The Rays*; Caufield, *Multiple Exposures*, 3–21; Eisenberg, *Radiology*.

3. Caufield, *Multiple Exposures*, 9–10. The best book explaining radiation for nonscientists is Gofman, *Radiation and Human Health*. This volume is highly controversial among scientists, as it contains extreme estimates for damage to human tissues by radioactivity; nonetheless, it is not outside the bounds of current scientific discourse. On X rays and radiation generally, see pp. 10–23.

4. Caufield, *Multiple Exposures*, 22–23.

5. Ibid., 23–24; Gofman, *Radiation and Human Health*, 23–30.

6. Bowing and Fricke, "Curie Therapy," 277; Martin and Martin, "History of Radium Therapy," in their *Low Intensity Radium Therapy*, 3–4, 6; Mould, *History of X-rays and Radium*, 6.

7. For the quotation, see "The Therapeutic Possibilities of Radium," 525.

8. Saubermann, "An Address on the Progress of Radium Therapy," 63–78, cited in Hewitt, "Rationalizing Radiotherapy," 918; Soddy, "Radium as Therapeutic Agent," 660.

9. Pifford, "A Few Words Concerning Radium," 1000.

10. Tracy, "Radium," 49. On early uses of radioactive solutions, see also Rollins, "Some Principles Involved in the Therapeutic Applications of Radioactivity," 542–44; Baskerville, *Radium and Radio-active Substances*, 139; and Wallian, "Radioactivity in Diabetes and Nephritis," 179–80. For claims of germicidal and antifermentive activity, see, for one, Tracy, "Radium," 50; for diseases treatable by radium, see Tracy, "Radium," 51; Wallian, "Radioactivity in Diabetes and Nephritis," 180; and Parkhurst, "Radium Therapy," 715. For lilac bud experiments, see Hewitt, "Rationalizing Radiotherapy," 917; for tadpole experiments, see Colwell and Russ, "Radium as a Pharmaceutical Poison," 221–23.

11. Lowe, "On Radium Emanations in Mineral Waters," 1051–53; Colwell and Russ, "Radium as a Pharmaceutical Poison," 221. For the "Berlin school," see Zueblin, "The Present Status of Radioactive Therapy," 112. For the springs at Gastein and Mache and the hydrotherapeutic clinic at the University of Berlin, see Fuerstenberg, "Radium Emanation Treatment for Rheumatism and Gout," 12–14; also Martin and Martin, "History of Radium Therapy," in their *Low Intensity Radium Therapy*, 7. For the Berlin clinic (run by Prof. His, with an emanatorium) and the Viennese medical clinic (run by von Noorden), see von Noorden, "Radium and Thorium-X Therapy." For Joachimsthal miners and inhabitants, see Zueblin, "The Present Status of Radioactive Therapy," 118, and Colwell and Russ, "Radium as a Pharmaceutical Poison," 221.

12. On the industrialist Armet de Lisle and the "French school," see Muir, "Story of Radium in Therapeutics," 605, and Colwell and Russ, "Radium as a Pharmaceutical Poison," 221. For radium institutes, see Case, "The Early History of Radium Therapy," 580–82.

13. Wallian, "Radioactivity in Diabetes and Nephritis," 179 (1st quotation); Sir Malcolm Norris, in introduction to Wickham and Degrais, *Radium Therapy*, iii. For the German influence on American radium therapy, see abstracts that appeared in various journals, especially *Radium*, and the published discussions following the presentation of papers, e.g., in Hinsdale, "Mineral Springs," 92–94, and Allen, "The Present Status of Radium Therapy," 23.

14. Landa, "From Buried Treasure to Buried Waste," 20, 22, 25 n. 15; Viol, "There Is No Radium Shortage," 7.

15. Landa, "From Buried Treasure to Buried Waste," 22; Cameron, "Mechanical and Physical Agents" 202, 196, and "Radium Emanation Therapy," 4.

16. Cameron, "A Treatment of Alveolar Pyorrhea," 41–42, "Mechanical and Physical Agents" (quotation, 198), and "Radium Emanation Therapy," 3–5; Proescher, "The Intravenous Injection of Soluble Radium Salts in Man," 9–10 (1st three quotations), "Intravenous Injection of Soluble Radium Salts," 45–53, 61–64, 77–87 (last quotation, 45), and "Influence of Intravenous Injections of Soluble Radium Salts in High Blood Pressure," 1–10, 17–21; Proescher and Almquest, "Contribution on the Biological and Pathological Action of Soluble Radium Salts," 65–71, 85–95. For Proescher's estimates on the number of patients, see his medical file, ANL. On Proescher, see Cameron, "Mechanical and Physical Agents," 202.

17. "Notes and Comments," *Radium* 2; Landa, "From Buried Treasure to Buried Waste," 22; "American Radium Society," 58–59.

18. Persson, "Treatment of Chronic Diseases at Spas," 1038 (statistic on spa users), 1038–39 (quotation of Mount Clemens physician); Lowe, "On Radium Emanations in Mineral Waters," 1053 (last quotation). For a brief history of spas, see Sigerist, "American Spas in Historical Perspective," esp. 71–73.

19. Engelmann, "Radium Emanation Therapy," 1225, 1227.

20. Moriarata, "Saratoga Springs," 16; "Saratoga Springs Reservation" and "Historical Sketch—Chronology of the Baths," mimeographed material supplied by Saratoga State Park.

21. "Saratoga Springs Reservation" and "Historical Sketch—Chronology of the Baths"; Thompson, "The Springs of Saratoga," 589–92 (quotations, 590, 592).

22. Moore and Whittemore, "The Radioactivity of the Waters of Saratoga Springs," 552; Moriarata, "Radium," 59–60, "Radium and Symptomatic Blood Pressure," 131–32, and "Pros and Cons of Radium," 86.

23. Memo, J. Lieben, June 22, 1972, in Irene LaPorte medical file, ANL (information on Moriarata); New York Conservation Commission, Division of Saratoga Springs, "Saratoga Mineral Springs and Baths," 15–16.

24. Saratoga Springs Commission, New York, Report . . . to the Legislature, 52; Albert Fall, Secretary, Department of the Interior, and Stephen Mather, Director of the National Park Service, "Hot Springs National Park, Arkansas," n.d. (ca. 1921), 8–9 (1st quotation), in FTCR; Reed, "Hot Springs National Park, Arkansas," U.S. Railroad Administration, n.d., 10–11, in FTCR.

25. Field, "Brief Historical Sketch," Robley Evans Papers, ANL, 36–39, 41–46, 58–60 (also available in Robley Evans Papers at MIT); Field, "The Efficiency of Radioactive Waters," 393, 392, and "Radium," 23–24; Bissel, "Some Radium Therapeutics," 58, "Intravenous Injection of Radium Element," 20, "Radium Therapeutics Otherwise Than for Malignancy," 107, and "The Medicolegal Aspects of Radium Therapy," 100.

26. [Bissel], "Joseph Biddleman Bissel," 63–64; Field, "Brief Historical Sketch," 61–64, 68–69.

27. Von Noorden, "Radium and Thorium-X Therapy," 95, 99–103.

28. Field, "Radium: Its Physio-Chemical Properties," as cited in American Institute of Medicine, *Radium: Abstracts*, 105–8; Saubermann, "Radium Emanation and Physiological Processes" ("disturbed metabolism," 298, and nerve endings theory, 299); Morton, "Radiochemicotherapy," 385 (aging). On radium as a catalyst, stimulating oxidation, see Tracy, "Recent Applications of Radium Emanations and Radium Water," 612–16.

29. Rowntree and Baetjer, "Radium in Internal Medicine," 1438–42.

30. Engelmann, "Radium Emanation Therapy," 1225, 1228.

31. Saubermann, "An Address on the Progress of Radium Therapy."

32. Zueblin, "The Present Status of Radioactive Therapy," 109, 115–16, 147.

33. Quimby, "The Background of Radium Therapy," 443; Hinsdale, "Mineral Springs," 92.

34. Starr, *The Social Transformation of American Medicine*, 131; "Notes and Comments," *Radium* 4.

35. Zueblin, "The Present Status of Radioactive Therapy," 109, 113–15.

36. Serwer, "The Rise of Radium Protection," chap. 2–3. The "Becquerel burn" (beginning of paragraph) refers to the burn Henri Becquerel received by carrying radium in his vest pocket.

37. Ibid., chap. 3 and 5; Gudzent and Halberstaedter, "Occupational Injuries Due to Radioactive Substances," 27–28.

38. Lazarus-Barlow, "On the Disappearance of Insoluble Radium Salts." The French work is by H. Dominici and is cited in Castle, Drinker, and Drinker, "Necrosis of the Jaw," 374, 381.

39. Proescher, "Intravenous Injection of Soluble Radium Salts," 45; Cameron and Viol, "Classification of the Various Methods Employed," 66; Seil, Viol, and Gordon, "The Elimination of Soluble Radium Salts."

40. Cameron and Viol, "Classification of the Various Methods Employed," 66.

41. Stedman, *Reference Handbook of the Medical Sciences*, 466, as cited in American Institute of Medicine, *Radium: Abstracts*, 26; Greenough, "The Value of Radium in the Treatment of Disease," 71; Simpson, *Radium Therapy*, 321; Pinch, "Report of the Work . . . at the Radium Institute . . . 1921," as reprinted in *Radium*, 125, 131–32.

42. Ordway, "Injuries Due to Radium," 254, and "Occupational Injuries Due to Radium," 122; Pinch, "Report of the Work . . . at the Radium Institute, London . . . 1919," 80–81, 85, 101.

43. Mottram and Clarke, "Leukocytic Blood Content"; Mottram, "The Red Blood Cell Count of Those Handling Radium," "Histological Changes in the Bone Marrow of Rats," and "The Effects of Increased Protection from Radiation."

44. Lazarus-Barlow, "Some Biological Effects of Small Quantities of Radium," 11.

45. Viol, "History and Development of Radium-Therapy," 33; Delano, "A Study in the Internal Therapeutics of Radium," 8; Fischel, "Radium and Internal Medicine."

46. Landa, "From Buried Treasure to Buried Waste," 22–23, 25 n. 15.

47. Ibid., 25 n. 15, 26 (quotation).

48. American Institute of Medicine, *Bibliography on Radium* and *Radium: Abstracts*.

49. Pfahler, "Effects of X-rays and Radium," 160; Price-Jones, "Observations on the Effects Produced," 146; Lazarus-Barlow, "On the Histological and Some Other Changes," 126–29; Williams, "Preliminary Note on Observations"; Ludwig and Lorensen, "Studies on the Content of Radium Emanation," 597.

50. Cameron and Viol, "Classification and Relative Value," 139, 144; "Effects of Radium Therapy upon Physiology and Metabolism"; "Radium Therapy in High Blood Pressure and Circulatory Diseases"; "Radium Therapy in Gout, Arthritis, and Rheumatism"; "An Outline of Radium and Its Emanations"; "The Spark of Life," 3–4.

51. Field, "Brief Historical Sketch," 61–64, 68–69; New Jersey Department of Public Health file on C. Everett Field, Radium Archives, ANL, including clippings from *New York American*, December 2, 1922 and *New York World*, December 10, 1922.

CHAPTER THREE

1. Production Manager Report, 1st quarter, April 17, 1924, reel 5, USRCR.

2. "Notes on Radium Cases," box 83, CLNJP; Leonore Young file, reel 3, RBP; Katherine Wiley, June 6, 1924, box 1, CLNJP.

3. The characterization of Wiley is by Cornelia Bradford, vice president of the Consumers' League of New Jersey (and head worker of a nearby settlement house), in "Historical Souvenir of the Consumers' League," 1926, box 5, CLNJP. Wiley's actions are detailed in "Notes on Cases," box 83, CLNJP. These notes exist with variations in reel 30, AALLP; in reel 3, RBP; and in reel 85, NCLR. The quotation is from Katherine Wiley, June 6, 1924, box 1, CLNJP.

4. Rosen, *Preventive Medicine*, 7–13 (quotation, 9); Rosner and Markowitz, *Dying for Work*, xii. For his early work on industrial health and suggestions on how further work might be shaped, see Rosen, "On the Historical Investigation of Occupational Disease," "The Medical Aspects of Controversy," A *History of Public Health*, 141–44, 272–75, 419–39, "Early Studies of Occupational Health," and "From Frontier Surgeon to Industrial Hygienist."

5. Sicherman, "Alice Hamilton" and *Alice Hamilton*; Nugent Young, "Interpreting the Dangerous Trades"; Hamilton, *Exploring the Dangerous Trades*; Taylor, *The Medical Profession and Social Reform*, 1–9. In Taylor's eyes, Hamilton brought the campaign for social medicine to Hull-House.

6. Sicherman, *Alice Hamilton*, 153–54, and "Alice Hamilton," 303–4; Hamilton, *Exploring the Dangerous Trades*, 114–18. The muckraking article was by William Hard. The British treatise was Oliver, *Dangerous Trades*. Hamilton's article was "Industrial Diseases."

7. Ramazzini, *De Morbis Artificium Diatriba* or *The Diseases of Workers*, discussed, e.g., in Kober, "Prefatory," xxii. On the origins of industrial health concerns, see Donegan, "For the Good of Us All," 17–21.

8. Examples of the nineteenth-century literature include J. T. Wilson, "Diseases Incident to Some Occupations" (1879–80), and George Rohe, "Hygiene of Occupations" (1884), and Dr. J. H. Ireland, "The Preventable Causes of Disease, Injury and Death in American Manufacture and Workshops and the Best Means for Preventing and Avoiding Them," Lomb Prize Essay (1886), all cited in Kober, "Prefatory," l. For other publications, see Kober, "Prefatory," l, and Nugent Young, "Interpreting the Dangerous Trades," 7–8. Specifically, see McReady, "On the Influence of Trades, Professions and Occupations," and James Henrie Lloyd, "The Diseases of Occupation," 354, cited in Nugent Young, "Interpreting the Dangerous Trades," 7–8, 17 n. 23.

9. This paragraph is drawn mostly from Donegan, "For the Good of Us All"; also from ideas in Marx, *The Machine in the Garden*; Rothman, *The Discovery of the Asylum*; and Dublin, *Women at Work*.

10. Craig Donegan ("For the Good of Us All," 362) writes that America was "not nearly so tardy in pursuing industrial reform as is often charged." Asger Braendgaard ("Occupational Health and Safety Legislation," 179) finds Pennsylvania to be "not substantially behind England" in regulating the problems of industrial development from 1870 to 1970; indeed, he claims that it "was close to getting a lead over England." For details, see Teleky, *History of Factory and Mine Hygiene*, chap. 2, and Kober, "Prefatory," xxx–xxxix.

11. Starr, *The Social Transformation of American Medicine*, 18–29, chap. 3.

12. Rosen, *Preventive Medicine*, 10, 16. On the history of public health, see esp. Rosen, *A History of Public Health*; Duffy, *A History of Public Health* and *The Sanitarians*; Rosenkrantz, *Public Health and the State*; and Leavitt, *The Healthiest City*.

13. Taylor, *The Medical Profession and Social Reform*; Sellers, "The Public Health Service's Office of Industrial Hygiene." Sellers's work has clarified this difference for me and helps to explain the relation among Progressive reformer-physicians, industrial medicine practitioners, and academic industrial health researchers.

14. Sicherman, *Alice Hamilton*, 136. On the concentration of capital and monopoly capitalism, see Braverman, *Labor and Monopoly Capital*, and Baran and Sweezy, *Monopoly Capital*; see also Chandler, *The Visible Hand*, and Porter, *Rise of Big Business*. The welfare capitalism thesis is best argued in Brody, *Steelworkers in America* and "Rise and Decline of Welfare Capitalism," and in Brandes, *American Welfare Capitalism*. On industrial medicine, see Nugent, "Fit for Work."

15. Frederick Hoffman deserves further study. He appears in Cattell, *American Men of Science*, 827, and, briefly, as a Prudential vice president and uncompromising foe of nationalized health insurance in Starr, *The Social Transformation of American Science*, 252, Lubove, *The Struggle for Social Security*, 83, 88, and Bale, "America's First Compensation Crisis," 37. See also Hoffman, "Mortality from Consumption in the Dusty Trades," November 1908 and May 1909.

16. Sigerist, "Historical Background of Industrial and Occupational Disease," 606.

17. Rosen, "From Frontier Surgeon to Industrial Hygienist," 641–42; Doehring, "Factory Sanitation and Labor Protection."

18. Kober, *Report of the Committee on Social Betterment*, 3; Rosen, "From Frontier Surgeon to Industrial Hygienist," 642; and Kober, *Industrial and Personal Hygiene*, 10 (Kober's last quotation), 16–17, 22–95, 126–69.

19. AALL Papers; also Domhoff, *The Higher Circle*, 170–73; Berman, *Death on the Job*; Starr, *The Social Transformation of American Medicine*, 243–49; Lubove, *The Struggle for Social Security*, 32–34; and Rosner and Markowitz, "The Early Movement for Occupational Safety and Health," 514. See also Pierce, "Activities of the American Association for Labor Legislation."

20. Sicherman, *Alice Hamilton*, 154–59, 161–62.

21. Good sources for details on the industrial hygiene movement are Kober, "Prefatory," esp. xli, xliii–xlix; Rosen, *A History of Public Health*, esp. 435; Lescohier, "The Campaigns for Health and Safety in Industry," esp. 363 on the AALL conferences; and Hazlett and Hummel, *Industrial Medicine in Western Pennsylvania*, esp. 48, 90–91. On the National Civic Federation and mercury poisoning, see Bates, *Mercury Poisoning*.

22. On medical-educational developments, see Kober, "Prefatory," xlvi–lvii; Curran, *Founders of the Harvard School of Public Health*, esp. 156–57; and Aub and Hapgood, *Pioneer in Modern Medicine*, esp. 249 on the *Journal of Industrial Hygiene*. Selleck and Whittaker's *Occupational Health in America* is especially useful on professional divisions among physicians interested in industrial health—see chap. 6.

23. Rosner and Markowitz, "Research or Advocacy?," 84–86.

24. On the Illinois occupational disease law, see chap. 8. On the compensation of industrial diseases, see Beers, "Compensation for Occupational Diseases"; Canan, "Industrial

Law," 594; Chamberlain, "Workmen's Compensation for Diseases"; Dodd, *Administration of Workmen's Compensation*, 757–72; and Reede, *Adequacy of Workmen's Compensation*, 45–53. On Hamilton and the AALL, see Sicherman, *Alice Hamilton*, 175.

25. Alice Hamilton (*Exploring the Dangerous Trades*, 183–99) discussed the desperate conditions of the munitions plants she studied in an amusing way; see also Sicherman, *Alice Hamilton*, 200–201, and Rosner and Markowitz, "Research or Advocacy?," 85.

26. Sicherman, *Alice Hamilton*, 1, 4–5.

27. Nugent, "Organizing Trade Unions"; Rosner and Markowitz, "Safety and Health as a Class Issue."

28. No good monograph on the Consumers' League yet exists, but information may be found in Nathan, *The Story of an Epoch-Making Movement*; Sara Evans, *Born for Liberty*, 148–52; Lemons, *The Woman Citizen*, esp. chap. 5; O'Neill, *Everyone Was Brave*, esp. 95–98, 105, 152, 218–19, 233, 237–38; Goldmark, *Impatient Crusader*; Athey, "The Consumers' League and Social Reform"; and Wolfe, "Women, Consumerism, and the National Consumers' League," 378–92. On Florence Kelley, see also Wade, "Florence Kelley," 1:317, and Sklar, *Florence Kelley and the Nation's Work*. On the founding of the Consumers' League of New Jersey, see Bradford, "Historical Souvenir of the Consumers' League," box 5, CLNJP.

29. Felice Gordon, *After Winning*, 59–60; Lemons, *The Woman Citizen*, 142–44.

30. Hill, "Protection of Women Workers and the Courts," 256; Lemons, *The Woman Citizen*, 181–208.

31. Felice Gordon, *After Winning*, 41–42, 59–60, 65, 216 n. 25; "Women in Industry, 1921–1923," Women in Industry Newsletter, December 28, 1922, and "Women in Industry, 1924–1927," box 2, LWVNJP. The woman linking New Jersey's League of Women Voters and Consumers' League was Juliet (Mrs. G. W. B.) Cushing.

32. Sicherman, *Alice Hamilton*, 175, 266; Hamilton to Robert Lovett, November 5, 12, 1925, and other papers, box 20, no. 12, AFPSR.

33. On Hamilton's talk before the National Conference of Charities and Correction and her influence in helping to pass Illinois's 1911 occupational disease law, see Sicherman, *Alice Hamilton*, 158, 161–62, and Nugent, "Workers' Health in Chicago." On Hamilton's effect on the Consumers' League, see Sicherman, *Alice Hamilton*, 241, 243, 312, 375, and Minutes, Board of Directors, December 15, 1922, January 24, 1923, NCLR; "Outline on Occupational Diseases," CLMP; Katherine Wiley, Report, December 7, 1923, box 1, CLNJP. See also Hamilton, "The Growing Menace of Benzene," 627–30.

34. Katherine Wiley, Reports, December 8, 1922, and February 20, 1923, box 1, CLNJP; "Lead Poisoning in New Jersey," press release, March 31, 1913, reel 70, AALLP.

35. Wiley to Hamilton, April 12, 1923, box 54, CLNJP.

36. "Women Who Win," feature article, Scrapbook (box 74), "Katherine G. T. Wiley—Biography and Writings" (box 53), and "Reports—K. G. T. Wiley, Executive Secretary, 1921–1924" (box 1), CLNJP. See also "Women in Industry, 1921–1923" and Women in Industry newsletter, December 28, 1922, box 2, LWVNJP.

37. Wiley to Hamilton, April 12, 1923, box 54, CLNJP.

38. Wiley to Roach, Bureau of Sanitation and Hygiene, April 23, 1923; Roach to Wiley, April 25, 1923; and Frederick Hoffman to Wiley, May 3, 1923, ibid.

39. Hamilton to Wiley, April 28, 1923, and Wiley to Roach, May 3, 1923, ibid.

40. Wiley, Report, May 11, 1923, box 1, CLNJP.

41. There are two exceptions to this general inattention: Roger Cloutier ("Florence Kelley and the Radium Dial Painters") noted Florence Kelley's work for the dialpainters, and Angela Nugent ("The Power to Define a New Disease," 183) observed that the Consumers' League involvement with the dialpainters "stemmed from a strong tradition of reform based on cooperation between women workers and middle-class women."

42. On Irene Chubb, see reel 29, AALLP; on Irene Osgood, see Sicherman, *Alice Hamilton*, 154–55; on Lillian Erskine, see Sicherman, *Alice Hamilton*, 177 n. a. See also Domhoff, *The Higher Circle*, 170, 178; Letterhead, n.d., CLMP.

43. Nugent, "Organizing Trade Unions," 427–28; Rosner and Markowitz, "Safety and Health as a Class Issue," 54.

44. Sklar, "The Historical Foundations of Women's Power," 74–75, 77–78, and "Two Political Cultures in the Progressive Era," 37, 45–46, 56.

45. Sklar, "The Historical Foundations of Women's Power," 44–45, 52–58, 61, 63–66, 72–74, and "Two Political Cultures in the Progressive Era," 37, 41, 59.

46. Muncy, *Creating a Female Dominion in American Reform*; Sklar, "The Historical Foundations of Women's Power," 44–45, 52–58, 61, 63–66, 72–74, and "Two Political Cultures in the Progressive Era," 37, 41, 59.

47. Wiley, "Notes on Cases," and Blum to Wiley, May 21, 1924, both in reel 30, AALLP; Schaub, "Radium," 139; Frederick Hoffman, affidavit, August 25, 1927, reel 1, RBP (charges regarding Blum).

48. Wiley, "Notes on Cases," reel 30, AALLP; Wiley to Josephine Goldmark, December 10, 1947, and Wiley to Mrs. [Susanna] Zwemer, January 27, 1958, reel 85, NCLR.

49. Wiley, "Notes on Cases," reel 30, AALLP; Wiley, "Notes on Cases," reel 3, RBP. There is some discrepancy in different records about when Martland first examined dialpainters. Wiley met with McBride on May 19, 1924, and Martland, according to Wiley's notes, had already told McBride that he could find no connection between the dialpainters' symptoms and their diseases. Thus Martland had presumably considered the matter before May 1924. Wiley called Martland in March 1925 and learned that his inquiry involved only the examination of a few women in Dr. Barry's office. But, according to the *New York Times* of June 20, 1925 (1:4, 6), Martland claimed that seven months earlier he had seen three women in the offices of Drs. Barry and Davidson—this would have been around November 1924, *after* Wiley's conversation with McBride in May 1924 about Martland but certainly before Martland spoke to Wiley in March 1925 about the dialpainter examinations. Also, in the *New York World* of June 19, 1925 (1), Martland was reported to have learned of the dialpainter deaths "before I took office a short time ago." The office referred to must have been that of the Essex County medical examiner, which title he assumed in June 1925 (*Newark Evening News*, June 2, 1925, 1).

Martland himself wrote to McBride about *two* meetings with dialpainters. The first one was a conversation with several women in the offices of the dentists Davidson and Barry. Martland could not "secure an autopsy" and so "lost interest in the matter temporarily." The second meeting was with Marguerite Carlough in May 1925, following which he was appointed county physician of Essex County, which office permitted Martland to insist on an autopsy of Carlough's sister, Sarah Maillefer, on her death in June 1925. See Martland to McBride, August 28, 1925, "Radium I: Scrapbook, 1925–1927," HMP. At any rate, we do not know exactly what Martland did about the first meeting that led McBride to inform Wiley that Martland could find no proof of any problem.

50. Wiley, "Notes on Cases," reel 30, AALLP; Wiley, "Notes on Cases," reel 3, RBP.

51. Wiley to John Andrews, June 19, 21, 1924, Hamilton to Andrews, and Ethelbert Stewart, Commissioner, USBOLS, to Andrews, June 24, 1924, reel 30, AALLP. For the fate of the Labor Department during the 1920s, I consulted Hawley, "Herbert Hoover, the Commerce Secretariat," 116–40, esp. 132 which discusses the department's atrophy, and Zieger, *The Republicans and Labor*, 87–106, 123, 131–43, 199–211, esp. 59 on the department's atrophy and 71–72 on "assaults" on the department in the guise of reorganizing the executive branch.

52. A. M. Stimson, Assistant Surgeon General, USPHS, to Wiley, April 18, 1924, and Ethel Johnson, Assistant Commissioner, Massachusetts Department of Labor and Industries, to Wiley, June 3, 1924, reel 30, AALLP; Wiley, Report to the National Consumers' League Council at the Consumers' League Annual Meeting, November 13, 1924, reel 4, NCLR.

53. Schaub, "Radium," 138–39; Blum, "Osteomyelitis of the Mandible and Maxilla," 802.

54. Wiley, Report to the National Consumers' League Council at the Consumers' League annual Meeting, November 13, 1924, reel 4, NCLR; Andrew McBride to Hoffman, January 1925, reel 2, RBP; Arthur Roeder to Wiley, January 9, 1925, reel 3, RBP.

CHAPTER FOUR

1. Gillespie, "Accounting for Lead Poisoning," esp. 305–6.

2. Hamilton to Wiley, January 30, 1925, box 84, CLNJP.

3. Hamilton to Wiley, January 30, 1925, February 2, 5, and 7, 1925, ibid.; Arthur Roeder to Drinker, Drinker testimony, March 12, 1924, HUPDR. On Drinker (and also for the history of industrial hygiene at Harvard), see Curran, *Founders of the Harvard School of Public Health*, 39, and Aub and Hapgood, *Pioneer in Modern Medicine*, 248, 262.

4. Cecil Drinker testimony, November 14, 1927, reel 1, RBP; Castle, Drinker, and Drinker, "Necrosis of the Jaw," 371.

5. Ibid., 373–75.

6. Ibid., 374–76. On the Drinkers' specialties, see Cecil Drinker testimony, November 14, 1927, reel 1, RBP, and Curran, *Founders of the Harvard School of Public Health*, 161–65, 230–31.

7. Castle, Drinker, and Drinker, "Necrosis of the Jaw," 374–75, 377–78, 380; interview with William Castle, March 23, 1984.

8. Castle, Drinker, and Drinker, "Necrosis of the Jaw," 376–77.

9. Drinker to Roeder, June 3, 1924, Roeder to Drinker, June 6, 18, 24, 1924, and Drinker to Roeder, June 27, 1924, all in Cecil Drinker testimony, November 14, 1927, reel 1, RBP. Most of these letters also appear in HUPDR.

10. Hamilton to Wiley, February 7, 1925, box 84, CLNJP; Wiley to Josephine Goldmark, December 10, 1947, reel 85, NCLR.

11. Wiley, "Notes on Cases," reel 3, RBP; Cecil Drinker to Arthur Roeder, February 17, 1925, HUPDR.

12. Drinker to Roeder, February 17, 1925, HUPDR; Alice Hamilton to Clara Landsberg, February 8, 1925, quoted in Sicherman, *Alice Hamilton*, 283; Hamilton to Wiley, February 27, 1925, box 85, CLNJP.

13. Wiley, "Notes on Cases," reel 3, AALLP; Hamilton to Wiley, March 16, 1925, box 84, CLNJP; *New York Times*, March 10, 1925, 2; Arthur Roeder to Cecil Drinker, April 9, 1925, Drinker testimony, November 14, 1927, reel 1, RBP.

14. Hamilton to Wiley, February 7, March 16, 1925, box 84, CLNJP; Hamilton to Florence Kelley, April 6, 1925, reel 26, NCLR; Hamilton to Katherine Drinker, April 4, 1925, HUPDR.

15. Hamilton to Katherine Drinker, April 4, 1925, and Drinker to Hamilton, April 17, 1925, HUPDR. More information on Katherine Drinker may be found in Short, "Katherine Rotan Drinker," and Bowen, *Family Portrait*.

16. John Roach, Deputy Commissioner, New Jersey Department of Labor, to Cecil Drinker, May 1, 1925, Drinker to Roach, May 29, 1925, Drinker to Arthur Roeder, June 18, 1925, and Andrew McBride to Drinker, June 25, 1925, HUPDR.

17. Josiah Stryker, lawyer for USRC, to Cecil Drinker, June 20, 1925, HUPDR; interview with Castle, November 16, 1984.

18. Curran, *Founders of the Harvard School of Public Health*, 4–5, 155–56; Aub and Hapgood, *Pioneer in Modern Medicine*, 182–93, 248–62. For a review of the founding of industrial hygiene educational programs in America, see Kober, "Prefatory," xlvi–xlviii.

19. Aub and Hapgood, *Pioneer in Modern Medicine*, 185; minutes of a meeting of those involved in the Industrial Clinic, Harry Linenthal correspondence file, HUIHDR.

20. Shattuck, "Industrial Medicine at Harvard," Shattuck Papers, Archives of the Countway Library, Harvard Medical School, Boston, 3–4; Cecil Drinker to Frederick Shattuck, November 28, 1922, HUIHDR.

21. Short, "Katherine Rotan Drinker"; Edsall to Wade Wright, n.d., in Shattuck, "Industrial Medicine at Harvard," 17; Minutes, Division of Industrial Hygiene, November 25, 1919, HUIHDR.

22. Shattuck to George C. Vincent of the Rockefeller Foundation, November 14, 1919, HUIHDR.

23. "Conference on Industrial Hygiene: Rockefeller Foundation, 1919," 7, 25, box 97, USPHSR.

24. Selby, "Studies of the Medical and Surgical Care of Industrial Workers," 22; Burlingame, "The Art, Not the Science, of Industrial Medicine," 368–69.

25. "Conference on Industrial Hygiene: Rockefeller Foundation, 1919," 19, 47, box 97, USPHSR; Selleck and Whittaker, *Occupational Health in America*, 61.

26. Aub and Hapgood, *Pioneer in Modern Medicine*, 256; Shattuck to "Jack," December 5, 1922, and to J. P. Morgan, October 4, 1922, in Shattuck, "Industrial Medicine at Harvard," 193–95, 201; Shattuck, "Five Years of the Harvard School of Industrial Hygiene," February 17, 1923, in Shattuck, "Industrial Medicine at Harvard," 171–75.

27. Wade Wright to Edsall, February 21, 1923, in Shattuck, "Industrial Medicine at Harvard," 48–49.

28. Wiley to Miss Dyckman, n.d., box 83, CLNJP.

29. Hoffman, "Radium (Mesothorium) Necrosis," 961–65; Castle, Drinker, and Drinker, "Necrosis of the Jaw," 374–75, 377–80.

30. Castle, Drinker, and Drinker, "Necrosis of the Jaw," 371–72, 377–78.

31. Hoffman, "Radium (Mesothorium) Necrosis," 961–65; Burricelli medical file, ANL ("blemishes" story); Clark, "Glowing in the Dark," 196 (progression of Grace Fryer's illness); "Radium Poisoning," USBOLS, 1222, 1229–30. Hoffman stated in a footnote that one of the Smith sisters had recently died from pernicious anemia, but other sources do not bear this out.

32. Hoffman, "Radium (Mesothorium) Necrosis," 962–65.

33. Ibid., 963; Arthur Roeder to William Castle, June 6, 1924, Drinker testimony, reel 1, RBP. The first objection I have found to the possibility of widespread radium contamination at the U.S. Radium plant is in a report by Swen Kjaer, April 9, 1925, SKP, although I have heard this position argued recently by people familiar with the dialpainting episodes. The argument suggests that dialpainters' "glowing in the dark" could be caused for brief periods by the action of zinc sulfide alone if it had been exposed to sunlight.

34. Kjaer, April 13, 1925, SKP.

35. Kjaer, April 4, 8, 15, 1925, SKP.

36. For the movements of the Radium Dial studio, see Kjaer, April 10, 13, 15, 1925, February 28, 1929, SKP; see also "Radium Poisoning," USBOLS, 39.

37. Clipping, March 17, 1936, vol. 1, Radium Archives, ANL; also reproduced in Mayo, "We Are All Guinea Pigs," *Village Voice*, December 25, 1978, 18ff.

38. On lippointing, see Catherine Donahue's testimony before the Illinois Industrial Commission, February 10, 1938, vol. 31, Radium Archives, ANL. On dust, compare Kjaer, April 20, 1925, SKP, to interview with Elizabeth Frenne, reported by A. M. Brues, April 7, 1972, vol. 2, Radium Archives, ANL. For an account of glowing in the dark, see Report on Doris Hedrick, vol. 2, Radium Archives, ANL; see also Conroy, "Radium City."

39. Memo, MIT Laboratory for Nuclear Science and Engineering, December 6, 1958, vol. 1, Radium Archives, ANL (for demonstration of radium eating); "Ranks of Living Dead Dwindle in 25 Years," clipping, *Chicago Daily News*, June 13, 1953, in vol. 1, Radium Archives, ANL ("glow in our cheeks" quotation). The "goodlooking" quotation is attributed to the dialpainter Margaret Glacinski in A. M. Brues, memo on contents of *Chicago Tribune* article, vol. 1, Radium Archives, ANL. See also "Radium Poisoned Workers Won Meager Compensation," clipping, *The Worker*, March 3, 1957, 7, box 83, CLNJP, and Swen Kjaer, April 10, 15, 17, and 20, 1925, SKP.

40. Kjaer, April 8, 1925, SKP.

41. Kjaer, April 17, 20, 1925, SKP.

42. "Radium Poisoning," USBOLS, 59 (Margaret Looney, case 38), 60 (Paul Hogue, case 40, and Charles Viol, case 41). On Catherine Wolfe, see A. M. Brues to file, December 29, 1971, memo regarding articles in the *Chicago Tribune*, February 11 (p. 1), 12 (n.p.), April 6 (p. 13), July 28 (p. 12), 1928, in vol. 31, Radium Archives, ANL.

43. Kjaer, "Occupational Disease—Radium Necrosis: Summary of Field Work" (dates of summary penciled in: April 4–25, 1925), SKP; Kjaer, April 9, 1925, SKP.

44. Kjaer, "Occupational Diseases—Radium Necrosis: Summary of Field Work."

45. On U.S. Radium "stonewalling," see Arthur Roeder to Frederick Hoffman, December 14, 1925, in Hoffman's affidavit, August 25, 1927, reel 2, RBP.

46. Schaub, "Radium," 138–39.

47. Martland to Andrew McBride, August 8, 1925, "Radium I: Scrapbook, 1925–1927," HMP (Martland's consultation for Davidson and Barry); Martland, Conlon, and Knef, "Some Unrecognized Dangers," 1769 (impetus from Hoffman's work and quotation about Hoffman's statistical work); Berg, *Harrison Stanford Martland*, 173–78 (opportunity presented by Lemen's death); Martland, "Occupational Poisoning," 1249–50; Reitter and Martland, "Leucopenic Anemia," 165.

48. Martland, "Occupational Poisoning," 1253–54.

49. Martland, Conlon, and Knef, "Some Unrecognized Dangers," 1770–74; "Radium Poisoning," USBOLS, 1224–25; Hoffman, "Radium (Mesothorium) Necrosis," 961–62; Sarah Maillefer and Marguerite Carlough medical files, ANL.

50. Martland, Conlon, and Knef, "Some Unrecognized Dangers," 1770–74.

51. Ibid., 1774.

52. Ibid., 1774–75.

53. Berg, *Harrison Stanford Martland*, 173–78; Martland, Conlon, and Knef, "Some Unrecognized Dangers," 88–92.

54. Martland, Conlon, and Knef, "Some Unrecognized Dangers," 89–90.

55. Ibid., 90.

56. Ibid., 91. The full paper is, of course, Martland, Conlon, and Knef, "Some Unrecognized Dangers in the Use and Handling of Radioactive Substances."

57. Ibid., 1774–75.

58. Schaub, "Radium," 139; Hoffman to Arthur Roeder, President of USRC, December 8, 1925 (reel 2), USRC to Hoffman, December 14, 1925 (reel 2), and Hoffman's affidavit, August 25, 1927 (reel 1), RBP.

59. Information on Flinn is from his daughter, vol. 28, Radium Archives, ANL; Cattell, "Frederick Flinn."

60. Rosner and Markowitz, "A 'Gift of God'?," 344–47.

61. Ibid., 347–49.

62. Ibid., 349–51.

63. Flinn to Alice Hamilton, November 28, 1927, AHP (cited in Baron, "Watches, Workers, and the Awakening World," 58 n. 167); Flinn and H. Barker correspondence, and H. Barker to H. Schlundt, n.d., reel 3, USRCR.

64. Flinn to Schaub, December 7, 1925, in "Stipulation," reel 1, RBP.

65. "Stipulation," ibid. It was Grace Fryer who was supposed to be healthier than Flinn.

66. Flinn to Drinker, January 16, 1926, "Stipulation," ibid. (also in HUPDR); Flinn, "Radioactive Material: An Industrial Hazard?."

67. Flinn, "A Case of Antral Sinusitis"; "Radium Poisoning," USBOLS, 1233; Flinn to Alice Hamilton, November 28, 1927, AHP.

68. C. H. Granger (Vice President, Waterbury Clock Co.), "History of Radium Painting and Radium Poisoning Cases Resulting therefrom at the Waterbury Clock Company," 1936, vol. 9, n.p., Radium Archives, ANL.

69. Federal Writers' Project, *Connecticut*, 304–5; Brass Workers' History Project, *Brass*

Valley, 71–77; Granger, "History of Radium Painting"; Swen Kjaer, December 8, 10, 1928, February 13, 1929, SKP.

70. Granger, "History of Radium Painting."

71. Swen Kjaer, December 12, 1928, February 14, 1929, SKP.

72. Granger, "History of Radium Painting"; clipping, *Waterbury Republican,* January 28, 1927, in Elizabeth Dunn medical file, ANL; Swen Kjaer, December 10, 11, 1928, SKP; Flinn, "A Case of Antral Sinusitis."

73. Flinn, "A Case of Antral Sinusitis," 347; letter (probably from Katherine Drinker) to Wiley, November 1, 1927, box 83, CLNJP; Raymond Berry to Hamilton, February 21, 1928, AHP.

74. Hamilton to Dr. Emerson of Columbia University, November 15, 1927, AHP.

75. Flinn to Hamilton, November 28, 1927, AHP.

76. Hamilton to Flinn, December 3, 1927, AHP.

77. Flinn to Hamilton, December 10, 1927, AHP; Flinn, "Some of the Newer Industrial Hazards"; Hamilton to Florence Kelley, n.d., and Wiley to Miss Dyckman, n.d., box 83, CLNJP; Wiley to Mrs. [Susanna] Zwemer, January 27, 1958, reel 85, NCLR; Raymond Berry to Hamilton, January 6, 1928, reel 3, RBP; "Frederick Flinn," in Cattell, *American Men of Science.*

78. Hoffman to Mrs. Francis J. Rigney, April 2, 1943, box 26, Frederick Hoffman Papers, Columbia University Rare Book and Manuscript Library, New York; Castle, Drinker, and Drinker, "Necrosis of the Jaw," 378; Martland, Conlon, and Knef, "Some Unrecognized Dangers," 1775.

CHAPTER FIVE

1. *New York Times,* March 19 (2:1), June 9 (14:3), June 19 (1:2), June 24 (21:3), 1925. On the Carlough suit, see also newspaper clipping, Carlough medical file, ANL.

2. The best sources for the history of tort law in industrial America are Friedman and Ladinsky, "Social Change and the Law of Industrial Accidents," esp. 51–52, and Friedman, *History of American Law,* 261–64, 409–27.

3. See, e.g., Canan, "Industrial Law," and Birnbaum and Oshinsky, *Occupational Disease Litigation,* 128–41.

4. Kuser, Carlough, and Maillefer medical files, ANL.

5. On settlements, see *New York Times,* May 5, 1926, 21:3; clippings, *Newark Ledger,* May 31, 1928, and *New York World,* n.d., in Kuser medical file, ANL; *New York World,* May 31, 1928, 1.

6. See Janet Jelinek, memo regarding a conversation with Sarah Maillefer's daughter, Margaret Ludwig, August 21, 1968, in Carlough medical file, ANL.

7. The New Jersey law is cited in *Cumulative Supplement to the Compiled Statutes of New Jersey, 1911–1924,* 230–33; it is discussed in Ritchie, "An Appraisal of Workmen's Compensation Legislation," 11–13, 27–31, 71; and it is criticized in Dodd, *Administration of Workmen's Compensation,* 765–66, and Beers, "Compensation for Occupational Diseases," 587. F. M. Wilcox charged that the radium victims illustrated the "perfidy" of New

Jersey's schedules, in "The Schedule Fraud in Occupational Disease Compensation," 120, cited in Dodd, *Administration of Workmen's Compensation*, 766. The radium cases are also considered in Reede, *Adequacy of Workmen's Compensation*, 49.

8. On women's cooperative political groups, see "Memos," October 30, 1924, "Legislation, 1926–1929" folder, LWVNJP; Felice Gordon, *After Winning*, 45–46.

9. Wiley to Lena Anthony Robbins, New Jersey Federation of Women's Clubs, March 22, 1926, "Correspondence 1927–1929" folder, Lena Robbins Papers, box 4, LWVNJP. On Agnes Jones, see Minutes, May 5, 1926, Business Minutes, 1922–25, Papers of the Woman's Club of Orange, From the Collections of the New Jersey Historical Society, Newark. On the radium necrosis bill, see Katherine G. T. Wiley, "Occupational Hazards—Whose the Responsibility?," clipping, *Civic Pilot*, ca. 1926, 7, CLNJP; "Committee on Women in Industry: A Statement of the Year's Program," and clipping, Lydia Sayre Walker, Chair, "Department of Women in Industry," *Civic Pilot* (March 1931), 11, both in "Women in Industry, 1924–1927" folder, LWVNJP; "Legislative Reports," February 3, March 3, March 12, 1926, in "Legislation, 1926–1929" folder, LWVNJP; and Felice Gordon, *After Winning*, 64.

10. "Minutes, Women's Organizations Interested in Legislation," "Legislation, 1926–1929" folder, LWVNJP (the failure of women in politics and Feickert quotation); Felice Gordon, *After Winning*, 36–38, 46–48, 78–79, 86–88, 90–97.

11. On the lack of opposition to the radium bill, see clipping, Mrs. F. W. Klaber, "Legislative Session—In Retrospect," *Civic Pilot*, and "Minutes, Women's Organizations Interested in Legislation," both in "Legislation, 1926–1929" folder, LWVNJP. On the silicosis bill, see clipping, *Ledger*, May 25, 1928, reel 2, RBP.

12. On Hoffman as the author of the radium necrosis bill, see Susanne Zwemer, "History of the Consumers' League of New Jersey," CLNJP, 27; on his support for blanket compensation of industrial diseases, see unidentified clipping, "Radium II: Scrapbook, 1928," HMP. On changing radium necrosis to radium poisoning, see clipping, *Civic Pilot*, March 1931, 11, "Legislation, 1930–1931," and Helen Paul, Legislative Report, March 31, 1931, all in box 2, LWVNJP.

13. Reitter and Martland, "Leucopenic Anemia" (Lemen); Martland, "Microscopic Changes of Certain Anemias."

14. Barker, "Radioactive Determination."

15. Martland, "Occupational Poisoning," 1256–58, 1264–65.

16. "Radium Victim #41," *Life*, December 17, 1951, 81ff; Martland, "The Occurrence of Malignancy in Radio-Active Persons," 2443.

17. Stocker and Eckert medical files, ANL; Martland and Humphries, "Osteogenic Sarcoma" (reported at New York Pathology Society conference, March 8, 1928). These early conclusions are also discussed in Martland, "The Occurrence of Malignancy in Radio-Active Persons"; on dosimetry and cancer causation, see esp. 2437, 2512–13.

18. Katherine Wiley to Mrs. Zwemer, January 27, 1958, reel 85, NCLR (comments on Berry); Raymond Berry, affidavit, May 1928, reel 1, RBP.

19. Hoffman, "Radium (Mesothorium) Necrosis," 961–62 (patient 8); "Stipulation" (Fryer's years of work) and "Statement of Case" (medical history), reel 1, RBP; Fryer medical file, ANL (date of birth and medical history). See also "Testimony, Grace Fryer," reel 1, RBP.

20. Clark, "Glowing in the Dark," 230–40; "Statement of Case," and Hussman, testimony in Chancery Court, January 1, 1928, both on reel 1, RBP.

21. Berry to Frederick Hoffman, July 22, 1927, and Berry to Vice Chancellor Backes, January 18, 1928, reel 3, RBP; Berry, affidavit, May 1928, reel 1, RBP.

22. For Berry's interpretation of *Rylands*, see "Memorandum of Law," reel 1, RBP; for *Rylands* history and controversy, see Friedman, *History of American Law*, 409, 425–26.

23. For Berry's research and reasoning, see "Memorandum of Law," reel 1, RBP. The cases cited are *Marshall v. Wellwood*, 38 NJL 339; *P. Ballantine & Sons v. The Public Service*, 76 NJL 358 (1908), 86 NJL 331 (1914); *Cuff's v. Newark and New York Railroads*, 35 NJL 17 and 35 NJL 574; and *McAndrews v. Collard*, 42 NJL 189.

24. Berry's "Memorandum of Law"; *Smith v. Oxford Iron Co.*, 42 NJL 467.

25. Berry's "Memorandum of Law."

26. For strategy, see "Proof of Knowledge of Danger by USRC" and "Stipulation," reel 1, RBP. Berry also relied on Frederick Hoffman to testify that von Sochocky knew of radium's dangers while an officer of the company and that the corporation denied the hazards of dialpainting after it had reasonable grounds for suspicion based on the Drinker team's report. See Berry to Hoffman, January 3, 1928, reel 3, RBP, and USRC to Hoffman regarding Katherine Schaub, December 14, 1925, in Hoffman's affidavit, August 25, 1927, reel 1, RBP. On June 14, 1924, Helen Kuser's physician, Dr. Blum, had also written to U.S. Radium about his concern with her health; this was more evidence that U.S. Radium knew of problems and covered them up. Hoffman's affidavit, August 25, 1927, reel 1, RBP (testimony regarding Blum's letter to U.S. Radium, June 14, 1924).

27. "Stipulation," reel 1, RBP, March 1928.

28. "Stipulation" (1st two quotations), "Proof of Knowledge of Danger by USRC" (on von Sochocky), and Thomas F. V. Curran to Raymond Berry, May 31, 1928, all in ibid.; von Sochocky testimony, April 26 or 27, 1928, reel 2, RBP (Berry's questions to von Sochocky); Grace Fryer testimony, January 12, April 26 or 27, 1928, reels 1–2, RBP. On U.S. Radium suspicions about von Sochocky, see "Report of Conversation with Miss Anna Rooney . . . July 20, 1927," vol. 5, Radium Archives, ANL.

29. "Stipulation," reel 1, RBP (Berry's reasoning and quotations); "Proof of Knowledge of Danger by USRC," reel 1, RBP. The publications Berry referred to were American Institute of Medicine, *Bibliography on Radium* and *Radium: Abstracts*.

30. Castle, Drinker, and Drinker, "Necrosis of the Jaw."

31. Drinker to J. A. Singmaster, July 14, 1924 (1st quotation), Drinker to O'Brien, December 19, 1932 (2d quotation), Drinker to Potter and Berry, Attorneys, July 7, 1927 (next quotation), Castle to Drinker, September 6, 1927, and Drinker to Castle, September 8, 1927 (block quotation), all in HUPDR.

32. Drinker testimony, November 14, 1927, Examination in Boston, reel 1, RBP.

33. Report, Wiley to Consumers' League of New Jersey, June 8, 1928, box 84, CLNJP.

34. Raymond Berry, affidavit, May 19, 1928 (date of transfer), and Fryer testimony, January 12, 1928, reel 1, RBP.

35. Von Sochocky testimony, April 26 or 27, 1928, reel 2, RBP.

36. Von Sochocky was "not friendly" to U.S. Radium's "controlling elements." Thomas Curran to Berry, May 31, 1928, reel 1, RBP. On von Sochocky as a hostile witness, see clipping, n.d., no identification, "Radium II: Scrapbook, 1928," HMP.

37. Hughes testimony, April 25 or 26, 1928, reel 2, RBP.

38. Ibid.; Armin St. George testimony, April 26 or 27, 1928, reel 2 RBP. The U.S. Radium attorneys drew their quotation from John, "Preliminary Report on the Therapeutic Use of Radium Salts."

39. See Schaub testimony, esp., reel 2, RBP; also the testimony of Quinta McDonald and Grace Fryer (reel 1) and Albina Larice (reel 2), RBP.

40. See, e.g., Schaub testimony, reel 2, RBP.

41. Clipping, n.d., no identification, "Radium II: Scrapbook, 1928," HMP; *New York Times*, April 27, 1928, 1:3; Martland testimony, reel 2, RBP.

42. Clippings from *Daily Courier*, April 30, 1928, and *Newark Ledger*, May 24, January 13, 1928, both in reel 2, RBP.

43. *New York World*, May 13, 1928, 1N–2N.

44. *New York World*, May 13, 1928, 1N–2N (McDonald's 1st quotation and characterization of Fryer); clippings, *New York Sun*, May 17, 1928 (McDonald's second quotation), and *Newark Sunday Call*, May 13, 1928 (Larice's quotation), reel 2, RBP.

45. Clippings, *Newark Sunday Call*, May 13, 1928 (Schaub's quotation), and *Newark Ledger*, January 13, 1928 (Hussman's 1st quotation), reel 2, RBP; *New York World*, May 13, 1928, 1N–2N.

46. Schaub, "Radium," 139–40; clipping, *New York Telegram*, May 15, 1928, reel 2, RBP.

47. Clark, "Glowing in the Dark," 239–40; "Statement of case," reel 1, RBP; Polednak, "Fertility of Women after Exposure."

48. Clipping, *Newark Ledger*, May 14, 1928 (1st quotation), and unidentified clipping (2d quotation), both in reel 2, RBP.

49. Goldmark, *Impatient Crusader*, chap. 16 (Florence Kelley's contributions); *New York World*, May 19, 1928, 14 (editorial), May 14, 1928 (cartoon, editorial page); clipping, *Newark [News?]*, May 21, 1928, reel 1, RBP.

50. *New York World*, May 18, 1928 (Flinn announcement); clippings, *Newark Star Ledger*, May 18, 1928, and *New York Sun*, May 17, 1928 (Hoffman responses), both in reel 2, RBP; editorial, *New York World*, May 19, 1928, 10. See also two unidentified clippings, "Radium II: Scrapbook, 1928," HMP. In "Some Precautions to Be Taken When Making Tests," Flinn explained the "mistakes" in his celebrated 1928 readings. Martland ("Occupational Poisoning in Manufacture of Luminous Watch Dials," 1265) later theorized that if a dialpainter had mostly mesothorium, which decays relatively quickly, deposited in her bones, and if she outlived the early years of her disease, she might then outlive the mesothorium in her bones and survive radium poisoning.

51. *New York World*, May 21 (pp. 1, 4), May 29 (p. 1), 1928, and *New York Times*, May 21, 1928, 2 (statute of limitations amendment); clipping dated May 20, 1928, Grace Fryer medical file, ANL, and *New York World*, May 25, 1928, 6 (message of sympathy); clipping, *Newark Ledger*, May 25, 1928, reel 2, RBP (Wiley's quotation); clippings from *Newark Star-Eagle*, June 15, 1928 (occupational disease commission), and June 5, 1928 (the fate of the legislation), vol. 7, Radium Archives, ANL.

52. *New York Times*, May 23 (p. 11), 30 (p. 7), June 1 (p. 27), and 5 (pp. 1, 8), 1928. The May 23 article is interpreted in Hilgartner, Bell, and O'Connor, *Nukespeak*, 11.

53. Clipping, reel 2, RBP; Berry to Goldmark, December 12, 1947, reel 3, RBP. Federal judge William Clark personally guaranteed the annuities and pensions granted to the

dialpainters in the settlement, in case of the dissolution or bankruptcy of U.S. Radium. See William Clark to U.S. District Court (New Jersey), June 4, 1928, reel 2, RBP. The bulk of the settlement was paid by insurance companies, however. See Berry to Krumbhaar, June 18, 1929, EKP.

54. Rosner and Markowitz, "A 'Gift of God'?." See also Chapter 6 of this book.

55. *New York Times*, June 5, 1928, 1, 8; *New York World*, June 3 (pp. 1, 4), 4 (p. 13), 1928; Berry to Collins and Corbin, June 2, 1928, reel 3, RBP (Berry's fee); *New York World*, June 4, 1928, 13 (Wiley's quotation); C. B. Lee, USRC, to Harris, June 18, 1928, Commissioner Harris Correspondence Files, ANL.

56. *New York Times*, June 5, 1928, 1, 8; Berry to Krumbhaar, October 18, 1930, EKP; Berry to Alice Hamilton, June 6, 1928, Fryer file, reel 3, RBP (quotation); C. B. Lee, USRC, to Harris, June 18, 1928, Commissioner Harris Correspondence Files, ANL.

57. On Ewing, see Charles Norris to Raymond Berry, June 7, 1928, Fryer files, reel 3, RBP, and Norris to Martland, July 2, 1928, and Martland's reply, July 3, 1928, both in "Radium II: Scrapbook, 1928," HMP. On Krumbhaar, see Martland to Krumbhaar, June 16, 1928, "Radium II: Scrapbook, 1928," HMP (also in EKP).

58. Krumbhaar to Raymond Berry, April 2, 1929, and Berry to Krumbhaar, October 18, 1930, EKP. See also these papers for further details on the functioning of the three-physician board. For the board's private conclusion that the dialpainters were radioactive and were suffering from the effects of radioactivity, see Krumbhaar, Craver, and Ewing to USRC and Berry, June 24, 1929, reel 2, RBP.

59. Berry to E. B. Krumbhaar, October 18, 1930, EKP; Rosner and Markowitz, *Deadly Dust*, 7, 10.

60. Clark, "Glowing in the Dark," 269; Canfield medical file, ANL; "Pre-trial Exam," January 17, 1929, reel 2, RBP.

61. Berry to Alice Hamilton, April 2, 1930, reel 1, RBP. Von Sochocky—even though a hostile witness—had been one link demonstrating early U.S. Radium knowledge of dialpainting dangers, but he had died, himself a victim of radiation.

62. On the Drinkers and Castle, see Catherine Drinker to Berry, March 3, 1930, Berry to Cecil Drinker, March 4, 1930, Berry to William Castle, March 4, 1930, Cecil Drinker to Berry, March 6, 1930, Castle to Berry, March 27, 1930, and Alice Hamilton to Berry, March 31, 1930, all in reel 1, RBP. On Kjaer, see Kjaer to Berry, February 29, 1930, and Charles Baldwin to Berry, February 19, 1930, both in reel 1, RBP. On Ewing and Krumbhaar, see testimony in Chancery Court of Dr. Edward Krumbhaar, Dr. James Ewing, Grace Fryer, Katherine Schaub, Edna Hussman, and Albina Larice, June 27, 1929, reel 2, RBP.

63. Clipping, ca. 1928, "Radium II: Scrapbook, 1928," HMP. Martland's views will be considered further in Chapter 7.

64. Hughes to Berry, October 11, 1926, reel 1, RBP.

65. Clipping, *Newark Star-Eagle*, March 10, 1930, Canfield medical file, ANL; "Statement—March 11, 1930," and Berry to Overseer of the Poor, Department of Public Welfare, February 5, 1932, both in reel 1, RBP.

66. Berry to Edwards and Smith, Attorneys for USRC, March 8, 1930, reel 1, RBP.

67. Federal Writers' Project, *Connecticut*, 29–31; Roth, *Connecticut*, 162.

68. Federal Writers' Project, *Connecticut*, 37–38, 64; Roth, *Connecticut*, 149–62, 172–85; Janick, *A Diverse People*, 1–16; Anderson, *From Yankee to American*, 55–63. For views

on reform impulses and Progressivism among Connecticut's immigrants, urbanites, and laborers, see Buenker, "Progressivism in Connecticut," and Heath, "Labor and the Progressive Movement."

69. Nareff, "Blue Collar, White Gloves"; Hirsch, *We've Come a Long Way*; Nichols, "A New Force in Politics," 2, 6, 12–13, 82, 84, 123, app. (item 24); Pendary, "Women's Secular Voluntary Organizations"; Lemons, *The Woman Citizen*, 93, 170–71.

70. Janick, *A Diverse People*, 24–26; Roth, *Connecticut*, 185, 188–89.

71. On the history of the compensation act and court interpretations, see Aspell, " 'Personal Injury,' " 142–49. On evaluations of why industrial diseases were made compensable, see Roth, *Connecticut*, 163–90; Janick, *A Diverse People*, 21–30; and Federal Writers' Project, *Connecticut*, 64.

72. Clippings, *Morning World*, May 18, 30, 1928, reel 2, RBP, and unidentified clippings, "Radium II: Scrapbook, 1928," HMP; Swen Kjaer, December 11, 1928, SKP (1st quotation, Dr. Arthur Holmes); clipping, "Dread Radium Disease Kills 4th Victim," *Waterbury Republican*, January 6, 1930, Marion Demolis medical file, ANL (2d quotation).

73. Granger, "History of Radium Painting"; Marjorie Dumschott medical file, ANL; "Radium Poisoning," USBOLS, 1237–38. Flinn wrote Alice Hamilton that the Connecticut cases were settled under his auspices, January 4, 1928, reel 3, RBP.

74. Kjaer, December 7, 10, 11, 1928, SKP. In Connecticut responsibility for industrial health lay with the health board, unlike in other states where Labor Departments were more active. Rosner and Markowitz, "Research or Advocacy?," 86.

75. Kjaer, February 9, 1929, SKP. See also Aspell, " 'Personal Injury,' " 148–49.

76. Cardow medical file, ANL, including clippings from *Waterbury Republican*, March 21–22, 1929, and *New York World*, March 20, 1929; clipping, *Waterbury Republican*, March 19, 1929, in Lena Wall medical file, ANL; "Radium," USBOLS, 1238; Minutes, Board of Directors, National Consumers' League, reel 2, NCLR.

77. Granger, "History of Radium Painting"; Clark, "Glowing in the Dark," 225 (details of dialpainting illnesses and deaths).

78. Granger, "History of Radium Painting"; Katherine Moore to Robley Evans, March 25, 1935, in Moore medical file, ANL; Clark, "Glowing in the Dark," 226–27.

79. Granger, "History of Radium Painting."

80. Clipping, *Newark Evening News*, December 26, [1925?], in Amelia Maggia medical file, ANL (Knef's thirty-seven dialpainter patients); Wiley's notes, December 10, 1924, reel 85, NCLR, and testimony of Clarence Lee, President of USRC, April 26–27, 1928, reel 2, RBP (Maggia suit); Knef to Waterbury Clock Co., September 10, 1925, vol. 4, Radium Archives, ANL.

81. Transcript of a conversation with Knef, vol. 4, Radium Archives, ANL; Wiley's notes, December 10, 1924, reel 3, RBP; unidentified clipping, October 19, 1927, Maggia medical file, ANL.

82. "Transcript of Conference, May 19, 1926," reel 2, RBP; Dean Edmonds (counsel to USRC), affidavit regarding conference with Knef, reel 3, RBP; Arthur Roeder (president of USRC), affidavit regarding conference with Knef, reel 3, RBP.

83. "Transcript of Conference, May 19, 1926," reel 2, RBP. On detectives, see Swen Kjaer report of conversation with J. S. Kelly, president of Radium Dial Co., March 9, 1929,

SKP. On children, see Elizabeth Lewis report of a conversation with Knef, March 22, 1930, vol. 5, Radium Archives, ANL.

84. Hamilton to Florence Kelley, January 9, 1929, in Barbara Sicherman, *Alice Hamilton*, 315.

85. Hamilton to Florence Kelley, April 6, 1925, Hamilton Correspondence, reel 26, NCLR.

CHAPTER SIX

1. On turf wars between the USPHS and the USDOL, see Nugent Young, "Interpreting the Dangerous Trades," 41–49, and Rosner and Markowitz, "Research or Advocacy?."

2. C. E. A. Winslow to Josephine Goldmark, June 27, 1928, and clippings, *New York Herald-Tribune* and *New York Times*, July 16, 1928, both in box 63, Florence Kelley Papers, NYPL.

3. Minutes, Board of Directors, May 25, October 18, 1928, NCLR; Proceedings of the National Consumers' League Annual Meeting, November 15–16, 1928, Organizational files, reel 5, NCLR; Sicherman, *Alice Hamilton*, 312; Goldmark, *Impatient Crusader*, 197–203; Nugent Young, "Interpreting the Dangerous Trades," 170–71. Nugent has considered the Radium Conference at length in her dissertation.

4. On the tetraethyl lead debates, see Rosner and Markowitz, "A 'Gift of God'?"; and Nugent Young, "Interpreting the Dangerous Trades," chap. 6. Concerning corporatism, see Weinstein, *The Corporate Ideal*, x; Wilson, *Herbert Hoover*, introduction, chap. 2–3; Hawley, "Herbert Hoover"; and Orloff, "The Political Origins," 62–63.

5. Kjaer's notes on the Radium Conference, SKP. The radium conference is also considered in Nugent Young, "Interpreting the Dangerous Trades," chap. 6, and Nugent, "The Power to Define a New Disease," 183–85.

6. Kjaer's notes on the Radium Conference, SKP; transcript, "Conference on Radium in Industry," vol. 28, Radium Archives, ANL—also in reel 85, NCLR, and reel 3, RBP.

7. Transcript of the Radium Conference, with Kjaer's notes, SKP.

8. Kjaer's notes on the Radium Conference, SKP (Granger and Wiley quotations); USRC to Commissioner of Health Harris, Harris Correspondence Files, ANL. For elaboration and an earlier history of this trend, see Clark, "The Menace of Benzene."

9. Kjaer's notes on the Radium Conference, SKP; transcript, "Conference on Radium in Industry," vol. 28, Radium Archives, ANL; Schwartz et al., "Health Aspects of Radium Dial Painting." Wiley indicated that neither of the two committees agreed upon during the conference ever met. Wiley to [Josephine?] Goldmark, December 10, 1947, reel 85, NCLR.

10. Montgomery, *Workers' Control in America*, esp. 1–6.

11. On the advisory committee, see Martland to Alice Hamilton, n.d. (ca. 1929), "Radium III: Scrapbook, January 1929–November 1930," HMP; *New York Times*, February 26, 1929, 21; Schwartz et al., "Health Aspects of Radium Dial Painting," 362–63.

12. All of these opinions are reported in Wiley to Florence Kelley, December 26, 1928, Radium files, reel 85, NCLR.

13. Hamilton to Florence Kelley, January 9, 1929, partially reproduced in box 83, CLNJP (original in NCLR and reproduced in Sicherman, *Alice Hamilton*, 314–15).

14. Nugent Young, "Interpreting the Dangerous Trades," 156–97. I have drawn heavily on Nugent's analysis.

15. Ibid.

16. Ibid. See also Nugent, "The Power to Define a New Disease."

17. For the fate of the federal Department of Labor during the 1920s, I consulted Hawley, "Herbert Hoover," esp. 132 (its atrophy), and Zieger, *The Republicans and Labor*, 87–106, 123, 131–43, 199–211, esp. 59 (its financial difficulties) and 71–72 ("assaults" on the department in the guise of reorganizing the executive branch).

18. Kjaer notes, SKP; "Necrosis of the Jaw among Workers Applying Luminous Paint"; "Necrosis of the Jaw among Workers Using Luminous Paint"; "Dangers in the Use and Handling of Radioactive Substances"; "Effects of Use of Radioactive Substances."

19. Kjaer, February 4–March 12, 1929, SKP; "Radium Poisoning," 1239–40.

20. Kjaer, February 26, 1929, SKP; "Statement by the Radium Dial Company," *Daily Republican Times*, June 7, 1928, 3, and Catherine Donahue testimony, February 10, 1938, both in vol. 31, Radium Archives, ANL.

21. "Radium Poisoning," 1239; Donahue testimony, February 10–11, 1938, vol. 31, Radium Archives, ANL; application to Industrial Commission of Illinois, Ella Cruse v. Radium Dial Co., filed June 8, 1928, in Ella Cruse medical file, ANL.

22. Kjaer, February 23–28, 1929, SKP; Donahue testimony, February 11, 1938, vol. 31, Radium Archives, ANL.

23. "Radium Poisoning," 1201–2, 1208, 1233.

24. Ibid., 1209–11.

25. Ibid., 1200–1201, 1206.

26. For the advisory committee, see Schwartz et al., "Health Aspects of Radium Dial Painting," 362–63.

27. Leake, "Radium Poisoning," 1078.

28. Ibid., 1078–79.

29. Ibid.

30. Barker to James Leake, June 5, 1931, vol. 28, Radium Archives, ANL (also in USRCR, reel 5, ANL); L. R. Thompson to C. H. Granger of Waterbury Clock Co., June 1, 1931, General File 0875-96-49, USPHSR.

31. Barker to Leake, June 5, 1931, vol. 28, Radium Archives, ANL; Thompson to Granger, June 1, 1931, General File 0875-96-49, USPHSR.

32. "Abstract of Discussion" following Leake, "Radium Poisoning," 1070–80.

33. Schwartz et al., "Health Aspects of Radium Dial Painting," 364, 376–81, 375.

34. Ibid., 365, 448.

35. Ibid., 365, 441–42, 448, 452.

36. Ibid., 376, 375, 365.

37. On the Tuck settlement, see unidentified clippings, "Radium VI: Scrapbook, August 1932–1938," HMP, and March 15, 1935, Radium files, NPL; *U.S. Radium Corporation v. Globe Indemnity et al.* On the Stasi settlement, see *New York Times*, March 6, 1931, and unidentified clipping, March 3, 1931, Radium files, NPL. On the various federal cases,

see clippings, March 13, July 8, December 18, 29, 1935, all in Radium files, NPL; see also *LaPorte v. United States Radium Corp.*, 13 F. Supp. 263.

38. *LaPorte v. U.S. Radium Corp.*, 264–65; Clark, "Glowing in the Dark," 290–91.

39. *LaPorte v. U.S. Radium Corp.*, 264, 271.

40. Ibid., 271–72.

41. Ibid., 266–68, 277, 269–70, 276, 271–72, 275–77.

42. On Berry's reasoning, see Chapter 5; on my arguments about the impact of finances on research, see Chapter 2.

43. For dismissal of the Tuck case and others, see unidentified clippings, May 9, 1936, December 18, 1935, Radium files, NPL.

44. *LaPorte v. U.S. Radium Corp.*, 277.

45. Ibid.

46. USRC to Martland, January 29, 1935, and Martland's reply, February 11, 1935, both in "Radium VI: Scrapbook," HMP; Martland, "The Occurrence of Malignancy," 2512–13; Roach, Deputy Commissioner of Labor, New Jersey Department of Labor, "Memorandum for the File" (regarding visit to Pioneer Instrument Company, Division of Bendix Aviation, May 9), May 12, 1941, "Radium IX: Scrapbook, 1939–1944," HMP.

CHAPTER SEVEN

1. Martland, Conlon, and Knef, "Some Unrecognized Dangers," 1776. Doubts about radium's efficacy and safety as an internal medicine are discussed at length in Chapter 2.

2. Stevens, "The Use of Intravenous Injections," 156.

3. Reitter and Martland, "Leucopenic Anemia."

4. On the AMA's attitudes about Bailey, see " 'Radithor' and William J. A. Bailey," 343; for details, see clipping, *New York Times*, May 8, 1915, Proudfoot Commercial Agency to the Department of Investigation, AMA, February 18, 1920, and Solicitor, Post Office Department, to Arthur Cramp, n.d., all in William Bailey files, AMADIR. Bailey's "consultant" status is from an unidentified clipping, March 28, 1925, "Advertisements," box 19, FDAR.

5. Hearing transcripts, August 22, 1930, FTCR.

6. Unidentified newspaper clipping, "Radium's Gamma Rays Sought," March 28, 1925, "Advertisements," box 19, FDAR; Morris, *Modern Rejuvenation Methods*, v, vi.

7. Morris, *Modern Rejuvenation Methods*, 218–19, 5–6, 9, 8, 69–70, 75–77, 80.

8. Ibid., 183–84, 196, 175.

9. Ibid., ix; pamphlets: "Neurasthenia: Its Treatment with Radium Water," 1926; "Modern Treatment of Arthritis and Kindred Conditions with Radium Water," 1927; "Radium in Rejuvenescence," 1929; "Modern Treatment of the Endocrine Glands with Radium Water: Radithor, the New Weapon of Modern Science," 1926; and others, in "Exhibits," FTCR; Morris testimony, September 4–5, 1930, hearing transcripts, pp. 50–53, FTCR.

10. Morris, *Modern Rejuvenation Methods*, 177.

11. Clipping, *Time*, April 1, 1932, Byers medical file, ANL; Eben MacBurney Byers and

Mary Jennings Hill medical files, ANL. Byers gave Radithor to one of his racehorses, too. See unidentified clipping, William Bailey files, AMADIR.

12. Byers and Hill medical files, ANL.

13. "Government Warns Public against Fraudulent Radioactive Water or Drugs," press release, July 13, 1926; W. S. Firsbie, chemist, to Dr. F. J. Osborne, health officer of East Orange, June 15, 1928; and notes on meeting between physicians at Howard Kelly Hospital, Baltimore, and Food, Drug, and Insecticide Administration officials regarding cooperative plan with the FTC to stop sales of Radithor through actions against false advertising, January 21, 1920—all in "Advertisements," box 19, FDAR.

14. "Complaint," Docket 1756, FTCR; "Complaint" in the matter of Bailey Radium Labs, Docket 1756, signed February 5, 1930, and "Answer," both in "Advertisements," box 19, FDAR.

15. Mr. Hulbert (Bailey's lawyer) in examination of Frederick Flinn, pp. 163–70, Albert Fall, Secretary of the Interior, and Stephen Mather, Director of the National Park Service, "Hot Springs National Park, Arkansas," n.d. (ca. 1921), p. 11, and Opie Reed, for U.S. Railroad Administration and National Park Service, "Hot Springs National Park, Arkansas," n.d. (ca. 1920s), pp. 10–11, all in Docket 1756, FTCR; New York, Conservation Commission, "Saratoga Mineral Springs and Baths," 15–16.

16. Flinn, "Effects of Radium and Radon Gas," "Dangers of Internal Radium Therapy," "Radium Salts and Emanation," and "Stimulating Action of Radio-active Deposits"; Flinn testimony, 92–96, 119–25, 152–81, 368–70, 414, Docket 1756, FTCR. On Flinn's continued work for the radium companies, see Flinn to Martland, January 7, 1930, "Radium IV: Scrapbook, July 1930–November 1931," and Flinn to Martland, September 27, 1932, "Radium VI: Scrapbook, August 1932–1938," both in HMP.

17. Martland testimony, Docket 1756, FTCR, 245–352.

18. Testimony of other physicians and dentists, Docket 1756, FTCR; Byers testimony, September 10, 1931, Docket 1756, FTCR, 490ff., including testimony about his operations by the physician Joseph Wheelwright (pp. 499–503) and the oral surgeon Harold Vaughn (p. 507); Flinn's testimony about Byers's radium burden, 557, Docket 1756, FTCR; Byers and Hill medical files, ANL.

19. Byers and Hill medical files, ANL; "Motion to Withdraw Answer," September 29, 1931, testimony of Mr. Hulbert (Bailey's lawyer), August 22, 1930, 4, and "Order to Cease and Desist," December 19, 1931, all in Docket 1756, FTCR.

20. Flinn to Martland, November 21, 1930, Martland's reply, n.d., typescript of Martland's remarks following presentation of the PHS paper on radium in workers, and Edwin Smith to Martland, July 31, 1930, all in "Radium IV: Scrapbook, July 1930–November, 1931," HMP; see also Flinn to Martland, October 26, 1931, "Radium V: Scrapbook, October 1931–July 1932," and Flinn to Martland, September 27, 1932, "Radium VI: Scrapbook, August 1932–1938," HMP.

21. Memo regarding Mrs. Robert Adler (Flinn's daughter) and her reminiscences about her father, June 3, 1958, vol. 28, Radium Archives, ANL.

22. See Starr, *The Social Transformation of American Medicine*.

23. "Beryllium Scrapbook" and "Benzol [benzene] Scrapbook," HMP; Ginley, "Harrison Stanford Martland," 204–13.

24. Clipping, n.d., "Radium II: Scrapbook, 1928" ("star witness"); Schaub to Mart-

land, June 28, 1928, "Radium II: Scrapbook, 1928"; Martland to Flinn, March 29, 1930, "Radium III: Scrapbook, January 1929–November 1930"; and Martland to Frances Perkins, September 3, 1947, "Beryllium Scrapbook" — all in HMP.

25. "Oaths and Bonds of Officers and Employees," in *Cumulative Supplement to the Compiled Statutes of New Jersey, 1911–1924*, 1:727; Martland, "Occupational Poisoning," 1235; Martland to James Cassidy (representing Eleanor Eckert's brother), November 3, 1927, "Radium III: Scrapbook, January 1929–November 1930," HMP.

26. "An Act . . . [concerning] chief medical examiners, coroners, and county morgue keepers," in *Supplement to the Compiled Statutes of New Jersey, 1925–1930*, 299–302; Martland, "Occupational Poisoning," 1273; Martland to Andrew McBride, New Jersey Labor Commissioner, July 28, 1925, "Radium I: Scrapbook, 1925–1927," HMP.

27. Martland, "Occupational Poisoning," 1274–75.

28. Ibid., 1274.

29. Ibid., 1275.

CHAPTER EIGHT

1. Two good popular treatments of the Illinois cases are Mayo, "We Are All Guinea Pigs," 18ff., and Conroy, "Radium City," *Illinois Times*, August 23–29, 1984, 4+.

2. Cruse's death certificate, her application to the Illinois Industrial Commission, and clipping, *Newark Evening News*, July 26, 1928, Cruse medical file, all in ANL. Looney's death certificate; Roberta Simon, memo reporting a conversation with Looney's aunt and fellow dialpainter Thelma McBriarly, July 9, 1981; and A. M. Brues, memo reporting on a conversation with Looney's sister and fellow dialpainter Catherine White, January 29, 1973, all in Looney medical file, ANL.

3. Purcell medical file, ANL; memo regarding conversation with Purcell, January 16, 1973, Purcell medical file, ANL; Donahue testimony, February 10, 1938, vol. 31, Radium Archives, ANL.

4. Donahue and Purcell testimony, February 10–11, 1938, vol. 31, Radium Archives, ANL.

5. Testimony of Charlotte Purcell, Catherine Donahue, and Tom Donahue, February 10–11, 1938, vol. 31, Radium Archives, ANL.

6. Vol. 31, Radium Archives, ANL (date of Vallat's filing); *Vallat v. The Radium Dial Co.*, 360 Ill. 407–9 (details of the case).

7. Although the discussion about health and safety reform is drawn from Beckner, *History of Labor Legislation in Illinois*, my interpretation of women reformers' role in labor legislation is reinforced by Tax, *The Rising of the Women*, esp. 82–84; Sklar, *Florence Kelley and the Nation's Work*, esp. 206–7, 236; Nugent, "Workers' Health in Chicago"; Buenker, "Urban Immigrant Lawmakers"; and Howard, *Illinois*, 502–19. Another good source on labor legislation is Staley, *A History of the Illinois State Federation of Labor*.

During these decades legislators were far more concerned with controlling labor radicalism than protecting workers' health. In response to an upsurge of labor radicalism (called forth by the hard times following the panic of 1873), an 1877 law made illegal the obstruction of railroads specifically and any business operation generally. The Armed Workmen

Law of 1879 attempted to secure a monopoly in firearms and military organization for the state militia but did not succeed in suppressing armed labor organizations. The 1880s success of the Knights of Labor and other labor groups ended with the Haymarket Riot in 1886, which led to enactment of the Merritt Conspiracy Law of 1887. This held guilty all members of a "conspiracy" for the actions of any one member, stipulating that to belong to such "conspiracy" one did not actually have to meet or communicate with one's fellow conspirators, but only write or speak anything encouraging them in their beliefs or actions. Labor had enough influence to prevent the statute from ever being enforced, and the Illinois State Federation of Labor won its repeal by 1891. More successfully, Illinois manufacturers began in 1886 to use the courts against unions. They pioneered in the use of injunctions to break strikes and boycotts. The Illinois Supreme Court invalidated labor legislation such as an 1893 law forbidding yellow dog contracts and an 1897 law protecting unions from prosecution under an Illinois antitrust law. Unions in Illinois were strong enough to win legislation but seldom saw it survive through court challenges. See Beckner, *History of Labor Legislation in Illinois*, 9–23, 53, 290–92, 225–26.

8. Beckner, *History of Labor Legislation in Illinois*, 254–68; Sklar, *Florence Kelley and the Nation's Work*, 206–7, 214, 216, 222, 234–45.

9. Beckner, *History of Labor Legislation in Illinois*, 188–89, 254–68; Sklar, *Florence Kelley and the Nation's Work*, esp. 234–36, 239, 254–55, 281; Sicherman, *Alice Hamilton*, 113; Wade, "Florence Kelley," 316–19. Ritchie I is more properly cited as *Ritchie v. People*, 155 Ill. 98.

10. Beckner, *History of Labor Legislation in Illinois*, 189–208. The decision is properly cited as *Ritchie & Co. v. Wayman*, 244 Ill. 509.

11. Ibid., 227–28, 250, 293–94.

12. Ibid., 188, 228–31.

13. Cohen, *Making a New Deal*, 82. Richard Jensen (*Illinois: A Bicentennial History*, 104) writes that in general social welfare legislation fared poorly in Illinois except mining and factory safety laws, which "appealed to the ethic of efficiency."

14. Nugent, "Workers' Health in Chicago," 1; Howard, *Illinois*, 424; Beckner, *History of Labor Legislation in Illinois*, 231–40, 296–301. See specifically Hurd, "Act to Provide for the Health, Safety, and Comfort of Employees," in *Revised Statutes of the State of Illinois, 1911*.

15. Beckner, *History of Labor Legislation in Illinois*, 272.

16. Ibid., 272–78; Sicherman, *Alice Hamilton*, 156–58; Nugent Young, "Interpreting the Dangerous Trades," 30–33.

17. Beckner, *History of Labor Legislation in Illinois*, 278–79.

18. Ibid., 279–82; Hurd, "Protection from Occupational Diseases," in *Revised Statutes of the State of Illinois, 1911*, 1145–48 (quotation, 1145). On the background of this act, see Staley, *History of the Illinois State Federation of Labor*, 277, and Bogart and Mathews, *The Modern Commonwealth*, 184; also Sicherman, *Alice Hamilton*, 156–58, and Nugent Young, "Interpreting the Dangerous Trades," 30–33.

19. Beckner, *History of Labor Legislation in Illinois*, 280–82; Cahill, "Section 199 of Occupational Disease Act," in *Revised Statutes of the State of Illinois, 1923*, 1664.

20. See *Burns v. Industrial Commission*, 356 Ill. 602, 191 N.E. 225.

21. Dodd, *Administration of Workmen's Compensation*, 761. Specifically, see *Raymond v. Industrial Commission*, 354 Ill. 586, 188 N.E. 861.

22. Amenta et al., "The Political Origins of Unemployment Insurance," 151.

23. Ibid., 150, 164–65, 169–73, 179.

24. *Vallat v. The Radium Dial Co.*, 407–10. Paul Clemens tells me that these were common grounds for criticizing much of the New Deal legislation of the 1930s.

25. *Vallat v. The Radium Dial Co.*, 410, 412–13.

26. For dismissal of the dialpainters' cases, see Donahue testimony, February 11, 1938, vol. 31, Radium Archives, ANL; for the new law, see clipping, March 17, 1936, as featured in Mayo, "We Are All Guinea Pigs"; on the new suits, see memo on news story in *Chicago Tribune*, February 11, 1938, vol. 1, Radium Archives, ANL; on Vallat, see in her medical file the report of a conversation with Frances O'Connel, recorded by Carol Croft, August 8, 1972, ANL.

27. There is a great deal of material on these hearings in vol. 31, Radium Archives, ANL, but see specifically Brues, notes on *Chicago Tribune* articles, esp. February 11, 1938; Thomas Donahue testimony, February 11, 1938; "Reply Brief for Petitioner"; and clipping, "Radium Victims Tells 'Living Death,'" *Chicago Herald Examiner*, February 11, 1938. See also memo, September 21, 1972, Catherine Donahue medical file, ANL.

28. Clippings, "Ottawa's Doomed Women" and "10 Gone! Victims Ask 'Who's Next?,'" vol. 1, Radium Archives, ANL; "Industrial martyr" photo and caption, reproduced in Mayo, "We Are All Guinea Pigs"; clipping, "Radium Dial Deals Death to Ninth of Suicide Club," March 14 (no year), vol. 1, Radium Archives, ANL.

29. The settlement figures are from the clipping, "Mrs. Donahue, Victim of Radium Poisoning, Dies," *Ottawa Daily Times*, July 27, 1938, vol. 1, Radium Archives, ANL, which also discusses her death and the Radium Dial appeal, as did the *Chicago Tribune* as reported by Brues, notes on an article published on July 28, 1938, vol. 31, Radium Archives, ANL. Different figures for Donahue's settlement are given in Brues's notes on a *Tribune* article dated April 5, 1938, vol. 31, Radium Archives, ANL: described there is a $3,740 total settlement that was based on an award of $11 weekly for missed work from 1934 to 1940, a $277 annual pension, and $2,500 in medical expenses.

30. On documenting the bond, see clipping, *Ottawa Daily Times*, July 27, 1938, Catherine Donahue medical file, ANL; on documenting self-insurance, see Joseph Kelly, President of Radium Dial, to Illinois Industrial Commission, November 2, 1928, read by the dialpainter attorney Leonard Grossman during hearings before the Illinois Industrial Commission, in vol. 31, Radium Archives, ANL.

31. For the dialpainters' version of the new company's start-up, see memo of conversation with Emma Peppler filed by C. D. Croft, Rufus Reed medical file, ANL; for the second version, see Carol Croft and Austin Brues, memo, September 27, 1972, vol. 1, Radium Archives, ANL; for the third version, see Mayo, "We Are All Guinea Pigs," and Conroy, "Radium City."

32. On Standard Chemical and Belgian radium, see Reference Department, Carnegie Library, Pittsburgh (a short article on the radium industry), in vol. 10, Radium Archives, ANL, and Swen Kjaer report, March 9, 1929, SKP. On the formation of the Radium Service Co. of America, see Kjaer report, February 25, 1929, SKP. "Chronological Summary,"

vol. 10, Radium Archives, ANL, also contains details on Radium Chemical and the Belgians and on Kelly as head of Luminous Processes. For formation of Radium Luminous and its relation to Radium Dial and Radium Chemical, see Kjaer report, March 9, 1929, SKP, and R. Evans, memo on conversation with Rufus Reed, May 13, 1959, Reed medical file, ANL. For the 1936 Radium Chemical consolidation, see "Radium Chemical Company, Certificate of Consolidation, July 28, 1936," vol. 29, Radium Archives, ANL, and "Chronological Summary," cited above. For interlocking directorships in 1973, see Dun and Bradstreet Information Report, July 19, 1973, vol. 1, Radium Archives, ANL.

33. *People ex rel. Radium Dial Co. v. Ryan*, 371 Ill. 597, 21 N.E. 2d 749.

34. "Court Upholds Award in Girl's Radium Death," *Chicago Tribune*, October 10, 1939, and "15 Walking Ghosts Jilted by Justice," *Chicago Daily Times*, July 7, 1937, clippings, vol. 1, Radium Archives, ANL; medical files of Charlotte Purcell (who received $300), Irene Politte, Catherine Touchette, and Helen (Munch) Albee (who withdrew their suits), and Romilda Bierman (whose file reports that she received "some settlement"), all at ANL.

35. Evans's story is from his "Inception of Standards," 442–43, and "Origins of Standards for Internal Emitters." This account is also outlined in Hacker, *The Dragon's Tail*, 24–25, and Stannard, *Radioactivity and Health*, 1413. For Evans on a safe level of deposited radium, see his "Protection of Radium Dial Workers," 256, where he cites as evidence of his earlier conclusions his "Quantitative Determination" and "Apparatus for the Determination of Minute Quantities," among other papers. Federal standards were recommended in Curtiss, "Safe Handling of Radioactive Luminous Compound," and in "Occupational Hazards to Young Workers, Radioactive Substances." See also "Safe Health Practices Relating to the Use and Handling of Radioactive Luminous Paints."

36. Evans, "Protection of Radium Dial Workers," 267; Morse and Kronenberg, "Radium Painting," 814–15. In the latter work, discrepancies in the number of workers in the study may be found (p. 815) by comparing the thirty-five workers described in the text (twenty-one above the tolerance level) to the thirty-six workers (sixteen or seventeen above the tolerance level) appearing in table III, "Breath Radon Samples," and to the thirty-nine workers (seventeen above the tolerance level) in the unlabeled table.

37. Morse and Kronenberg, "Radium Painting," 810. On benzene poisoning, see Clark, "The Menace of Benzene."

38. Morse and Kronenberg, "Radium Painting," 814–16.

39. Ibid., 810–11.

40. Godfrey Schmidt to Evans, December 6, 1943; Evans to Martland, December 14, 1943; Evans to George Keogh, October 16, 1943; and Martland to Evans, November 22, 1943 — all in "Radium IX: Scrapbook, 1939–1944," HMP. For further information on Martland's changing ideas about a safe level of radium deposition, see Martland to Dr. Edwin Ziegler, n.d. (ca. 1940), "Radium VIII: Scrapbook, 1941–1944," HMP, and data on the struggle over the American Standards Association subcommittee on standards for radium, radon, and gamma radiation, "Radium IX: Scrapbook, 1939–1944," HMP.

41. Morse and Kronenberg, "Radium Painting," 814–15, 817.

42. Ibid., 815, 817.

43. Ibid., 812.

44. Anecdotes from Mayo, "We Are All Guinea Pigs"; Conroy, "Radium City"; "The

Dial Painters: The Live and the Dead Still Raise Questions about Job Radiation," clipping, *Wall Street Journal*, September 19, 1983, vols. 1, 52, Radium Archives, ANL.

45. Most of this narrative is from Conroy, "Radium City," but see also clipping, "Ottawa Firm Cited for Radioactivity Level," March 11, 1977, vol. 1, Radium Archives, ANL.

46. The suit was eventually withdrawn; the former dialpainters continued to seek evidence that they were injured by their exposure to radium. "The Dial Painters: The Live and the Dead Still Raise Questions about Job Radiation," clipping, *Wall Street Journal*, September 19, 1983, vols. 1, 52, Radium Archives, ANL.

47. Again, I rely on Conroy, "Radium City," but see also clippings, December 1980, June 1981, vol. 1, Radium Archives, ANL.

48. Conroy, "Radium City."

49. "Evidence Suggests Contaminated Material Remains at Clean Site," clippings, *Banner-Herald*, August 1, 5, 1982, vol. 1, Radium Archives, ANL.

50. Conroy, "Radium City."

51. Ibid.; clipping, *Ottawa Daily Times*, November 4, 1983, vol. 1, Radium Archives, ANL.

52. Rowland and Lucas, "Radium-Dial Workers," 235–36 (table 2); Stehney et al., "Survival Times of Women Radium Dial Workers," 344.

CONCLUSION

1. Clippings: unidentified, August 17, 1985, and *Newark Star-Ledger*, December 22, 1983, February 3, March 4, 1984, all in vol. 52, Radium Archives, ANL; "Sellers of Toxic Land Ruled 'Forever' Liable," *Newark Star-Ledger*, March 28, 1991, 1, 40.

2. "Sellers of Toxic Land Ruled 'Forever' Liable," *Newark Star-Ledger*, March 28, 1991, 1, 40; Arthur Roeder to Cecil Drinker, June 6, 1924, Drinker testimony, reel 1, RBP.

3. "Sellers of Toxic Land Ruled 'Forever' Liable," *Newark Star-Ledger*, March 28, 1991, 40.

4. For this insight into the motivations of radium industry officials I must thank Dr. James Stebbings, an epidemiologist involved in studying the health effects of deposited radionuclides.

5. Medical files of Viol, von Sochocky, Roth, Michael Lemen, and Barker, all at ANL; interview with Florence Wall; J. Lieben, memo, June 22, 1972, in Irene LaPorte medical file, ANL.

6. On industrial safety, employers' liability, and workers' compensation, see Asher, "Business and Workers' Welfare"; Lubove, "Workmen's Compensation"; and Weinstein, "Big Business."

7. Martland, Conlon, and Knef, "Some Unrecognized Dangers," 1770; Martland, "Occupational Poisoning," 1245–47; Hoffman, "Radium (Mesothorium) Necrosis," 963–64; Flinn, "Radioactive Material: An Industrial Hazard?," 2079.

8. Curran, *Founders of the Harvard School of Public Health*, 53, 55, 165, 230–34. This transformation can be traced in the records of the Industrial Hygiene and Physiology Departments, Archives of the Countway Library of Medicine, Harvard Medical School, Boston.

9. On conflict between Labor Departments and Health Departments, see Nugent Young, "Interpreting the Dangerous Trades," esp. chap. 3; Rosner and Markowitz, "Research or Advocacy?."

10. Zwemer, "History of the Consumers' League of New Jersey," 29, box 5, CLNJP; documents from Scrapbook, box 4, CLNJP: clippings, April 1932, February 19, 1933; CLNJP Bulletin, November 1929, May 1930; and letter soliciting money, n.d. (ca. 1931–32).

11. "The Dial Painters' Story," transcript of taped interview, "Radium Cases—Dyckman Reminiscences" folder, box 84, CLNJP; unidentified clipping, n.d. (ca. 1933), and flyer, n.d. (ca. 1937), both in Scrapbook, box 4, CLNJP; Zwemer, "History of the Consumers' League of New Jersey," 29–30.

12. Zwemer, "History of the Consumers' League of New Jersey," 30.

13. Ibid., 26, 30–32.

14. National Consumers' League, "State Regulation of Radioactive Material," report, 1958, and Susanna Zwemer to Leo Goodman, atomic energy adviser, United Auto Workers, May 28, 1962, both in reel 84, NCLR (see other related documents on reels 84–85); C. W. Wallhausen, Vice President of USRC, to U.S. Atomic Energy Commission, October 16, 1961, box 83, CLNJP. One wonders what role the Consumers' League might have played in the legal suit.

15. On continuing efforts to protect workers and the public from radioactivity, see documents on reels 84–85, NCLR.

16. Berry to Mrs. G. W. B. Cushing, President of the Consumers' League of New Jersey, June 18, 1928, box 84, CLNJP.

17. Schaub, "Radium," 140–41. The rest of the autobiography was destroyed by her family. See Schaub medical file, ANL.

18. Clipping, *Newark Sunday Call*, May 13, 1928, reel 2, RBP; Schaub, "Radium," 140–41; Schaub to Harrison Martland, December 5, 1930, "Radium IV: Scrapbook, July 1930–November 1931," HMP.

19. Unidentified newspaper clipping, reel 2, RBP; clipping, *Evening Graphic*, May 28, 1928, Schaub medical file, ANL.

20. Schaub, "Radium," 141; Schaub to James Ewing, April 4, 1930, in Sharpe, "Radium Osteitis," 1071–72; Martland, "The Occurrence of Malignancy," Case 6, 2467–70.

21. Sharpe, "Radium Osteitis," 1073–75, 1077–78; Schaub to Dr. Robert Humphries, October 28, 1931, "Radium V: Scrapbook, October 1931–July 1932," HMP; Schaub, "Radium," 141.

22. Sharpe, "Radium Osteitis," 1075–77, 1079–82; letters among Lloyd Craver, Robert Humphries, and James Ewing, 1932, Schaub's medical file, ANL. Schaub's burial site is given in her medical file.

BIBLIOGRAPHY

This bibliography is organized as follows:
    Manuscript Collections
    Interviews
    Autobiographies
    Newspapers
    Published Legal Decisions
    Contemporary Sources (including government documents)
    Secondary Sources

MANUSCRIPT COLLECTIONS

Argonne, Illinois
Center for Human Radiobiology, Argonne National Laboratory
    Robley Evans Papers
    Commissioner Harris (New York City Health Department) Correspondence Files
    Swen Kjaer (Department of Labor investigator) Papers (also filed as part of
        Radium Archives)
    Medical Files
    New Jersey Department of Public Health Files on Everett Field
    Radium Archives (looseleaf notebooks)
    U.S. Radium Corporation Records

Boston, Massachusetts
Harvard Medical School, Harvard University
    Archives of the Francis A. Countway Library of Medicine
        Industrial Hygiene Department Records
        Physiology Department Records
        Frederick Shattuck Papers

Cambridge, Massachusetts
Schlesinger Library, Harvard University
    Consumers' League of Massachusetts Papers
    Alice Hamilton Papers
    Hamilton Family Papers

Chicago, Illinois
American Medical Association Headquarters
    Department of Investigation Records

Ithaca, New York
Labor Management Documentation Center, New York State School of Industrial and

Labor Relations, Cornell University
American Association for Labor Legislation Papers

Newark, New Jersey
Special Collections and Archives, University Libraries, University of Medicine and
      Dentistry of New Jersey
   Harrison Martland Papers
Newark Public Library
   Radiation and Radium Clipping File
New Jersey Historical Society
   Women's Club of Orange Records

New Brunswick, New Jersey
Rutgers University Libraries
   Special Collections and University Archives
      Consumers' League of New Jersey Papers
      League of Women Voters of New Jersey Papers

New York, New York
Columbia University Rare Book and Manuscript Library
   Frederick Hoffman Papers
The New York Public Library
   Astor, Lenox and Tilden Foundations, Rare Books and Manuscripts Division
      American Fund for Public Service Records
      Florence Kelley Papers (filed with Nicholas Kelley Papers)

Philadelphia, Pennsylvania
Library of the College of Physicians of Philadelphia
   Edward Bell Krumbhaar Papers

Saratoga Springs, New York
Saratoga Springs State Park
   Saratoga Springs State Park Records

College Park, Maryland
National Archives, Archives II
   Federal Trade Commission Records, Docket 1756

Washington, D.C.
Library of Congress
   Raymond Berry Papers (filed as Records of the National Consumers' League)
   National Consumers' League Records
National Archives
   Record Group 62: U.S. Council of National Defense
   U.S. Department of Labor

Record Group 86: Women's Bureau Papers
Record Group 100: Bureau of Labor Standards
Record Group 174: Commission on Industrial Relations, 1912–25
Record Group 257: Bureau of Labor Statistics
Record Group 88: U.S. Food and Drug Administration
Record Group 90: U.S. Public Health Service

INTERVIEWS

Dr. William Castle, Boston, March 23, November 16, 1984
Grace Swanner, Saratoga Springs, New York, 1985
Florence Wall, New York, New York, 1986

AUTOBIOGRAPHIES

Hamilton, Alice. *Exploring the Dangerous Trades: The Autobiography of Alice Hamilton,*
    M.D. 1943. Reprint, Boston: Northeastern University Press, 1985.
Hardy, Harriet. *Challenging Man-Made Disease.* New York: Praeger, 1983.
Schaub, Katherine. "Radium." *Survey Graphic* 68 (May 1, 1932): 138–41.

NEWSPAPERS

American Weekly
Banner-Herald (Athens, Ga.)
Chicago Tribune
Daily Republican Times (Ottawa, Ill.)
Newark Evening News
Newark Star-Eagle
Newark Star-Ledger
New York American
New York Sun
New York Times
New York World
Ottawa Daily Times
Wall Street Journal
Waterbury Republican

PUBLISHED LEGAL DECISIONS

Burns v. Industrial Commission, 356 Ill. 602, 191 N.E. 225 (1934).
Cuff's v. Newark and New York Railroads, 35 NJL 17 (1870), 35 NJL 574 (1871).

LaPorte v. United States Radium Corp., 13 F. Supp. 263 (1935).

McAndrews v. Collard, 42 NJL 189 (1880).

Marshall v. Wellwood, 38 NJL 339 (1876).

Muller v. Oregon, 208 U.S. 412 (1908).

P. Ballantine & Sons v. The Public Service, 76 NJL 358 (1908), 86 NJL 331 (1914).

People ex rel. Radium Dial Co. v. Ryan, 371 Ill. 597, 21 N.E.2d 749 (1939).

Raymond v. Industrial Commission, 354 Ill. 586, 188 N.E. 861 (1933).

Ritchie v. People, 155 Ill. 98 (1895) [Ritchie I].

Ritchie & Co. v. Wayman, 244 Ill. 509 (1910) [Ritchie II].

Rylands v. Fletcher (England), L.R. 3 H.L. 330 (1868).

Smith v. Oxford Iron Co., 42 NJL 467 (1880).

United States Radium Corp. v. Globe Indemnity Co., New Amsterdam Casualty Co., Aetna Life Insurance Co., and American Mutual Liability Insurance Co., 13 N.J. Mis. R. 316, 116 NJL 90 (1935).

Vallat v. The Radium Dial Co., 360 Ill. 407 (1935).

CONTEMPORARY SOURCES (INCLUDING GOVERNMENT DOCUMENTS)

Allen, John. "The Present Status of Radium Therapy." *Colorado Medicine* 11 (1914): 18–24.

American Institute of Medicine. *Bibliography on Radium*. Published for U.S. Radium Corp., 1922.

———. *Radium: Abstracts*. Published for U.S. Radium Corp., 1922.

"The American Radium Society." *Radium* 8 (December 1916): 58–59.

Andrews, John. "Phosphorous Poisoning in the Match Industry in the United States." U.S. Bureau of Labor Statistics Bulletin 86. 1910.

Arlidge, J. T. *The Hygiene, Diseases, and Mortality of Occupations*. London: Percival and Co., 1892.

Aub, Joseph, Robley Evans, D. M. Gallagher, and D. M. Tibbets. "Effects of Treatment on Radium and Calcium Metabolism in the Human Body." *Annals of Internal Medicine* 11 (February 1938): 1443–63.

Aub, Joseph, Robley Evans, Louis Hemplemann, and Harrison Martland. "The Late Effects of Internally Deposited Radioactive Materials in Man." *Medicine* 31 (1952): 221–329.

Barker, Howard. "Radioactive Determination on Post-Mortem Specimens Taken from One Exposed to Radioactive Material over a Prolonged Period." *American Journal of Roentgenology* 21 (January 1929): 31–37.

Baskerville, Charles. *Radium and Radio-active Substances: Their Application Especially to Medicine*. Philadelphia: Williams, Brown and Earle, 1905.

Bates, Lindon (Mrs.). "Mercury Poisoning in the Industries of New York City and Vicinity." Women's Welfare Department, New York and New Jersey section of the National Civic Federation, 1911.

Beers, George. "Compensation for Occupational Diseases." *Yale Law Journal* 37 (1928): 583–87.

Bissel, Joseph. "Intravenous Injection of Radium Element." *Pennsylvania Medical Journal* 18 (November 1914): 129–33. Abstracted in *Radium* 5 (April 1915): 19–20.

———. "The Medicolegal Aspects of Radium Therapy." *Medical Record* 89 (July 1917): 102–4. Reprinted in *Radium* 9 (September 1917): 97–101.

———. "Radioactivity in Therapeutics." *Medical Record* 93 (1918), 142–44. Cited in *Radium: Abstracts*, American Institute of Medicine, 1922.

———. "Radium Therapeutics Otherwise Than for Malignancy." *Medical Record* 87 (June 1915): 1023–24. Reprinted in *Radium* 5 (August 1915): 105–7.

———. "Some Radium Therapeutics." *Medical Record* 86 (July 1914): 55–58.

[Bissel]. "Joseph Biddleman Bissel." Obituary. *Radium* 12 (January 1919): 63–64.

Blum, Theodor. "Osteomyelitis of the Mandible and Maxilla." *Journal of the American Dental Association* 11 (September 1924): 802–5.

Bogart, Ernest, and John Mathews. *The Modern Commonwealth, 1893–1918.* Chicago: McClury, 1922.

Bowing, Harry, and Robert Fricke. "Curie Therapy." In *The Science of Radiology*, edited by Otto Glasser, pp. 267–90. Springfield, Ill.: Thomas, 1933.

Bremner, Robert. "Brief." In "Creation of a Bureau of Labor Safety," p. 11. U.S. Senate Report, July 27, 1914.

Brues, A. M. "The Long-Term Follow-Up of Radium Dial Painters and Thorium Workers." In *The Medical Basis for Radiation Accident Preparedness*, edited by K. F. Hubner and S. A. Fry, pp. 441–50. New York: Elsevier North Holland, 1980.

Burlingame, C. C. "The Art, Not the Science, of Industrial Medicine." *Journal of Industrial Hygiene* 2 (February 1921): 368–73.

Cahill, James. *Revised Statutes of the State of Illinois, 1923.* Chicago: Collagan and Co., 1923.

Cameron, William. "Mechanical and Physical Agents: Special Reference to Radium." *Pennsylvania Medical Journal* 17 (December 1913): 194–202.

———. "Radium Emanation Therapy in Arthritis Deformans." *Radium* 1 (August 1913): 3–5.

———. "A Treatment of Alveolar Pyorrhea with Radium Emanation: A Review." *Radium* 2 (December 1913): 41–43.

Cameron, William, and Charles Viol. "Classification and Relative Value of the Various Methods Employed for the Internal Administration of Radium Emanation and Radium Salts." *Radium*, n.s., 2 (July 1923): 136–48.

Cameron, William, and Charles Viol. "Classification of the Various Methods Employed in the Internal Administration of Radium Emanation and Radium Salts." *Radium* 4 (January 1915): 57–68.

Canan, Robert. "Industrial Law: Workmen's Compensation, Occupational Diseases and the Statute of Limitations." *Columbia Law Review* 24 (July 1936): 594–604.

Castle, William B., Katherine Drinker, and Cecil Drinker. "Necrosis of the Jaw in Workers Employed in Applying a Luminous Paint Containing Radium." *Journal of Industrial Hygiene* 7 (August 1925): 371–82.

Chamberlain, J. P. "Workmen's Compensation for Diseases Due to Employment." *American Bar Association Journal* 10 (September 1924): 647–51.

Colwell, J., and Sidney Russ. "Radium as a Pharmaceutical Poison." *Lancet* 2 (July 23, 1932): 221–23.

Compiled Statutes of New Jersey. Newark: Soney and Sage Co., 1911.

Craver, Lloyd, and Herman Schlundt. "Parathyroid Extract and Viosterol Treatment of Radium Poisoning." *JAMA* 105 (September 21, 1935): 959–60.

Cumulative Supplement to the Compiled Statutes of New Jersey, 1911–1924. Newark: Soney and Sage Co., 1925.

Curtiss, L. F., et al. "Safe Handling of Radioactive Luminous Compound." Handbook H27. U.S. Department of the Interior, National Bureau of Standards, 1941.

"Dangers in the Use and Handling of Radioactive Substances." U.S. Bureau of Labor Statistics. *Monthly Labor Review* 22 (January 1926): 171–74.

Delano, Samuel. "A Study in the Internal Therapeutics of Radium." *Radium* 6 (October 1915): 1–15.

Dodd, Walter. *Administration of Workmen's Compensation*. New York: The Commonwealth Fund, 1936.

Doehring, C. F. W. "Factory Sanitation and Labor Protection." U.S. Bureau of Labor Bulletin 44. 1903.

Dublin, Louis, and P. Leiboff. "Occupation Hazards and Diagnostic Signs." U.S. Bureau of Labor Statistics Bulletin 306. 1922.

Edsall, David. "Industrial Poisoning." *American Labor Legislation Review* 2 (1912): 231–34.

———. "The Prevention of Disease in War Industries: Extent and Importance of the Problem." *Medical Record* 93 (1918): 611.

———. "The Study of Occupational Diseases in Hospitals." U.S. Bureau of Labor Statistics. *Monthly Labor Review* 5 (1917): 169–85.

———. "Supposed Physical Effects of the Pneumatic Hammer on Workers in Indiana Limestone." *Public Health Reports* 33 (March 22, 1918): 394–403.

Edsall, David, F. P. Wilbur, and Cecil Drinker. "The Occurrence, Course, and Prevention of Chronic Manganese Poisoning." *Journal of Industrial Hygiene* 1 (August 1919): 183–93.

"Effects of Radium Therapy upon Physiology and Metabolism." National Radium Products Co., 1924.

"Effects of Use of Radioactive Substances on the Health of Workers." U.S. Bureau of Labor Statistics. *Monthly Labor Review* 22 (May 1926): 19–31.

Engelmann, W. "Radium Emanation Therapy." *Lancet* 1 (May 3, 1913): 1225–58.

Evans, Robley. "Apparatus for the Determination of Minute Quantities of Radium." *Review of Scientific Instruments* 6 (1935): 99–102.

———. "The Effect of Skeletally Deposited Alpha-ray Emitters in Man." *British Journal of Radiology* 39 (1966): 881–95.

———. "Inception of Standards for Internal Emitters, Radon, and Radium." *Health Physics* 41 (1981): 437–48.

———. "Origins of Standards for Internal Emitters." In *Health Physics: A Backward Glance*, edited by R. Kathren and P. Ziemer, pp. 141–57. New York: Pergamon Press, 1980.

———. "Protection of Radium Dial Workers and Radiologists from Injury by Radium." *Journal of Industrial Hygiene and Toxicology* 25 (September 1943): 253–69.

———. "The Quantitative Determination of the Radium Content of Living Persons." *American Journal of Roentgenology and Radium Therapy* 17 (1937): 368–78.

———. "Radium Poisoning: A Review of Present Knowledge." *American Journal of Public Health* 23 (October 1933): 1017–23.

———. "Radium Poisoning, II: The Quantitative Determination of the Radium Content and Radium Elimination Rate of Living Persons." *American Journal of Roentgenology* 37 (March 1937): 368–78.

Field, C. E. "The Efficiency of Radioactive Waters for the Control of Faulty Elimination." *Medical Record* 87 (1915): 390–94.

———. "Radium: Its Physio-Chemical Properties Considered with Relation to High Blood Pressure." *Radium* 7 (April 1916): 23–29.

"Fifty Years of Occupational Health." Pamphlet. U.S. Public Health Service. Division of Occupational Health. 1964.

Fischel, Ellis. "Radium and Internal Medicine." *Annals of Clinical Medicine* 3 (September 1924): 245–48.

Fishbein, M. "Radioactive Waters." *Scientific American* 141 (November 1929): 442.

Fisher, Irving. *Report on National Vitality, Its Wastes and Conservation.* Committee of One Hundred on National Health Bulletin 30. Washington, D.C.: GPO, 1909.

Flinn, Frederick. "A Case of Antral Sinusitis Complicated by Radium Poisoning." *Laryngoscope* 37 (May 1927): 341–49.

———. "Dangers of Internal Radium Therapy." *American Journal of Physical Therapy* 9 (June 1932): 65–70. Also in *Archives of Physical Therapy, X-ray, and Radium* 8 (August 1932): 476–83.

———. "Effects of Radium and Radon Gas Taken Internally" (a response to a letter). *JAMA* 98 (June 25, 1932): 2306.

———. "Elimination of Radium Salts from the Human Body." *JAMA* 96 (May 23, 1931): 1763–65.

———. "Parathormone in the Treatment of Radium Poisoning." *Johns Hopkins Hospital Bulletin* 45 (1929): 269.

———. "Radioactive Material: An Industrial Hazard?" *JAMA* 87 (December 18, 1926): 2078–81.

———. "Radium Salts and Emanation." *American Journal of Roentgenology* 31 (June 1934): 830–36.

———. "Some of the Newer Industrial Hazards." *Boston Medical and Surgical Journal* 197 (January 12, 1928): 1309–14.

———. "Some Precautions to Be Taken When Making Tests for Radioactivity in the Living Body." *American Journal of Roentgenology* 22 (December 1929): 554–56.

———. "Stimulating Action of Radio-active Deposits in the Body." *Radiology* 23 (September 1934): 331–38.

Florin, A. A., et al. "Epidemiological Followup of New Jersey Radium Cases, Progress Report to April, 1964." New Jersey State Department of Health, Radium Research Project, NYO-2181-2, 1964.

Fuerstenberg, A. "Radium Emanation Treatment for Rheumatism and Gout." *Radium* 1 (April 1913): 12–14.

Gettler, Alexander. "Radio-activity in the Human and Its Detection." *Archives of Physical Therapy, X-ray, and Radium* 8 (August 1932): 473–76.

Gettler, Alexander, and Charles Norris. "Poisoning from Drinking Radium Water." *JAMA* 100 (February 11, 1933): 400–402.

Greenough, R. B. "The Value of Radium in the Treatment of Disease." *Rhode Island Medical Journal* 4 (1921): 71.

Gudzent, F., and L. Halberstaedter. "Occupational Injuries Due to Radioactive Substances." *Deutsche medizinische Wochenschrift* 40 (1914): 633–35. Reprinted in *Radium* 3 (May 1914): 27–28.

Hamilton, Alice. "A Discussion of the Etiology of So-Called Aniline Tumors of the Bladder." *Journal of Industrial Hygiene* 3 (May 1921): 16–28.

———. "Ether Poisoning in the Manufacture of Smokeless Powder." *Journal of Industrial Hygiene* 2 (June 1920): 41–49.

———. "The Growing Menace of Benzene (Benzol) Poisoning in American Industry." *JAMA* 78 (March 4, 1922): 627–30.

———. "Industrial Diseases: With Special Reference to the Trades in Which Women Are Employed." *Charities and the Commons* 20 (September 5, 1908): 655–58.

———. "Industrial Poisoning by Compounds of the Aromatic Series." *Journal of Industrial Hygiene* 1 (August 1919): 200–12.

———. *Industrial Poisons in the United States.* New York: Macmillan, 1925.

———. *Industrial Toxicology.* New York: Harper and Brothers, 1934 (2d and 3d editions with Harriet Hardy, 1949, 1974; revised by Asher Finkel as *Hamilton and Hardy's Industrial Toxicology,* 1983).

———. "Inorganic Poisons, Other Than Lead, in American Industry." *Journal of Industrial Hygiene* 1 (June 1919): 89–102.

———. "Lead Poisoning in American Industry." *Journal of Industrial Hygiene* 1 (May 1919): 8–21.

———. "Lead Poisoning in Illinois." *JAMA* 61 (April 29, 1911): 1240–44.

———. "The Lessening Menace of Benzol Poisoning in American Industry." *Journal of Industrial Hygiene* 10 (1928): 227–33.

———. "Occupational Poisoning in the Viscose Rayon Industry." U.S. Department of Labor. Washington, D.C.: GPO, 1940.

———. *Report of the Illinois Commission on Occupational Diseases.* Springfield, 1911.

———. "Some New and Unfamiliar Poisons." *New England Journal of Medicine* 215 (September 3, 1936): 425–30.

———. "TNT as an Industrial Poison." *Journal of Industrial Hygiene* 3 (July 1921): 16–28.

———. "The Toxicity of the Chlorinated Hydrocarbons." *Yale Journal of Biology and Medicine* 15 (July 1943): 787–800.

———. "Women in the Lead Industries." U.S. Bureau of Labor Statistics Bulletin 253. 1919.

———. "Women Workers and Industrial Poisons." U.S. Department of Labor, Women's Bureau, Bulletin 57. 1926.

Hammond, Roland. "A Case of Industrial Radium Poisoning with Bony Changes." *Rhode Island Medical Journal* 15 (January 1932): 9–11.

Hayhurst, Emery. *A Survey of Industrial Health-Hazards and Occupational Disease in Ohio.* Columbus: F. J. Hear Print Co., 1915.

"Health Problems of Women in Industry." U.S. Department of Labor, Women's Bureau, Bulletin 18 (revised as Bulletin 136 in 1935). 1931.

Hinsdale, Guy. "Mineral Springs: Their Analyses, Classification, Therapeutic Uses, Radioactivity, and Newer Methods of Application, with Special Reference to American Springs." *Transactions of the American Climatological Association* 29 (1913): 64–94.

Hoffman, Frederick. "Mortality from Consumption in Dusty Trades." U.S. Bureau of Labor Bulletin 79 (pp. 633–875). November 1908.

———. Mortality from Consumption in Dusty Trades." U.S. Bureau of Labor Bulletin 82 (pp. 471–638). May 1909.

———. "Radium (Mesothorium) Necrosis." *JAMA* 85 (September 26, 1925): 961–65.

Hurd, Henry, ed. *Revised Statutes of the State of Illinois, 1911.* Chicago: Chicago Legal News Co., 1912.

"Industrial Hazards of Radioactive Material." *Scientific American* 136 (April 1927): 240.

Ingram, Marylou. *Biological Effects of Ionizing Radiation: An Annotated Bibliography Covering the Years 1898–1957.* 2 vols. U.S. Atomic Energy Commission, Division of Technical Information, 1966.

John, Findley. "Preliminary Report on the Therapeutic Use of Radium Salts." *Illinois Medical Journal* 51 (May 1927): 379–83.

Kober, George M. *Diseases of Occupational Vocational Hygiene.* Philadelphia: P. Blakiston's Son and Co., 1916.

———. *Industrial and Personal Hygiene.* A Report of the Commission on Social Betterment. President's Home Commission, 1908.

———. "Industrial Hygiene." U.S. Bureau of Labor Bulletin 75 (pp. 472–591). 1908.

———. "Prefatory: Historical Review of Industrial Hygiene and Its Effects on Public Health." In *Industrial Health,* edited by George Kober and Emery Hayhurst, pp. xii–lviii. Philadelphia: P. Blakiston's Son and Co., 1924.

———. *Report of the Committee on Social Betterment.* President's Home Commission, 1908.

Kober, George M., and Emery Hayhurst, eds. *Industrial Health.* Philadelphia: P. Blakiston's Son and Co., 1924.

Lazarus-Barlow, W. S. "On the Disappearance of Insoluble Radium Salts from the Bodies of Mice after Subcutaneous Injection." *Archives of the Middlesex Hospital* 30 (1913): 92–94.

———. "On the Histological and Some Other Changes Produced in Animals by Exposure to the Gamma Rays of Radium." In *Medical Uses of Radium: Studies of Effects of Gamma Rays from a Large Quantity of Radium,* edited by the [British] Medical Research Committee, part 4, pp. 33–129. London: H.M. Stationary Office, 1922.

———. "Some Biological Effects of Small Quantities of Radium." *Archives of Radiology and Electrotherapy* 24 (June 1919): 1–12.

Leake, James. "Radium Poisoning." *JAMA* 98 (March 26, 1932): 1077–80.

Legge, Thomas M. "Charles Turner Thackrah: A Pioneer in Industrial Hygiene." *Journal of Industrial Hygiene* 1 (1919–20): 578–81.

———. "Industrial Diseases in the Middle Ages." *Journal of Industrial Hygiene* 1 (1919–20): 550–56.

———. "Medical Supervision in Factories." *Journal of Industrial Hygiene* 2 (1920–21): 66–71.

———. *Shaw Lectures on Thirty Years' Experience of Industrial Maladies*. London, 1929.

———. "Twenty Years' Experience of the Notification of Industrial Diseases." *Journal of Industrial Hygiene* 1 (1919–20): 590–96.

"List of Industrial Poisons and Other Substances Injurious to Health Found in Industrial Processes." U.S. Bureau of Labor Bulletin 100. 1912.

Lloyd, James Henrie. "The Diseases of Occupation." In *Twentieth-Century Practice: An International Encyclopedia of Modern Medical Science*, edited by Thomas L. Stedman, vol. 3. New York: William Wood and Co., 1895.

Lowe, T. Pagan. "On Radium Emanations in Mineral Waters." *Lancet* 1 (1912): 1051–53. Also in *British Medical Journal* 1 (1912): 884–86.

Ludwig, P., and E. Lorensen. "Studies on the Content of Radium Emanation in the Air of the Schneeberger Mines." *Strahlentherapie* 16 (1924): 428–35. Abstracted in *American Journal of Roentgenology* 13 (1925): 597.

McReady, Benjamin. "On the Influence of Trades, Professions, and Occupations." Prize dissertation for 1837 of the New York State Medical Society. Reprinted, Baltimore: Johns Hopkins University Press, 1943.

Martin, Charles, and James Martin. *Low Intensity Radium Therapy*. Boston: Little, Brown, 1957.

Martland, Harrison. "The Danger of Increasing the Normal Radioactivity of the Human Body." *Emanuel Libman Anniversary Volume* (New York: International Press, 1932): 2:781–91.

———. "Microscopic Changes of Certain Anemias Due to Radioactivity." *Archives of Pathology and Laboratory Medicine* 2 (1926): 465–72.

———. "Occupational Poisoning in Manufacture of Luminous Watch Dials." *Monthly Labor Review* 28 (June 1929): 1242–75. Reprinted from *JAMA* 92 (February 9, 1929): 466–73, and (February 16, 1929): 552–59.

———. "The Occurrence of Malignancy in Radio-Active Persons." *American Journal of Cancer* 15 (October 1931): 2435–516.

Martland, Harrison, Philip Conlon, and Joseph Knef. "Some Unrecognized Dangers in the Use and Handling of Radioactive Substances." *JAMA* 85 (December 5, 1925): 1769–76. Presented to the New York Pathological Society by Martland and published in part with a report of the discussion that followed in *Proceedings of the New York Pathological Society*, n.s., 25 (1925): 78–92.

Martland, Harrison, and R. E. Humphries. "Osteogenic Sarcoma in Dial Painters Using Luminous Paint." *Archives of Pathology* 7 (1929): 406–17.

Martland, Harrison, and Harrison Martland Jr. "Placental Barrier in Carbon Monoxide, Barbiturate, and Radium Poisoning." *American Journal of Surgery* 80 (1950): 270–79.

"Memorial on Occupational Diseases." Letter sent from the First National Conference on Industrial Disease to the president of the United States, 1911.

Moore, Richard, and C. F. Whittemore. "The Radioactivity of the Waters of Saratoga Springs." *American Journal of Industrial and Engineering Chemistry* 6 (July 1914): 552–53.

Moriarata, Douglas. "Pros and Cons of Radium." *Albany Medical Annual* 41 (1920): 73–87.

———. "Radium: A Recognition of Its Efficiency and a Plea for More Thorough Investigation." *Radium* 7 (May 1916): 55–62.

———. "Radium and Symptomatic Blood Pressure." *Radium* 7 (September 1916): 129–36.

———. "Saratoga Springs—History, Origin, Chemical Constituents, General Therapeutic Action." *New York Medical Journal* 95 (January 6, 1912): 16–18.

Morris, Charles Evans. *Modern Rejuvenation Methods*. New York: Scientific Medical Publishing Co., 1926.

Morse, Kenneth, and Milton Kronenberg. "Radium Painting." *Industrial Medicine* 12 (December 1943): 810–21.

Morton, William J. "Radiochemicotherapy: The Internal Therapeutics of the Radio-elements." *Medical Record* 87 (1915): 381–90.

Mottram, J. C. "The Effect of Increased Protection from Radiation upon the Blood Condition of Radium Workers." *Archives of Radiology and Electrotherapy* 25 (1920–21): 368–72. Condensed in *Radium* 18 (December 1921): 44–46.

———. "Histological Changes in the Bone Marrow of Rats Exposed to the Gamma Radiations from Radium." *Archives of Radiology and Electrotherapy* 25 (1920–21): 197–99. Reprinted in *Radium* 17 (June 1921): 65–67.

———. "The Red Blood Cell Count of Those Handling Radium for Therapeutic Purposes." *Archives of Radiology and Electrotherapy* 25 (1920): 194–97. Reprinted in *Radium* 17 (June 1921): 62–65.

Mottram, J. C., and J. R. Clarke. "The Leukocytic Blood Content of Those Handling Radium for Therapeutic Purposes." *Archives of Radiology and Electrotherapy* 24 (1919–20): 345–50. Reprinted in *Radium* 15 (June 1920): 40–44. Also in *Proceedings of the Royal Society of Medicine [London]* 9 (1919–20): 25–32 and *American Journal of Roentgenology*, n.s., 7 (1920): 355.

Muir, J. "Story of Radium in Therapeutics: The Curies' Gift to Humanity." In *Physicians Throughout the Ages*, edited by A. S. Brown, 2:601–6. Capehart-Brown Co., 1928.

"Necrosis of the Jaw among Workers Applying Luminous Paint on Watch Dials." U.S. Bureau of Labor Statistics. *Monthly Labor Review* 21 (November 1925): 181–84.

"Necrosis of the Jaw among Workers Using Luminous Paint." U.S. Bureau of Labor Statistics. *Monthly Labor Review* 21 (November 1925): 185–87.

New Jersey Bureau of Statistics of Labor and Industry. *Annual Reports*. 1881–1906.

New Jersey Department of Labor. *Annual Report*. July 1, 1923–June 30, 1924.

———. *Annual Report*. July 1, 1924–June 30, 1925.

New Jersey State Library, Trenton. "Legislative History of the Workers' Compensation Act."

New Jersey Statutes Annotated. St. Paul, Minn: West Publishing Co., 1986.

New York Conservation Commission, Division of Saratoga Springs. "Saratoga Mineral
Springs and Baths: Their Value as Therapeutic Agents." 1917.

"Notes and Comments." *Radium* 2 (October 1913): 9–10.

"Notes and Comments." *Radium* 4 (February 1915): 104.

"Occupational Hazards to Young Workers: Radioactive Substances." U.S. Department of
Labor, Children's Bureau, Publication 286. 1942.

Oliver, Thomas, ed., *Dangerous Trades: The Historical, Social, and Legal Aspects of
Industrial Occupations as Affecting Health*. London: J. Murray, 1902.

———. *Diseases of Occupation from the Legislative, Social, and Medical Points of View*.
London: Methuen and Co., 1908.

Ordway, T. "Injuries Due to Radium: Reports of Typical Cases and Recommendations
for Protection." *Scientific American*, supp. 81 (1916): 254–56. Reprinted in *Radium:
Abstracts*, American Institute of Medicine, pp. 163–64. Published for U.S. Radium
Corp., 1922.

———. "Occupational Injuries Due to Radium: Report of Cases." *JAMA* 66 (1916):1–6.
Reprinted in *Radium* 6 (March 1916): 121–34.

"An Outline of Radium and Its Emanations." National Radium Products Co., 1925.

Parkhurst, Burleigh. "Radium Therapy: Experiences and Suggestions." *North American
Journal of Homeopathy*, 3d s., 20 (November 1905): 713–17. Also in *Transactions of the
American Institute of Homeopathy* 61 (1905): 792–95.

Persson, G. A. "Treatment of Chronic Diseases at Spas, Particularly with Reference to
the Modern Conception of Radium Emanation." *New York Medical Journal* 95 (1912):
1038–42.

Pfahler, G. "Effects of X-rays and Radium on the Blood and General Health of
Radiologists." *American Journal of Roentgenology* 9 (October 1922): 647–56.
Reprinted in *Radium*, n.s., 2 (July 1923): 159–68.

Pifford, Henry. "A Few Words Concerning Radium." *Medical Record* 65 (June 8, 1904):
999–1004.

Pinch, A. E. Hayward. "A Report of the Work Carried Out at the Radium Institute,
London, from January 1st, 1919 to December 31st, 1919." Reprinted in *Radium* 15
(April 1920): 69–84 (September 1920): 85–101.

———. "A Report of the Work Carried Out at the Radium Institute from January 1, 1921
to December 31, 1921." Condensed in *Radium*, n.s., 1 (July 1922): 124–48.

"Poison Paintbrush." *Time*, September 4, 1928, pp. 40–41.

Price, George. *The Modern Factory; Safety, Sanitation, and Welfare*. New York: J. Wiley
and Sons, 1914.

Price-Jones, Cecil. "Observations on the Effects Produced on the Bone Marrow by
Exposing Rabbits to 5 Grams of Radium Bromide." In *Medical Uses of Radium:
Studies of Effects of Gamma Rays from a Large Quantity of Radium*, edited by the
[British] Medical Research Committee, pp. 130–46. London: H.M. Stationary Office,
1922.

Proescher, Frederick. "Contribution on the Therapeutic Value of the Intravenous
Injection of Soluble Radium Salts in the Treatment of Pernicious Anemia and
Leukemia." *Radium* 7 (June 1916): 71–77 and (July 1916): 102–12.

———. "Influence of Intravenous Injections of Soluble Radium Salts in High Blood Pressure." *Radium* 3 (April 1914): 1–10 and (May 1914): 17–21.

———. "Intravenous Injection of Soluble Radium Salts." *Radium* 2 (January 1914): 45–53, (February 1914): 61–64, and (March 1914): 77–87.

———. "The Intravenous Injection of Soluble Radium Salts in Man." *Radium* 1 (July 1913): 9–10.

Proescher, Frederick, and M. R. Almquest. "Contribution on the Biological and Pathological Action of Soluble Radium Salts—With Special Reference to Its Therapeutic Value in Pernicious Anaemia and Leukemia." *Radium* 3 (August 1914): 65–71 and (September 1914): 85–95.

Proescher, Frederick, and M. R. Almquest. "Contribution on the Therapeutic Value of the Intravenous Injection of Soluble Radium Salts in the Treatment of Pernicious Anemia and Leukemia." *Radium* 6 (January 1916): 85–96.

"Radioactivity after Death." *Scientific American* 141 (July 1929): 75.

"'Radithor' and William J. A. Bailey." *JAMA* 88 (January 29, 1927): 343.

"Radium Drinks." *Time*, April 1, 1932; *New York Times*, April 1, 1932; *New York American*, April 1, 1932.

"Radium Poisoning." U.S. Bureau of Labor Statistics. *Monthly Labor Review* 28 (June 1929): 1200–75.

"Radium Salts and Emanation." *American Journal of Roentgenology* 31 (June 1934): 830–36.

"Radium Therapy in Gout, Arthritis, and Rheumatism." National Radium Products Co., 1925.

"Radium Therapy in High Blood Pressure and Circulatory Diseases." National Radium Products Co., 1925.

"Radium Victim #41." *Life*, 1951, pp. 81–85.

Ramazzini, B. *The Diseases of Workers*. Translated by W. C. Wright. 1713. Reprint, New York: Hafner Publishing Co., 1964.

Reitter, G. S., and H. S. Martland. "Leucopenic Anemia of the Regenerative Type Due to Exposure to Radium and Mesothorium: Report of a Case." *American Journal of Roentgenology* 16 (1926): 161–67.

Rolleston, Humphry. "Critical Review: The Harmful Effects of Irradiation (X-rays and Radium)." *Quarterly Journal of Medicine* 24 (October 1930): 101–31.

Rollins, William. "Some Principles Involved in the Therapeutic Applications of Radioactivity." *Boston Medical and Surgical Journal* 149 (November 12, 1903): 542–44.

Rowntree, L. G., and W. A. Baetjer. "Radium in Internal Medicine: Its Physiologic and Pharmacologic Effects." *JAMA* 61 (October 18, 1913): 1438–42.

"Safe Health Practices Relating to the Use and Handling of Radioactive Luminous Paints." Massachusetts Department of Labor and Industries, Division of Occupational Hygiene, Publication 290. December 1942.

St. George, A. V., Alexander Gettler, and Ralph Muller. "Radioactive Substances in a Body Five Years after Death." *Archives of Pathology* 7 (March 1929): 397–405.

Saratoga Springs Commission, New York. "Report of the Saratoga Springs Commission to the Legislature," Bernard Baruch, chair. Document 70. 1930.

Saubermann, S. "An Address on the Progress of Radium Therapy." *Journal of the Roentgen Society* 9 (July 1913): 63–78. Reprinted in *Archives of the Roentgen Ray* 18 (August 1913): 99–116.

———. "Radium Emanation and Physiological Processes." *Archives of the Roentgen Ray* 16 (January 1912): 293–317.

Schlundt, Herman, H. H. Barker, and Frederick Flinn. "The Detection and Estimation of Radium and Mesothorium in Living Persons." *American Journal of Roentgenology and Radium Therapy* 21 (April 1929): 345–54, 24 (October 1930): 418–23, 26 (August 1931): 265–71, and 30 (October 1933): 515–22.

Schwartz, Louis, Fred Knowles, Rollo Britten, and Lewis Thompson. "Health Aspects of Radium Dial Painting." *Journal of Industrial Hygiene* 15 (September 1933): 362–82 and 15 (November 1933): 433–55.

Seil, H. A., C. H. Viol, and M. A. Gordon. "The Elimination of Soluble Radium Salts Taken Intravenously and Per Os." *New York Medical Journal* 101 (May 1915): 896–98. Reprinted in *Radium* 5 (1915): 40–44.

Selby, Clarence D. "Studies of the Medical and Surgical Care of Industrial Workers." U.S. Public Health Service Bulletin 99. 1919.

"Sellers of Toxic Land Ruled 'Forever' Liable." *Newark Star-Ledger*, March 28, 1991, pp. 1, 40.

Simpson, Frank Edward. *Radium Therapy*. St. Louis, Mo.: C. V. Mosby Co., 1922.

Soddy, Frederick. "Radium as Therapeutic Agent." *Medical Record* 64 (1903): 660.

"The Spark of Life." National Radium Products Co., 1925.

Stannard, J. Newell. *Radioactivity and Health: A History*. U.S. Department of Energy, Office of Health and Environmental Research, 1988.

"State Reporting of Occupational Diseases, Including a Survey of Legislation Applying to Women." U.S. Department of Labor, Women's Bureau, Bulletin 114. 1934.

Stevens, Rollin. "The Use of Intravenous Injections of Radium Chloride in Some of the Malignant Lymphomata." *American Journal of Roentgenology* 16 (August 1926): 155–61.

Subcommittee on the Handling of Radioactive Isotopes and Fission Products, U.S. National Committee on Radiation Protection and Measurements, National Bureau of Standards, U.S. Department of the Interior. "Safe Handling of Radioactive Isotopes." Handbook 42, 1949.

Supplement to the Compiled Statutes of New Jersey, 1925–1930. Newark Soney and Sage, 1931.

"The Therapeutic Possibilities of Radium." *Boston Medical and Surgical Journal* 149 (November 5, 1903): 524–25.

Thompson, W. Gilman. *The Occupational Diseases: Their Causation, Symptoms, Treatment, and Prevention*. New York: D. Appleton and Co., 1914.

———. "The Springs of Saratoga—Their Value in the Treatment of Disease." *Medical Record* 89 (April 1916): 589–92.

Tracy, Samuel. "Radium: Induced Radio-activity and Its Therapeutical Possibilities." *New York Medical Journal* 79 (January 9, 1904): 49–52.

———. "Recent Applications of Radium Emanations and Radium Water." *New York Medical Journal* 101 (1915): 612–16.

U.S. National Institute of Occupational Safety and Health. President's Report on Occupational Safety and Health. 1972.

Viol, Charles. "History and Development of Radium-Therapy." *Journal of Radiology* 2 (September 1921): 29–34.

———. "There Is No Radium Shortage." *Radium* 13 (April 1919): 6–10.

Von Noorden, Carl. "Radium and Thorium-X Therapy." *Medical Record* 83 (January 18, 1913): 95–103.

Von Sochocky, S. A. "Can't You Find the Keyhole?" *American Magazine* 91 (January 1921): 24–27, 106–8.

Wallian, Samuel. "Radioactivity in Diabetes and Nephritis." *St. Louis Medical Record* 52 (August 26, 1905): 179–80.

Ward, Emma F. "Phosphorous Necrosis in the Manufacture of Fireworks and in the Preparation of Phosphorous." U.S. Bureau of Labor Statistics Bulletin 405. 1926.

"Whose Responsibility?" *Scientific American* 139 (August 1928): 108.

Wickham, Louis F., and Paul Degrais. *Radium Therapy*. Translated by S. Ernest Dore, with an introduction by Sir Malcolm Norris. London: Cassell and Co., 1910.

Williams, R. C. "Preliminary Note on Observations Made on Physical Condition of Persons Engaged in Measuring Radium Preparations." *Public Health Reports* 38 (December 21, 1923): 3007–32. Reprinted in *Radium*, n.s., 3 (April 1924): 43–64.

"Women in New Jersey Industry." U.S. Department of Labor, Women's Bureau, Bulletin 37. 1924.

Zueblin, Ernst. "The Present Status of Radioactive Therapy in Medicine." *Maryland Medical Journal* 57 (May–June 1914): 108–24, 141–51.

SECONDARY SOURCES

Allison, Malorye. "The Radioactive Elixir." *Harvard Magazine* 94 (January–February 1992): 72–75.

Amenta, Edwin, Elisabeth Clemens, Jefren Olsen, Sunita Parikh, and Theda Skocpol. "The Political Origins of Unemployment Insurance in Five American States." *Studies in American Political Development* 2 (1987): 137–82.

Anderson, Ruth. *From Yankee to American: Connecticut, 1865–1914*. Bridgeport, Conn.: Pequot Press, 1975.

Asher, Robert. "Business and Workers' Welfare in the Progressive Era: Workmen's Compensation Reform in Massachusetts, 1880–1911." *Business History Review* 43 (1969): 452–75.

———. "Failure and Fulfillment: Agitation for Employers' Liability Legislation and the Origins of Workmen's Compensation in New York State, 1876–1910." *Labor History* 24 (Spring 1983): 198–222.

———. "Industrial Safety and Labor Relations in the United States, 1865–1917." In *Life and Labor: Dimensions of American Working Class History*, edited by Charles Stephenson and Robert Asher, pp. 115–30. Albany: SUNY Press, 1986.

———. "The Limits of Big Business Paternalism: Relief for Injured Workers in the Years before Workmen's Compensation." In *Dying for Work: Workers' Safety and Health in*

Twentieth-Century America, edited by David Rosner and Gerald Markowitz, pp. 19–33. Bloomington: Indiana University Press, 1987.

Ashford, Nicholas. *Crisis in the Workplace: Occupational Diseases and Injuries.* Cambridge: MIT Press, 1975.

Aspell, William. " 'Personal Injury' and the Connecticut Workmen's Compensation Act, 1913–1951." *Connecticut Bar Journal* 26 (1952): 142–55.

Assennato, Giorgio, and Vicente Navarro. "Workers' Participation and Control in Italy: The Case of Occupational Medicine." *International Journal of Health Services* 10 (1980): 217–32.

Athey, Louis. "The Consumers' League and Social Reform, 1890–1923." Ph.D. dissertation, University of Delaware, 1965.

Aub, Joseph, and Ruth Hapgood. *Pioneer in Modern Medicine: David Linn Edsall of Harvard.* Cambridge: Harvard Medical Alumni Association, 1970.

Badash, Lawrence. *Radioactivity in America: Growth and Decay of a Science.* Baltimore: Johns Hopkins University Press, 1979.

Bale, Anthony. "America's First Compensation Crisis: Conflict over the Value and Meaning of Workplace Injuries under the Employers' Liability System." In *Dying for Work: Workers' Safety and Health in Twentieth-Century America*, edited by David Rosner and Gerald Markowitz, pp. 34–52. Bloomington: Indiana University Press, 1987.

———. "Assuming the Risks: Occupational Disease in the Years before Workers' Compensation." *American Journal of Industrial Medicine* 13 (1988): 499–514.

———. "A Brush with Justice: The New Jersey Radium Painters and the Courts." *Health/PAC Bulletin* 17 (Fall 1987): 18–21.

———. " 'Hope in Another Direction': Compensation for Work-Related Illness among Women, 1900–1960—Part I." *Women and Health* 15 (1989): 81–102.

———. " 'Hope in Another Direction': Compensation for Work-Related Illness among Women, 1900–1960—Part II." *Women and Health* 15 (1989): 99–115.

———. "Women's Toxic Experience." In *Women, Health, and Medicine in America*, edited by Rima Apple, pp. 411–39. New York: Garland, 1990.

Baran, Paul, and Paul Sweezy. *Monopoly Capital: An Essay on the American Economic and Social Order.* New York: Modern Reader Paperbacks, 1966.

Baron, Sherry Lee. "Watches, Workers, and the Awakening World: A Case Study in the History of Occupational Medicine in America." Senior thesis, Harvard University, 1977.

Beckner, Earl. *A History of Labor Legislation in Illinois.* Chicago: University of Chicago Press, 1929.

Bell, Carolyn. "Implementing Safety and Health Regulations for Women in the Workplace." *Feminist Studies* 5 (1979): 286–301.

Berg, Samuel. *Harrison Stanford Martland, M.D.: The Story of a Physician, a Hospital, and an Era.* New York: Vantage Press, 1978.

Berkowitz, Monroe. *Workmen's Compensation: The New Jersey Experience.* New Brunswick, N.J.: Rutgers University Press, 1960.

Berman, Daniel. *Death on the Job: Occupational Health and Safety Struggles in the United States.* New York: Monthly Review Press, 1978.

Bernstein, Irving. *The Lean Years, 1920–1933*. Boston: Houghton Mifflin, 1960.

Birnbaum, Sheila, and Jerold Oshinsky. *Occupational Disease Litigation*. New York: Practicing Law Institute, 1986.

Bowen, Catherine Drinker. *Family Portrait*. New York: Little, Brown, 1970.

Braendgaard, Asger. "Occupational Health and Safety Legislation and Working-Class Political Action: A Historical and Comparative Analysis." Ph.D. diss., University of North Carolina at Chapel Hill, 1974.

Brandes, Stuart. *American Welfare Capitalism, 1880–1940*. Chicago: University of Chicago Press, 1976.

Brass Workers' History Project. *Brass Valley: The Story of Working People's Lives and Struggles in an American Industrial Region*. Philadelphia: Temple University Press, 1982.

Braverman, Harry. *Labor and Monopoly Capital: The Degradation of Work in the Twentieth Century*. New York: Monthly Review Press, 1974.

Brecher, Ruth, and Edward Brecher. *The Rays: A History of Radiology in the United States and Canada*. Baltimore: Williams and Williams, 1969.

Brodeur, Paul. *Expendable Americans*. New York: Viking, 1974.

———. *Outrageous Misconduct: The Asbestos Industry on Trial*. New York: Pantheon Books, 1985.

Brody, David. "The Rise and Decline of Welfare Capitalism." In *Change and Continuity in Twentieth-Century America: The 1920s*, edited by David Brody, John Braeman, and Robert Bremner, pp. 147–78. Columbus: Ohio State University Press, 1968.

———. *Steelworkers in America: The Nonunion Era*. New York: Harper and Row, 1960.

Buder, Stanley. *Pullman: An Experiment in Industrial Order and Community Planning, 1880–1930*. New York: Oxford University Press, 1967.

Buenker, John D. "Progressivism in Connecticut: The Thrust of the Urban, New Stock Democrats." *Connecticut Historical Society Bulletin* 35 (October 1970): 97–109.

———. "Urban Immigrant Lawmakers and Progressive Reform in Illinois." In *Essays in Illinois History: In Honor of Glenn Huron Seymour*, edited by Donald Tingley, pp. 52–74. Carbondale: Southern Illinois University Press, 1968.

Burch, Philip, Jr. *Industrial Safety Legislation in New Jersey*. New Brunswick, N.J.: Bureau of Government Research, Rutgers University, 1960.

Case, James. "The Early History of Radium Therapy and the American Radium Society." *American Journal of Roentgenology* 82 (1959): 574–85.

Cattell, Jacques. "Frederick Flinn." In *American Men of Science*, 9th ed., edited by Jacques Cattell, p. 356. Lancaster, Pa.: Science Press, 1955.

———. "Frederick Hoffman." In *American Men of Science*. 7th ed., edited by Jacques Cattell, p. 827. Lancaster, Pa.: Science Press, 1944.

Caufield, Catherine. *Multiple Exposures: Chronicles of the Radiation Age*. Chicago: University of Chicago Press, 1989.

Chandler, Alfred. *Strategy and Structure: Chapters in the History of the Industrial Enterprise*. Cambridge: MIT Press, 1962.

———. *The Visible Hand: The Managerial Revolution in American Business*. Cambridge, Mass.: Belknap Press, 1977.

Chapman, Carleton. *Physicians, Laws, and Ethics*. New York: New York University Press, 1984.

Chavkin, Wendy. *Double Exposure: Women's Health Hazards on the Job and at Home*. New York: Monthly Review Press, 1984.

———. "Occupational Hazards to Reproduction: A Review Essay and Annotated Bibliography." *Feminist Studies* 5 (1979): 310–25.

Cherniack, Martin. *The Hawk's Nest Incident: America's Worst Industrial Disaster*. New Haven: Yale University Press, 1986.

Clague, Ewan. *The Bureau of Labor Statistics*. New York: Praeger, 1968.

Clark, Claudia. "Glowing in the Dark: The Radium Dialpainters, the Consumers' League, and Industrial Health Reform in the United States, 1910–1935." Ph.D. dissertation, Rutgers University, 1991.

———. "The Menace of Benzene: Alice Hamilton and the Health of American Workers." Paper in the author's possession.

Cloutier, Richard. "Florence Kelley and the Radium Dial Painters." *Health Physics Journal* 39 (November 1980): 711–17.

Cohen, Lizbeth. *Making a New Deal: Industrial Workers in Chicago, 1919–1939*. New York: Cambridge University Press, 1990.

Conroy, John. "Radium City." *Chicago Free Weekly*, June 22, 1984, and *Illinois Times*, August 23–29, 1984. Clippings available at Argonne National Laboratory, Center for Human Radiobiology.

Corn, Jacqueline. "Protective Legislation for Coal Miners, 1870–1900: Responses to Safety and Health Hazards." In *Dying for Work: Workers' Safety and Health in Twentieth-Century America*, edited by David Rosner and Gerald Markowitz, pp. 67–82. Bloomington: Indiana University Press, 1987.

Curran, Jean Alonzo. *Founders of the Harvard School of Public Health with Biographical Notes, 1909–1946*. New York: Josiah Macy Jr. Foundation, 1970.

Derickson, Alan. " 'To Be His Own Benefactor': The Founding of the Coeur d'Alene Miners' Union Hospital, 1891." In *Dying for Work: Workers' Safety and Health in Twentieth-Century America*, edited by David Rosner and Gerald Markowitz, pp. 3–18. Bloomington: Indiana University Press, 1987.

———. *Workers' Health, Workers' Democracy: The Western Miners' Struggle, 1891–1925*. Ithaca, N.Y.: Cornell University Press, 1988.

Dewing, Stephen. *Modern Radiology in Historical Perspective*. Springfield, Ill.: Charles C. Thomas, 1962.

Dodyk, Delight. "Winders, Warpers, and the Girls on the Loom: A Study of Women in the Paterson Silk Industry and Their Participation in the General Strike of 1913." M.A. thesis, Sarah Lawrence College, 1979.

Domhoff, William. *The Higher Circle*. New York: Random House, 1970.

Donegan, Craig. "For the Good of Us All: Early Attitudes Toward Occupational Health with Emphasis on the Northern United States from 1787 to 1870." Ph.D. dissertation, University of Maryland, 1984.

Donizetti, Pino. *Shadow and Substance: The Story of Medical Radiography*. Oxford: Pergamon Press, 1967.

Dublin, Thomas. *Women at Work: The Transformation of Work and Community at Lowell, Massachusetts, 1826–1860.* New York: Columbia University Press, 1979.

Dubofsky, Melvyn. *When Workers Organize: New York City in the Progressive Era.* Amherst: University of Massachusetts Press, 1968.

Duffy, John. *A History of Public Health in New York City, 1866–1966.* New York: Russel Sage Foundation, 1968.

———. *The Sanitarians: A History of American Public Health.* Chicago: University of Illinois Press, 1990.

Edwards, Richard. *Contested Terrain: The Transformation of the Workplace in the Twentieth Century.* New York: Basic Books, 1979.

Evans, Sara. *Born for Liberty: A History of Women in America.* New York: The Free Press, 1989.

Federal Writers' Project. *Connecticut: A Guide to Its Roads, Lore, and People.* Boston: Houghton Mifflin, 1938.

"Fifty Years of Radium." *American Journal of Roentgenology* 60 (December 1948): 7–26.

Fones-Wolf, Kenneth. "Class, Professionalism, and the Early Bureaus of Labor Statistics." *Insurgent Sociologist* 10 (Summer 1980): 38–45.

Foster, James. "The *Western Dilemma*: Miners, Silicosis, and Compensation." *Labor History* 26 (Spring 1985): 268–87.

Fox, David, and Judith Stone. "Black Lung: Miners' Militancy and Medical Uncertainty, 1968–1972." *Bulletin of the History of Medicine* 54 (1980): 43–63.

Frank, Nancy Kay. "From Criminal Law to Regulation: A Historical Analysis of Health and Safety Law." Ph.D. dissertation, State University of New York at Albany, 1982.

Friedman, Lawrence. *A History of American Law.* New York: Simon and Schuster, 1973.

Friedman, Lawrence, and Jack Ladinsky. "Social Change and the Law of Industrial Accidents." *Columbia Law Review* 67 (1967): 50–82.

Galishoff, Stuart. *Safeguarding the Public Health: Newark, 1895–1918.* Westport, Conn.: Greenwood Press, 1975.

Gillespie, Richard. "Accounting for Lead Poisoning: The Medical Politics of Occupational Health." *Journal of Social History* 15 (October 1990): 303–32.

Ginley, Francis. "Harrison Stanford Martland, 1883–1954: Personal Recollections." *Academy of Medicine of New Jersey Bulletin* 15 (December 1969): 204–13.

Gofman, John. *Radiation and Human Health: A Comprehensive Investigation of the Evidence Relating Low-Level Radiation to Cancer and Other Diseases.* San Francisco: Sierra Books, 1981.

Goldmark, Josephine. *Impatient Crusader: Florence Kelley's Life Story.* 1953. Reprint, Westport, Conn.: Greenwood Press, 1976.

Golin, Steven B. "Bimson's Mistake, or How the Paterson Police Helped to Spread the 1913 Strike." *New Jersey History* 100 (Spring/Summer, 1982): 57–86.

Gordon, Bonnie. "Phossy-Jaw and the French Match Workers: Occupational Health and Women in the Third Republic." Ph.D. dissertation, University of Wisconsin-Madison, 1985.

Gordon, David, Richard Edwards, and Michael Reich. *Segmented Work, Divided Workers: The Historical Transformation of Labor in the United States.* Cambridge: Cambridge University Press, 1982.

Gordon, Felice. *After Winning: The Legacy of the New Jersey Suffragists, 1920–1947.* New Brunswick, N.J.: Rutgers University Press, 1986.

Gordon, Jerome, Allan Akman, and Michael Brooks. *Industrial Safety Statistics: A Reexamination.* New York: Praeger, 1971.

Graebner, William. *Coal Mining Safety in the Progressive Period: The Political Economy of Reform.* Lexington: University Press of Kentucky, 1976.

———. "Hegemony through Science: Information Engineering and Lead Toxicology, 1925–1965." In *Dying for Work: Workers' Safety and Health in Twentieth-Century America,* edited by David Rosner and Gerald Markowitz, pp. 140–59. Bloomington: Indiana University Press, 1987.

Green, Marguerite. *The National Civic Foundation and the American Labor Movement, 1900–1925.* Westport, Conn.: Greenwood Press, 1956.

Greenbaum, Fred. "Social Ideas of Samuel Gompers." *Labor History* 7 (Winter 1966): 35–61.

Grigg, Emmanuel. *The Trail of Invisible Light, from X-Strahlen to Radio(bio)logy.* Springfield, Ill.: Charles C. Thomas, 1965.

Grob, Gerald. *Workers and Utopia: A Study of Ideological Conflict in the American Labor Movement, 1865–1900.* Chicago: Northwestern University Press, 1961.

Grossman, Jonathon. *The Department of Labor.* New York: Praeger, 1973.

Hacker, Barton. *The Dragon's Tail: Radiation Safety in the Manhattan Project, 1942–1946.* Berkeley: University of California Press, 1987.

———. *Elements of Controversy: The Atomic Energy Commission and Radiation Safety in Nuclear Weapons Testing, 1947–1974.* Berkeley: University of California Press, 1994.

Hawley, Ellis. "Herbert Hoover, the Commerce Secretariat, and the Vision of the 'Associative State,' 1921–1928." *Journal of American History* 61 (1974): 116–40.

Haynes, William. *The American Chemical Industry: A History.* 3 vols. New York: Van Nostrand Co., 1954.

Hays, Samuel. *Beauty, Health, and Permanence: Environmental Politics in the United States, 1955–1985.* Cambridge: Cambridge University Press, 1987.

Hazlett, T. Lyle, and William Hummel. *Industrial Medicine in Western Pennsylvania, 1850–1950.* Pittsburgh: University of Pittsburgh Press, 1957.

Healy, J. W. "The Origin of Current Standards." *Health Physics* 29 (1967): 489–94.

Heath, Frederick. "Labor and the Progressive Movement in Connecticut." *Labor History* 12 (Winter 1971): 52–67.

Hewitt, Harold. "Rationalizing Radiotherapy: Some Historical Aspects of the Endeavor." *British Journal of Radiology* 46 (October 1973): 917–26.

Hilgartner, Stephen, Richard Bell, and Rory O'Connor. *Nukespeak: Nuclear Language, Vision, and Mindset.* San Francisco: Sierra Club, 1992.

Hill, Ann Corinne. "Protection of Women Workers and the Courts: A Legal Case History." *Feminist Studies* 5 (1979): 247–73.

Hirsch, Adelaide Morgan. *We've Come a Long Way: A History of the League of Women Voters in Connecticut, 1921–1933* (pamphlet, 52 pp.). New Haven: John Corbett Press, 1953.

Hirst, David. "James Fairman Fielder." In *The Governors of New Jersey, 1664–1974,* edited by Paul Stellhorn and Michael Birkner, pp. 183–86. Trenton: New Jersey Historical Commission, 1982.

"Historical Sketch—Chronology of the Baths." Mimeographed material supplied by Saratoga Springs State Park, New York.

Howard, Robert. *Illinois: A History of the Prairie State.* Grand Rapids: Eerdmans, 1972.

Hricko, Andrea, and Melanie Brunt. *Working for Your Life: A Woman's Guide to Job Health and Hazards.* University of California, Labor Occupational Health Program, 1976.

Hunt, Vilma. "A Brief History of Women Workers and Hazards in the Workplace." *Feminist Studies* 5 (1979): 274–85.

Hunt, Vilma, Kathleen Lucas-Wallace, and Jeanne Manson. *Work and the Health of Women.* Boca Rota, Fla.: CRC Press, 1979.

Ineson, Antonia, and Deborah Thom. "T.N.T. Poisoning and the Employment of Women Workers in the First World War." In *The Social History of Occupational Health*, edited by Paul Weindling, pp. 89–107. London: Croom Helm, 1985.

Janick, Herbert, Jr. *A Diverse People: Connecticut, 1914 to the Present.* Chester, Conn.: Pequot Press, 1975.

Jensen, Richard. *Illinois: A Bicentennial History.* New York: Norton, 1978.

Judkins, Bennet. *We Offer Ourselves as Evidence: Toward Workers' Control of Occupational Health.* New York: Greenwood Press, 1986.

Katzman, David. *Seven Days a Week: Women and Domestic Service in Industrializing American.* Urbana: University of Illinois Press, 1981.

Kazis, Richard, and Richard Grossman. *Fear at Work: Job Blackmail, Labor, and the Environment.* New York: Pilgrim Press, 1982.

Kessler-Harris, Alice. *Out to Work: A History of Wage-Earning Women in the United States.* New York: Oxford University Press, 1982.

Kotelchuck, David. "Asbestos: 'The Funeral Dress of Kings'—and Others." In *Dying for Work: Workers' Safety and Health in Twentieth-Century America*, edited by David Rosner and Gerald Markowitz, pp. 192–207. Bloomington: Indiana University Press, 1987.

Koven, Seth, and Sonya Michel. "Gender and the Origins of the Welfare State." *Radical History Review* 43 (1989): 112–20.

Krabbenhoft, Kenneth L. "History of Radiologic Journals." In *Classic Descriptions in Diagnostic Roentgenology*, edited by Andre Bruwer, 2:1991–2030. 2 vols. Springfield, Ill.: Charles C. Thomas, 1964.

Landa, Edward. "From Buried Treasure to Buried Waste: The Rise and Fall of the Radium Industry." *Colorado School of Mines* 8, no. 2 (1987): 1–77 (special issue).

Lang, Daniel. "A Most Valuable Accident." *New Yorker*, May 2, 1959, pp. 49–92.

Leavitt, Judith Walzer. *The Healthiest City: Milwaukee and the Politics of Health Reform.* Princeton, N.J.: Princeton University Press, 1982.

Lebsock, Suzanne. "Women and American Politics, 1880–1920." In *Women, Politics, and Change*, edited by Louise Tilly and Patricia Gurin, pp. 35–62. New York: Russell Sage Foundation, 1990.

Lee, R. Alton. "The Eradication of Phossy Jaw: A Unique Development of Federal Police Power." *The Historian* 20 (1966): 1–21.

Leiby, James. *Carroll Wright and Labor Reform*. Cambridge: Harvard University Press, 1960.

Lemons, Stanley. *The Woman Citizen: Social Feminism in the 1920s*. Urbana: University of Illinois Press, 1975.

Lescohier, Don. "The Campaigns for Health and Safety in Industry." In *History of Labor in the United States, 1896–1932*, vol. 3, edited by John Commons, pp. 359–70. New York: Macmillan, 1935.

Levenstein, Charles. "A Brief History of Occupational Health in the United States." In *Occupational Health: Recognizing and Preventing Work-Related Diseases*, edited by Barry Levy and David Wegman, pp. 11–12. 2d ed. Boston: Little, Brown, 1988.

Levenstein, Charles, Dianne Plantamura, and William Mass. "Labor and Byssinosis, 1941–1969." In *Dying for Work: Workers' Safety and Health in Twentieth-Century America*, edited by David Rosner and Gerald Markowitz, pp. 208–23. Bloomington: Indiana University Press, 1987.

Link, Arthur. "What Happened to the Progressive Movement in the 1920s?" *American Historical Review* 64 (1959): 833–51.

Lubove, Roy. *The Struggle for Social Security*. Cambridge: Harvard University Press, 1968.

———. "Workmen's Compensation and the Prerogatives of Voluntarism." *Labor History* 8 (Fall 1967): 254–79.

McCord, Carey. "The Beginnings of Occupational Health in the United States." *Industrial Medicine and Surgery* 21 (1952): 361–62.

———. "Occupational Health Publications in the United States Prior to 1900." *Industrial Medicine and Surgery* 24 (1955): 363–68.

Macklis, Roger. "The Great Radium Scandal." *Scientific American* (August 1993): 94–99.

———. "Radithor and the Era of Mild Radium Therapy." *JAMA* 264 (August 1, 1990): 614–21.

Mahoney, Joseph Francis. "George Sebastian Silzer." In *The Governors of New Jersey*, edited by Paul Stellhorn and Michael Birkner, pp. 194–96. Trenton: New Jersey Historical Commission, 1982.

———. "New Jersey Politics after Wilson: Progressivism in Decline." Ph.D. dissertation, Columbia University, 1964.

Markowitz, Gerald, and David Rosner. "Death and Disease in the House of Labor." *Labor History* 30 (1989): 113–17.

Marx, Leo. *The Machine in the Garden: Technology and the Pastoral Idea in America*. London: Oxford University Press, 1972.

Mayo, Ann. "We Are All Guinea Pigs." *Village Voice*, December 25, 1978, p. 18ff. Clipping available at Argonne National Laboratory, Center for Human Radiobiology.

Mays, C. W., ed. *Delayed Effects of Bone-seeking Radionuclides*. Salt Lake City: University of Utah Press, 1969.

Milles, Dietrich. "From Workers' Diseases to Occupational Diseases: The Impact of Experts' Concepts on Workers' Attitudes." In *The Social History of Occupational Health*, edited by Paul Weindling, pp. 55–77. London: Croom Helm, 1985.

Mock, Harry. "Industrial Medicine and Surgery—A Resume of Its Development and Scope." *Journal of Industrial Hygiene* 1 (1919): 1–7 and 2 (1919): 251–54.

Montgomery, David. *The Fall of the House of Labor: The Workplace, the State, and American Labor Activism, 1865–1925.* New York: Cambridge University Press, 1987.

———. *Workers' Control in America: Studies in the History of Work, Technology, and Labor Struggles.* New York: Cambridge University Press, 1979.

Mould, Richard. *A History of X-rays and Radium: With a Chapter on Radiation Units, 1895–1937.* Sutton, England: IPC Building and Contract Journals, 1980.

Muncy, Robyn. *Creating a Female Dominion in American Reform, 1890–1935.* New York: Oxford University Press, 1991.

Myers, W. S. "Andrew McBride." In *The Story of New Jersey,* edited by W. S. Myers, 4:436. New York: Lewis Historical Publishing Co., 1945.

Nareff, Margaret. "Blue Collar, White Gloves: The History of the Consumers' League in Connecticut, 1902–1930." M.A. thesis, Trinity College, 1980.

Nathan, Maud. *The Story of an Epoch-Making Movement.* Garden City, N.Y.: Doubleday, 1926.

Navarro, Vicente. "The Labour Process and Health: A Historical Materialist Interpretation." *International Journal of Health Services* 12 (1982): 5–29.

Nelkin, Dorothy, and Michael Brown. "Knowing about Workplace Risks: Workers Speak Out about the Safety of Their Jobs." *Science for the People* 16 (1984): 17–22.

Nelson, William, and Charles Shriner, eds. *History of Paterson and Its Environs.* New York: Lewis Historical Publishing Co., 1920.

Newman, Philip. *The Labor Legislation of New Jersey.* Washington, D.C.: American Council on Public Affairs, 1943.

Nichols, Carole Artigiani. "A New Force in Politics: The Suffragists' Experience in Connecticut." M.A. thesis, Sarah Lawrence College, 1979.

Nugent, Angela. "Fit for Work: The Introduction of Physical Examinations in Industry." *Bulletin of the History of Medicine* 57 (1983): 578–95.

———. "Organizing Trade Unions to Combat Disease: The Workers' Health Bureau, 1921–1928." *Labor History* 26 (1985): 423–46.

———. "The Power to Define a New Disease: Epidemiological Politics and Radium Poisoning." In *Dying for Work: Workers' Safety and Health in Twentieth-Century America,* edited by David Rosner and Gerald Markowitz, pp. 177–91. Bloomington: Indiana University Press, 1987.

———. "Workers' Health in Chicago, 1905–1911." Paper presented at the 77th Annual Meeting of the Organization of American Historians, 1984.

Nugent Young, Angela. "Interpreting the Dangerous Trades: Workers' Health in America and the Career of Alice Hamilton, 1910–1935." Ph.D. dissertation, Brown University, 1982.

Oliver, R. A. "Seventy-five Years of Radiation Protection." *British Journal of Radiology* 46 (October 1973): 854–60.

O'Neill, William. *Everyone Was Brave: The Rise and Fall of Feminism in America.* Chicago: Quadrangle Books, 1969.

Orloff, Anna Shola. "The Political Origins of America's Belated Welfare State." In *The Politics of Social Policy in the United States,* edited by Margaret Weir, Anna Shola

Orloff, and Theda Skocpol, pp. 37–80. Princeton, N.J.: Princeton University Press 1988.

Page, Joseph, and Mary-Win O'Brien. *Bitter Wages: Ralph Nader's Study Group Report on Disease and Injury on the Job.* New York: Grossman Publishers, 1973.

Pendary, Joyce. "Women's Secular Voluntary Organizations and Their Leadership: Stanford, Connecticut, 1860–1900." M.A. thesis, Sarah Lawrence College, 1978.

Petchesky, Rosalind. "Workers, Reproductive Hazards, and the Politics of Protection: An Introduction." *Feminist Studies* 5 (1979): 233–46.

Pierce, Lloyd. "Activities of the American Association for Labor Legislation on Behalf of Social Security and Protective Labor Legislation." Ph.D. dissertation, University of Wisconsin, 1953.

Polednak, A. "Fertility of Women after Exposure to Internal and External Radiation." *Journal of Environmental Pathology and Toxicology* 4 (1980): 457–70.

Porter, Glenn. *Rise of Big Business, 1860–1910.* New York: Crowell, 1973.

Price, George. "Union Health Center: The International Ladies' Garment Workers' Union." *Industrial Medicine and Surgery* 22 (1953): 489–97.

Price, George. *The Modern Factory: Safety, Sanitation, and Welfare.* New York: Arno, 1969.

Proskaver, Curt. "A Civil Ordinance of the Year 1846 to Combat Phosphorous Necrosis." *Bulletin of the History of Medicine* 11 (1942): 561–69.

Quimby, Edith. "The Background of Radium Therapy in the United States, 1906–1956." *American Journal of Roentgenology* 75 (March 1956): 443–56.

Reede, Arthur. *Adequacy of Workmen's Compensation.* Cambridge: Harvard University Press, 1947.

Ritchie, Fred. "An Appraisal of Workmen's Compensation Legislation in New Jersey." Ph.D. dissertation, Princeton University, 1952.

Rosen, George. "Charles Turner Thackrah in the Agitation for Factory Reform." *British Journal of Industrial Medicine* 10 (1953): 285–87.

———. "Early Studies of Occupational Health in New York City in the 1870s." *American Journal of Public Health* 67 (1977): 1100–1102.

———. "From Frontier Surgeon to Industrial Hygienist: The Strange Career of George M. Kober." *American Journal of Public Health* 65 (1975): 638–43.

———. *The History of Miners' Diseases: A Medical and Social Interpretation.* New York: Schuman's, 1943.

———. *A History of Public Health.* New York: M. D. Publications, 1958.

———. "The Medical Aspects of Controversy over Factory Conditions in New England, 1840–1850." *Bulletin of the History of Medicine* 15 (May 1944): 488–94.

———. "Occupational Health Problems of English Painters and Varnishers in 1825." *British Journal of Industrial Medicine* 10 (1953): 195–99.

———. "On the Historical Investigation of Occupational Disease: An Aperçu." *Bulletin of the History of Medicine* 5 (1937): 941–46.

———. *Preventive Medicine in the United States.* New York: Prodist, 1977.

Rosenkrantz, Barbara. *Public Health and the State: Changing Views in Massachusetts, 1842–1972.* Cambridge: Harvard University Press, 1972.

Rosner, David, and Gerald Markowitz. *Deadly Dust: Silicosis and the Politics of*

Occupational Disease in Twentieth-Century America. Princeton, N.J.: Princeton University Press, 1991.

———. "The Early Movement for Occupational Safety and Health, 1900–1917." In *Sickness and Health in America: Readings in the History of Medicine and Public Health,* edited by Judith Walzer Leavitt and Ronald L. Numbers, pp. 507–24. 2d ed. Madison: University of Wisconsin Press, 1985.

———. "A 'Gift of God'?: The Public Health Controversy over Leaded Gasoline during the 1920s." In *Dying for Work,* edited by Rosner and Markowitz, pp. 121–39. Also in *American Journal of Public Health* 95 (1985): 344–52.

———. "Introduction: Workers' Health and Safety—Some Historical Notes." In *Dying for Work,* edited by Rosner and Markowitz, pp. ix–xx.

———. "Research or Advocacy?: Federal Occupational Safety and Health Policies during the New Deal." In *Dying for Work,* edited by Rosner and Markowitz, pp. 83–103. Also in *Journal of Social History* 18 (Spring 1985): 365–82.

———. "Safety and Health as a Class Issue: The Workers' Health Bureau of America during the 1920s." In *Dying for Work,* edited by Rosner and Markowitz, pp. 53–64. Also in *Science and Society* 48 (1984–85): 466–82.

———, eds. *Dying for Work: Workers' Safety and Health in Twentieth-Century America.* Bloomington: Indiana University Press, 1987.

Roth, David. *Connecticut: A Bicentennial History.* Norton, N.Y.: American Association of State and Local History, 1979.

Rothman, David. *The Discovery of the Asylum: Social Order and Disorder in the New Republic.* Boston: Johns Hopkins University Press, 1972.

Rowland, R. E., et al. "Current Status of the Study of 226-Ra and 228-Ra in Humans at the Center for Human Radiobiology." *Health Physics* 35 (July 1978): 159–66.

Rowland, Robert, and Henry Lucas Jr. "Radium-Dial Workers." In *Radiation Carcinogenesis: Epidemiology and Biological Significance,* edited by J. D. Brice Jr. and J. F. Fraumeni Jr., pp. 231–40. New York: Raven Press, 1984.

"Saratoga Springs." Mimeographed material supplied by Saratoga Springs State Park, New York.

Satre, Lowell J. "After the Match Girls' Strike: Bryant and May in the 1890s." *Victorian Studies* 26 (1982): 7–31.

Scott, Rachel. *Muscle and Blood: The Massive Hidden Agony of Industrial Slaughter in America.* New York: Dutton, 1974.

Sealander, Judith. *As Minority Becomes Majority: Federal Reaction to the Phenomenon of Women in the Work Force, 1920–1963.* Westport, Conn.: Greenwood Press, 1983.

Selleck, Henry B., and Alfred Whittaker, *Occupational Health in America.* Detroit: Wayne State University Press, 1962.

Sellers, Christopher. "Manufacturing Disease: Experts and the Ailing American Worker." Ph.D. dissertation, Yale University, 1991.

———. "The Public Health Service's Office of Industrial Hygiene and the Transformation of Industrial Medicine." *Bulletin of the History of Medicine* 65 (1991): 42–73.

Serwer, Daniel. "The Rise of Radiation Protection: Science, Medicine, and Technology in Society, 1896–1935." Ph.D. dissertation, Princeton University, 1977.

Sharpe, William. "Chronic Radium Intoxication—Clinical and Autopsy Findings in Long-Term New Jersey Survivors." *Environmental Research* 8 (1974): 243–383.

———. "Harrison D. Martland and the New Jersey Radium Dial Painters." *Academy of Medicine of New Jersey Bulletin* 16 (1970): 55–62.

———. "The New Jersey Radium Dial Painters: A Classic in Occupational Carcinogenesis." *Bulletin of the History of Medicine* 52 (Winter 1978): 560–70.

———. "Radium Osteitis with Osteogenic Sarcoma: The Chronology and Natural History of a Fatal Case." *Bulletin of the New York Academy of Medicine* 47 (September 1971): 1059–82.

Short, Louise J. "Katherine Rotan Drinker: Sketch of a Semiprivate Life." Paper written for the Harvard University History of Science Program, January 12, 1982, available at the Francis A. Countway Library of Medicine, Harvard University, Cambridge.

Sicherman, Barbara. "Alice Hamilton." In *Notable American Women: The Modern Period*, edited by Barbara Sicherman and Carol Hurd Green, pp. 303–6. Cambridge: Harvard University Press, 1980.

———. *Alice Hamilton: A Life in Letters*. Cambridge: Harvard University Press, 1984.

Sigerist, Henry. "American Spas in Historical Perspective." In *Henry E. Sigerist on the History of Medicine*, edited by Felix Marti-Ibañez, pp. 66–79. New York: M. D. Publications, 1960.

———. "Historical Background of Industrial and Occupational Disease." *Bulletin of the New York Academy of Medicine*, 2d s., 12 (1936): 597–609.

Sklar, Kathryn Kish. *Florence Kelley and the Nation's Work: The Rise of Women's Political Culture*. New Haven: Yale University Press, 1995.

———. "The Historical Foundations of Women's Power in the Creation of the American Welfare State, 1830–1930." In *Mothers of a New World: Maternalist Politics and the Origins of Welfare States*, edited by Seth Koven and Sonya Michel, pp. 43–93. New York: Routledge, 1993.

———. "Two Political Cultures in the Progressive Era: The National Consumers' League and the American Association for Labor Legislation." In *U.S. History as Women's History: New Feminist Essays*, edited by Linda K. Kerber, Alice Kessler-Harris, and Kathryn Kish Sklar, pp. 36–62, 356–64. Chapel Hill: University of North Carolina Press, 1995.

Smith, Barbara Ellen. *Digging Our Own Graves: Coal Miners and the Struggle over Black Lung Disease*. Philadelphia: Temple University Press, 1987.

———. "The History and Politics of the Black Lung Movement." *Radical America* 17 (1983): 89–109.

Spear, Frederick, and K. Griffiths. *The Radium Commission: A Short History of Its Origin and Work, 1929–1948*. London: H.M. Stationary Office, 1951.

Staley, Eugene. *A History of the Illinois State Federation of Labor*. Chicago: University of Chicago Press, 1930.

Starr, Paul. *The Social Transformation of American Medicine*. New York: Basic Books, 1982.

Stebbings, J. H. "Human Health Effects of Radium: An Epidemiological Perspective of Research at Argonne National Laboratory." Typescript report from the DOE Radiation Epidemiology Contractors' Workshop, Rockville, Md., 1982.

Stebbings, James, Henry Lucas, and Andrew Stehney. "Mortality from Cancers of Major Sites in Female Radium Dial Workers." *American Journal of Industrial Medicine* 5 (1984): 435–59.

Stehney, A. F., H. F. Lucas Jr., and R. E. Rowland. "Survival Times of Women Radium Dial Workers First Exposed before 1930." In *Late Biological Effects of Ionizing Radiation: Proceedings of the Symposium on the Late Biological Effects of Ionizing Radiation, Held by the International Atomic Energy Agency in Vienna, March 13–17, 1978*, pp. 333–51. Vienna: International Atomic Energy Agency, 1978.

Stellhorn, Paul, and Michael Birkner. *The Governors of New Jersey, 1664–1974*. Trenton: New Jersey Historical Commission, 1982.

Stellman, Jeanne. *Women's Work, Women's Health: Myth and Realities*. New York: Pantheon, 1977.

Straight, Wilma. "Alice Hamilton: First Lady of Industrial Medicine." Ph.D. dissertation, Case Western Reserve, 1974.

Swartz, Joel. "Silent Killers at Work." *Crime and Social Justice* 3 (1975): 15–20.

Tax, Meredith. *The Rising of the Women: Feminist Solidarity and Class Conflict, 1880–1917*. New York: Monthly Review Press, 1980.

Taylor, Lloyd, Jr. *The Medical Profession and Social Reform, 1885–1945*. New York: St. Martin's Press, 1974.

Teleky, Ludwig. *History of Factory and Mine Hygiene*. New York: Columbia University Press, 1948.

———. "Lessons from the History of Lead Poisoning." *Industrial Medicine* 9 (1940): 17–20.

Tentler, Leslie Woodcock. *Wage-Earning Women: Industrial Work and Family Life in the United States, 1900–1930*. New York: Oxford University Press, 1979.

Wade, Louise. "Florence Kelley." In *Notable American Women*, edited by Edward James, 2:316–19. 3 vols. Boston: Harvard University Press, 1971.

Wall, Florence. "Radioactivity in Industry." *Chemistry* 42 (April 1969): 17–19.

Weindling, Paul. "Linking Self-Help and Medical Science: The Social History of Occupational Health." In *The Social History of Occupational Health*, edited by Weindling, pp. 1–31.

———, ed. *The Social History of Occupational Health*. London: Croom Helm, 1985.

Weinstein, James. "Big Business and the Origins of Workmen's Compensation." *Labor History* 8 (Spring 1967): 156–74.

———. *The Corporate Ideal in the Liberal State*. Boston: Beacon Press, 1968.

White, Lawrence. *Human Debris: The Injured Worker in America*. New York: Seaview/Putnam, 1983.

Whorton, James. *Before Silent Spring: Pesticides and Public Health in Pre-DDT America*. Princeton, N.J.: Princeton University Press, 1974.

Wilentz, Sean. *Chants Democratic: New York City and the Rise of the American Working Class, 1788–1850*. New York: Oxford University Press, 1984.

Wilson, Joan Hoff. *Herbert Hoover: Forgotten President*. Boston: Little, Brown, 1975.

Wolfe, Allis Rosenberg. "Women, Consumerism, and the National Consumers' League in the Progressive Era, 1900–1923." *Labor History* 16 (1975): 378–92.

Wright, Michael. "Reproductive Hazards and 'Protective' Discrimination." *Feminist Studies* 5 (1979): 302–9.

Yellowitz, Irwin. *Labor and the Progressive Movement in New York, 1897–1916.* Ithaca, N.Y.: Cornell University Press, 1965.

Young, Angela Nugent. "Interpreting the Dangerous Trades; Workers' Health in America and the Career of Alice Hamilton, 1910–1935." Ph.D. dissertation, Brown University, 1976.

Young, James Harvey. *The Medical Messiahs: A Social History of Health Quackery in Twentieth-Century America.* Princeton, N.J: Princeton University Press, 1967.

———. *Toadstool Millionaires: A Social History of Patent Medicines in America before Federal Regulation.* Princeton, N.J.: Princeton University Press, 1961.

Zieger, Robert. *The Republicans and Labor, 1919–1929.* Lexington: University of Kentucky Press, 1969.

Zwerling, Craig. "Salem Sarcoid: The Origins of Beryllium Disease." In *Dying for Work: Workers' Safety and Health in Twentieth-Century America,* edited by David Rosner and Gerald Markowitz, pp. 103–18. Bloomington: Indiana University Press, 1987.

INDEX